Lecture Notes in Electrical Engineering

Volume 1072

The book series *Lecture Notes in Electrical Engineering* (LNEE) publishes the latest developments in Electrical Engineering—quickly, informally and in high quality. While original research reported in proceedings and monographs has traditionally formed the core of LNEE, we also encourage authors to submit books devoted to supporting student education and professional training in the various fields and applications areas of electrical engineering. The series cover classical and emerging topics concerning:

- Communication Engineering, Information Theory and Networks
- Electronics Engineering and Microelectronics
- Signal, Image and Speech Processing
- Wireless and Mobile Communication
- Circuits and Systems
- Energy Systems, Power Electronics and Electrical Machines
- Electro-optical Engineering
- Instrumentation Engineering
- Avionics Engineering
- Control Systems
- Internet-of-Things and Cybersecurity
- Biomedical Devices, MEMS and NEMS

For general information about this book series, comments or suggestions, please contact leontina.dicecco@springer.com.

To submit a proposal or request further information, please contact the Publishing Editor in your country:

China

Jasmine Dou, Editor (jasmine.dou@springer.com)

India, Japan, Rest of Asia

Swati Meherishi, Editorial Director (Swati.Meherishi@springer.com)

Southeast Asia, Australia, New Zealand

Ramesh Nath Premnath, Editor (ramesh.premnath@springernature.com)

USA, Canada

Michael Luby, Senior Editor (michael.luby@springer.com)

All other Countries

Leontina Di Cecco, Senior Editor (leontina.dicecco@springer.com)

**** This series is indexed by EI Compendex and Scopus databases. ****

Chung Siung Choo · Basil T. Wong ·
Khairul Hafiz Bin Sharkawi · Daniel Kong
Editors

Proceedings of ASEAN-Australian Engineering Congress (AAEC2022)

Engineering Solutions in the Age of Digital Disruption

Springer

Editors
Chung Siung Choo
Swinburne University of Technology
Sarawak Campus
Kuching, Malaysia

Basil T. Wong
Swinburne University of Technology
Sarawak Campus
Kuching, Malaysia

Khairul Hafiz Bin Sharkawi
Sarawak Digital Economy Corporation
Berhad
Kuching, Malaysia

Daniel Kong
Monash University Malaysia
Subang Jaya, Malaysia

ISSN 1876-1100 ISSN 1876-1119 (electronic)
Lecture Notes in Electrical Engineering
ISBN 978-981-99-5546-6 ISBN 978-981-99-5547-3 (eBook)
https://doi.org/10.1007/978-981-99-5547-3

This Springer imprint is published by the registered company Springer Nature Singapore Pte Ltd.
The registered company address is: 152 Beach Road, #21-01/04 Gateway East, Singapore 189721, Singapore

Paper in this product is recyclable.

Foreword by Ir. Dennis Ong

The ASEAN Australian Engineering Congress, AAEC 2022, jointly organised by Engineers Australia, Swinburne University of Technology, Sarawak Campus and Sarawak Digital Economy Corporation Berhad (SDEC) in partnership with TEGAS Digital Village and Sarawak Development Institute (SDI) was held in Kuching, Sarawak, Malaysia, on 12–14 July 2022.

AAEC2022 was also supported by the Sarawak state government, the Australian Trade Commission (Austrade) and a number of engineering institutions and companies.

The Theme for this Congress is on 'Engineering Solutions in the Age of Digital Disruption', centred around digital transformation and Industry 4.0 in engineering practice.

Innovative engineering projects and potential solutions were presented in the form of technical presentations and forum discussions. This congress also provided networking opportunities amongst university lecturers, students, staff of government implementing agencies and practicing engineers to have constructive discussions and propose solutions to problem statements presented during the congress.

The Congress covered six Plenary Sessions on (1) Industry 4.0 and Future Technologies; (2) The Internet of Things, Network Communication, and Cybersecurity; (3) Artificial Intelligence and Automation in IR4.0; (4) Smart Energy, Cities, and Factories; (5) BIM and Construction for IR4.0; and (6) Education 4.0.

There were three Technical Sessions where 43 papers were accepted for Congress and presented.

AAEC 2022 also proudly presented three keynote speakers: (1) Mr. Paul Sanda, Minister-Counsellor (Commercial) and, Senior Trade and Investment Commissioner, Malaysia and Brunei, Australian Trade and Investment Commission (Austrade), who gave a keynote on 'Growing the "Digital Economy" Agenda in Australia and ASEAN'. (2) Dr. Nick Fleming, National President and Chair of Board Engineers Australia, Australia, who gave a keynote on 'Engineering Future-fit Solutions: Digital and Beyond'. (3) Professor Tan Sri Dato Ir. Dr. Chuah Hean Teik, President, ASEAN Academy of Engineering and Technology, Malaysia, who gave a keynote on 'Educating Future Engineers in the Age of Digital Discruption'.

We would like to thank all those involved in organising this congress as well as all the sponsors.

On behalf of the AAEC 2022 organizing committee.

Kuching, Malaysia

Ir. Dennis Ong
Organising Chairman, AAEC 2022

Foreword by Prof. Jugdutt Singh

The global economy has been severely affected by the COVID-19 pandemic, geopolitical issues and Russia–Ukraine war, and these will have a long-term effect on how economics and financial systems operate in the future. In this challenging environment, Sarawak Government launched the Post-COVID-19 Development Strategy 2030 (PCDS 2030) in 2021, outlining its plan to accelerate Sarawak's progress to be a developed State with a thriving economy driven by data and innovation where everyone enjoys economic prosperity, social inclusivity and a sustainable environment.

Many great engineering achievements have become commonplace that we now take them mostly for granted, for example, the widespread development and distribution of electricity and clean water, ease and efficiency of global travel, spacecraft and lasers, medical imaging and the Internet are just some of the examples in which engineering transformed and enhanced every aspect of humanity. As significant as these engineering achievements are, there are many more great challenges that remain to be fulfilled in areas of sustainability, health, security and the joy of living.

In this age of technological transformation, digitalisation has positioned the engineering industry for prolific growth. Industry 4.0 is revolutionising the way companies manufacture, improve and distribute their products. Businesses are integrating new technologies, including Internet of Things (IoT), cloud computing and analytics, artificial intelligence (AI) and machine learning into their production facilities and throughout their operations.

Globally, we are seeing major infrastructure initiatives largely supported by the increased use of technology. For example, digital technology is one of the critical levers for accelerating the transition to green hydrogen, especially artificial intelligence and IoT technologies that enables the optimisation and automation of systems through enhanced data management and analytics. Digital twins can model multiple designs and scenarios, optimising each design to maximise return on investment and minimise risk. Digitally monitored hydrogen plants will leverage near real-time data to ensure end-to-end traceability along the entire life cycle and automate input to certification process, offering more confidence and reliability and increasing future-proofing. Subsequently, industries such as agriculture and mining have opportunities to adopt smart and efficient practices aided by autonomous systems, IoT and data analytics to improve productivity, production efficiency and safety.

With technology underpinning sustainable global transformation, it is imperative that we look into nations' readiness in particular in its human capital, digital infrastructure and connectivity, technology adoption, innovation and start-up ecosystem, business and government investments, ease of doing business, cybersecurity and basic infrastructure to achieve sustainable growth.

As the nations explore impact metrics beyond financial measures, businesses are scrutinised for their societal responsibilities. Central to this is what is our role in the economy and our commitment to the society. This requires a holistic strategy around environmental, social and governance (ESG), creating a sustainability culture, where everyone feels a sense of responsibility and for sustainability.

Kuching, Malaysia

Prof. Jugdutt Singh
Chief Scientist and Chief Advisor for
Digital Economy

Preface

This book comprises the papers that were presented during the ASEAN Australian Engineering Congress 2022 (AAEC2022) which took place from 12 to 14 July, 2022. The conference was conducted in a hybrid mode, given the world's emergence from the COVID-19 pandemic. Held in Kuching, Malaysia, under the theme of 'Engineering Solutions in the Age of Digital Disruption', AAEC2022 provided a platform for engineers, researchers, academics and government officials to exchange the latest knowledge on Industry 4.0 technologies. The papers included in this compilation cover a broad range of topics, including artificial intelligence and machine learning, automation and sensors, engineering computing, materials science and manufacturing, simulation and optimisation, smart energy and cities and other related Industry 4.0 technologies.

The conference was launched by the Premier of the state of Sarawak, Malaysia, The Right Honourable Abang Abdul Rahman Zohari Abang Openg, who welcomed all to AAEC2022. This was followed by addresses by Dennis Lee Khian Ong as President of Engineers Australia Malaysia Chapter, followed by Prof. Hieng Ho Lau as Pro Vice-Chancellor and Chief Executive Officer of Swinburne University of Technology Sarawak Campus. This was followed by keynote speeches by Paul Sanda, Minister-Counsellor (Commercial) and Senior Trade and Investment Commissioner, Malaysia and Brunei from Australian Trade and Investment Commission, Dr. Nick Fleming, National President and Chair of Engineers Australia, and Dr. Hean Teik Chuah, President of ASEAN Academy of Engineering and Technology.

This book features selected papers from AAEC2022 having undergone a peer-reviewed process to ensure quality and relevance. We believe that this conference proceeding will be a valuable resource for academics, researchers and professionals who are interested in Industry 4.0 technologies and their applications.

Kuching, Malaysia — Chung Siung Choo
Kuching, Malaysia — Basil T. Wong
Bandar Sunway, Malaysia — Daniel Kong
Kuching, Malaysia — Khairul Hafiz Bin Sharkawi

Acknowledgements

The editors are indebted to the partnership between Sarawak Digital Economy Corporation, Sarawak Multimedia Authority, Engineers Australia (Malaysia Chapter), Swinburne University of Technology Sarawak Campus, and Sarawak Development Institute, as well as the conference committee members like Dennis Lee Khian Ong, Prof. Hieng Ho Lau, Prof. Hieng Tiong Su, Prof. Wallace Wong, Dr. Sing Muk Ng, Hartini Hamdan, Rosalind Wong, Evelette Robin, Lelia Sim and Astley Pung.

Special thanks are also offered to the partners of AAEC2022, namely, the Sarawak State Ministry of Education, Innovation and Talent Development, Digital Nasional, Sarawak Metro, TEGAS Digital Village, SAINS, Gamuda and Sarawak Energy. We also acknowledge the support from the Australian Trade and Investment Commission, Institution of Civil Engineers, ASEAN Academy of Engineering and Technology, Association of Consulting Engineers Malaysia, Malaysia Australia Business Council, Sarawak Business Federation, Business Events Sarawak, Sarawak Multimedia Authority, Malaysia Robotics and Automation Society, The Institution of Engineering and Technology, The Institution of Engineers Malaysia and Malaysian Technology Development Corporation.

We appreciate the knowledge shared by the authors during AAEC2022, along with the efforts of the Technical Program Committee in making the conference a success. We would like to acknowledge the support of the panel chairs, namely Hazwan Razak, Prof. Patrick Then, Dr. Elammaran Jayamani, Dr. Chang Saar Chai, Dr. Almon Chai, Dr. Daniel Looi and Dr. Victor Bong. A special mention also goes to the student helpers, Afzol Husain, Edwin Mun Chuen Thou, Jasman Yi Hao Chai, Jit Boon Bong, Lit Yen Yeo, Melissa Wen Wen Sia, Nicholas Lee Tien Lau, Pei Gee Kueh, Phil Siew Thong Chan and Zahara Rashini Amarasiri Gunawardana. The editors would like to acknowledge the tremendous efforts rendered by Jit Boon Bong and Jason Shun Fui Pei who have stood by the team so steadfastly and readily to contribute their assistance in proofreading and formatting all the materials presented in this book. We also extend our gratitude to the publishing team at Springer, and to the series editors for the *Lecture Notes in Electrical Engineering*.

None of this would have been possible without the support of our partners, keynote speakers, plenary panel members, presenters and authors, expert reviewers, panel chairs and participants.

This book would not have been possible without the contributions of everyone.

List of Reviewers

Basil Tung Liong Wong
Chang Saar Chai
Chung Siung Choo
Hong Siang Chua
Daniel Kong
Bing Shen How
Hudyjaya Siswoyo Jo
Jiuan Jing Chew
Kelvin Yong
Mark Kit Tsun Tee
Miko Chang May Lee
Ming Fai Chow
Norazah Abdul Aziz
Riady Siswoyo Jo
Kok Heng Soon
Yie Hua Tan
Victor Bong
Vivi Anggraini
Wan Tze Vong

Contents

Education 4.0

Engineering Computing

Smart Energy and Cities

Artificial Intelligence and Machine Learning

A Systematic Literature Review on Determining the Effectiveness of Short-Term COVID-19 Prediction Models

T. Ramraj and Valliappan Raman

Abstract Scientific communities are working toward mitigating the repercussions of the SARS-CoV-2 infectious virus and its spread toward the community. The dynamics of the spreading nature can be determined by prediction models. Various prediction models are devised and/or used to know the disease dynamics and the existing ones based on statistical models are being developed for single or multiple countries. Dynamic prediction of new variants of SARS-CoV-2 infections during the present pandemic scenario is critical and herculean and this prediction is cardinal for preparing strategic maneuvers. Many review articles commonly address the statistical models adopted, whereas the studies indicating effective models that address disease dynamics and forecast potential contagion scenarios, viz., Holt's method, Wright's modified Holt's method, and un-replicated linear functional relationship model (ULFR) for new variants are found scarce. This work aims at collating the basic working philosophies of most cited COVID-19 dynamic prediction model reports by a systematic literature study. The study is reported, by formulating a search strategy that is refined in terms of geographical purview, time period according to a predefined protocol and scientific reports presented across the globe. The outcome of this review shows that the study findings are critically relevant to front-line healthcare professionals and government agencies for disease prognostication, effective risk stratification, and adopting strategic maneuvers in policies. Additionally, this works aids in identifying the lacuna in the existing and or customized prediction models so as to improvise their data-driven approaches effectively.

Keywords Forecasting · COVID-19 · Statistical models · Machine learning methods · Deep learning approaches

T. Ramraj (✉) · V. Raman
Department of Artificial Intelligence and Data Science, Coimbatore Institute of Technology, Coimbatore, India
e-mail: ramraj@cit.edu.in

C. S. Choo et al. (eds.), *Proceedings of ASEAN-Australian Engineering Congress (AAEC2022)*, Lecture Notes in Electrical Engineering 1072,
https://doi.org/10.1007/978-981-99-5547-3_1

1 Introduction

Rapid spreading of COVID-19 disease is an unprecedented experience for humankind around the world. This disease has changed the nature of human society, it has hugely affected the healthcare systems, tourism, education, manufacturing, food and beverage, and political and legal systems. Despite actions taken around the globe by World Health Organization (WHO), government sectors, and healthcare departments, still the virus mutates and keeps spreading faster due to its high infectious power. Government agencies across the globe collect and analyze COVID-19 patient's data that challenges healthcare facilities. Apart from medical research studies, in specific, data science tools are more valuable to fight against this pandemic situation by predicting earlier COVID-19 cases, these predictions are helpful for government around the globe to prepare for the upcoming situations. Many research studies are being carried out to forecast COVID-19 dynamics across the communities to take strategic maneuvers and modify policies [1]. Effective sentiment analysis and data mining are carried out in COVID-19 datasets for clinical risk stratification [2]. Machine learning and statistical modeling are the state-of-the-art prediction models to forecast incidence cases [3–10] Multi-criteria decision framework, viz., Technique for Order of Preference by Similarity to Ideal Solution (TOPSIS) method, are adopted for forecasting purposes [11–13]. In forecasting, the popular methods are the average approach, time-series methods, econometric forecasting methods, and artificial neural networks. In economics, the widely applied methods fall in time-series methods and econometric forecasting methods. However, the performance of this approach has been tested only on a relatively geographically medium-sized dataset; research findings of the systematic literature review (SLR) recommend that healthcare systems and stakeholders should use the best prediction model to forecast incidence cases and adapt effective risk stratification strategies by imposing social distancing and lockdown measures during the pandemic. The present review gives an insight into various prediction models and how to choose the best model in terms of maximizing accuracy and minimizing errors. This information will be vital in decision-making and/or policy changing for government, the healthcare sector, and other stakeholders. The literature review applied various inclusion and exclusion criteria to generate comprehensive tables reviewing the literature that contain information about various COVID-19 prediction models, sample sizes, model accuracy, and prediction model characteristics.

In this paper, different machine learning models were analyzed for short-term prediction of the COVID-19 outbreak and summary of existing works conducted by different researchers were highlighted. The paper is organized as follows: Sect. 2 briefs the objectives, research questions, inclusion and exclusion criteria behind the study. Section 3 explains the various machine learning models used for short-term COVID-19 prediction. Discussion about the study is presented in Sect. 4. Finally, the conclusion was made in Sect. 5.

2 Objectives

Owing to the proliferation of COVID-19 incidence cases and its repercussions on social behaviors, several research groups have investigated various aspects of the virus, such as its epidemiological characteristics, socio-economic effects, and factors and parameters aiding the spread of the virus. The present work is an SLR with the following objectives: (1) To systematically review the prediction models that have been developed for COVID-19; (2) To collate the basic working philosophies of the most cited COVID-19 statistical/dynamic prediction model; (3) To synthesize and extract useful results and conclusions about the COVID-19 prediction models. A systematic literature review is a supplementary methodology used to help evaluate studies by capturing principal analyses and its purpose is to summarize the studies carried out and to identify gaps between previous studies and current studies. For the purpose of this review process, the following research questions (RQs) are articulated that focusses on the cardinal issues. The motivation and RQs of this study are as follows:

Motivation: To define the working philosophies of short-term prediction models of COVID-19 infections.

RQ1: Identify the cardinal factors that support the prediction of COVID-19 infections.

RQ2: Discern the methods that are abided in data-driven modeling for predicting COVID-19 infections.

2.1 Inclusion and Exclusion Criteria

i. As search engines provide a high level of recall, which leads to a large number of irrelevant resources being retrieved, the following inclusion and exclusion criteria are adopted for an effective review process.
ii. The type of document: only published documents were included and all other manuscripts are excluded.
iii. Specific domains (i.e., the subject area identified for the study) were the second screening criterion.
iv. Effective prediction models developed for or used in the COVID-19 domain are included, while other documents were excluded.
v. Language criterion: The technical reports and manuscripts that are released in English are prevailed. Other language documents are excluded in this study.

2.2 Quality Assessment and Coding

Quality evaluation of a phenomenon is conducted as a systematic way to avoid biases and errors. Thereby, an SLR includes quality assessment as an essential step. In this study, in the initial phase, nearly one thousand documents were chosen. Based on their titles, these documents were further analyzed and 114 documents were screened among which 28 are found duplicates. Based on the scrutinization of the title, content, abstract, and conclusion, 86 studies were finally selected for the review.

3 Methods

3.1 Machine Learning Models

Based on the COVID-19 time-series data [14], numerous studies in the literature emphasized on modeling, predicting, and forecasting COVID-19 spread. In Sarkar et al. [15] proposed a mathematical model—SARIIqSq, to forecast the transmission dynamics of COVID-19. This model is an extension of the traditional SEIR (Susceptible, Exposed, Infectious, Removed) model and is based on monitoring the dynamics of six behaviors, viz., infected (I), recovered (R), susceptible (S), asymptomatic (A), isolated infected (Iq), and quarantined susceptible (Sq). In Abbasi et al. [16] presented an alternative SIER approach—SQEIAR, that discerns the working philosophy of the COVID-19 epidemical dynamics by considering Quarantined individuals (Q) and Asymptomatic (A) parameters. The proposed method by Hameni et al. in [17] agglutinates the benefits of the ensemble of Kalman filter and the SIR (Susceptible, Infected, Recovered) model for modeling COVID-19 transmission. This hybrid approach enables estimating the unknown transmission rates of COVID-19 to reconstruct the basic reproduction number R_0. The forecasting performance of this approach has been verified based on real data from March 6 to May 17, 2020, of the current COVID-19 pandemic outbreak in Cameroon. The method proposed in [18] applied regression models for forecasting COVID-19 cases in Brazil by considering Seasonal Auto-Regressive Integrated Moving Average (SARIMA) and Auto-Regressive Integrated Moving Average (ARIMA) [13], cubist regression, random forest, ridge regression, SVR, and stacking-ensemble learning for one-, three-, and six-step ahead forecasting. The outcome of the study showed that the SVM regression and stacking-ensemble learning achieved better forecasting than the other models. The SVR approach coupled with different nonlinearity structures is used to predict COVID-19 cases and is presented in [19]. This statistical study implied that SVR with Gaussian kernel gave better prediction outcomes compared to SVR with other Kernel functions. In [20], for the global forecast of COVID-19, a nonlinear machine learning approach with partial derivative regression to build an epidemiological prediction method. For more computation efficiency and better dataset parameters, the progressive partial derivative linear regression is used. The study by Nasution et al. [21] employed an autoregression model utilizing Poisson distribution known as Poisson Autoregression (PAR) to predict the confirmed and recovered cases of COVID-19 specific to a particular demographic area.

Computational studies [22–24] deployed logistic regression to develop prediction-based models. As the Random Forest (RF) repeatedly builds regression trees based on the training data, research findings [3, 25–28] indicate that prediction of incidence COVID-19 is effective by selecting optimized feature subsets. Other prediction models deployed in research findings using ANN [25], KNN [7, 29–33] used a supervised learning ML approach. In [34, 35], the authors proposed genetic programming and analytics for forecasting cases in voluminous data. Other forecasting variants, viz., a stochastic-based SEIQHRF model [36], used to simulate disease prevalence in the susceptible, infected, recovered, and fatal compartments. Hybrid models [37] and MCMD [13] facilitate healthcare authorities in decision-making, especially for short-term prediction.

3.2 Deep Learning Models

With outstanding generalization capability and superior nonlinear approximation, Deep Learning (DL) methods have gained attention in time-series modeling and analysis [38, 39]. As the DL models can be obtained by concatenating several layers into the neural network structures [40], the cardinal DL model feature focuses on their capacity to automatically extract relevant information from large data [41]. They have attracted attention in the machine learning community and investigated a wide range of COVID forecasting applications [42–45].

For instance, in [43], an long short-term memory (LSTM) model is applied to forecast new confirmed cases of COVID-19 in the Canadian dataset between January 22 and March 31, 2020. A similar approach—an improvised LSTM model is used Harrou et al. [40], to predict the epidemic trends of COVID-19 in Russia, Peru, and Iran. In the seminal work by Shahid et al. [45] the support vector regression (SVR), LSTM, bidirectional LSTM (BiLSTM), and GRU are employed to forecast COVID-19 time-series data in ten countries most affected by COVID-19. The inference of the presented study revealed the superior performance of the BiLSTM based on COVID-19 data available until June 27, 2020. As Recurrent Neural Network (RNN) possesses the ability to process variable-length sequential data and can store only short-term memory, a few findings [46–48] used RNN to estimate the number of patients on a specific day. The below tabulation shows the major benchmark DL approaches used in prediction models apart from baseline methods, namely, logistic regression (LR) and support vector regression (SVR), for forecasting confirmed and recovered COVID-19 cases time-series data (Table 1).

3.3 Predictive Analysis

For forecasting time-series data, statistical methods [2, 50] and ARIMA [5, 51–55] are used. Other predictive analysis methods, viz., Holt's method, Wright modified

Table 1 The considered benchmark DL methods for forecasting time-series data [49]

Model	Description	Key points
GRU model	The major demarcation of GRU from LSTM is that only one unit is used to control both the forgetting factor and the decision to update the state unit [42] Gate mechanism is used to memorize the historical data features The GRU is trained in a supervised way The GRU has been used in time-series data sequences	It requires shorter training time when compared to LSTM and takes fewer parameters for processing [42] The scope for improvisation is low learning efficiency, slower convergence, and under-fitting
Restricted Boltzmann Machine (RBM)	RBMs are stochastic and generative neural networks consisting of visible and hidden units with strong connections between them RBMs are trained in an unsupervised way based on the contrastive divegence learning methods	They are good for data distribution approximation They can be stacked to build DL models The scope for improvisation lies in slow training phase due to the contrastive divergence approach
CNN	They are trained in a supervised way using backpropagation with gradient descent	Efficient to learn relevant features The scope for improvisation is a requirement of large data storage for storing labeled data
GAN	Complete generative model [43]	Has good data capability for data distribution approximation The scope for improvisation lies in the problem of convergence and vanishing gradient

Holt's method, and unreplicated linear functional relationship model (ULFR), are deployed to increase accuracy in forecasting the incidence cases. Holt's method [56], is defined as linear-exponential smoothing. This smoothing model is well known for forecasting data with trends. Wright [57] introduced a modification of Holt's method for the data with irregularity in time spacing. However, the modified Holt's method possesses a few unique characteristics, namely, greater computational efficiency and flexibility in terms of having more smoothing constants. Besides, the modified model outperforms SOTA methods with empirical, missing, or zero data. The ULFR [58], is a variant of linear regression model for analyzing the dependent and independent variables' relationship. Nevertheless, the relationship between the variables becomes fuzzy due to spikes and unusual fluctuations with defined variables. Moreover, it is unlikely to measure precisely independent variables in all circumstances. Based on the above-mentioned methods, various research studies [5, 51–55] are carried out for effective prediction.

3.4 GLEM and Hybrid EAMA Models

Global epidemic and mobility (GLEM) models are being used in COVID-19-related studies. These models involve a stochastic computational framework that combines demographic and mobility data so as to predict epidemic distribution across the globe. The objective of the GLEM model is to optimize versatility in specifying the disease compartment model and configuring the simulation scenario. It allows the user to set a number of criteria, including compartment-specific features, transition values, and environmental effects [59]. Similarly, hybrid EAMA model [60] uses iterative methods to predict the epidemic distribution.

3.5 Other Methods

Other models include PRISMA [61, 62], Singular Spectrum Analysis (SSA) Model [63], and Vikor method [64] forecasts incidence cases that can be used for solving prioritization problems, effective clinical prognostication, and to assess the effects of SDMs (e.g., isolation, quarantine) for reducing the transmission of COVID-19. The authors in [65] proposed a cloud-based short-term forecasting model to predict the number of COVID-19 confirmed cases for the next 7 days. The results of this study indicate the importance of the cloud-based short-term forecasting model in decision-making to prepare the needed medical resources.

4 Discussion

Various ML and DL techniques are carried out to provide suitable forecasting accuracy. Toward this purpose, the LSTM-CNN model has been reviewed for COVID-19 time-series forecasting. The overall comparison of the hybrid models (i.e., LSTM-CNN, GAN-DNN, and GAN-GRU) and three DL models (standalone), viz., CNN, LSTM, and RBM reveal the domination of the LSTM-CNN model due to its capability to learn higher level features that permit good forecasting precision as prescribed by Dairi et al. [66]. Essentially, the performance of DL models when applied to a small amount of time-series data and compare their forecasting accuracy with the traditional time-series models indicate that the DL models provide satisfying results, even with relatively small-sized data. On the other end, ML models exclude the time dependency in the COVID-19 data series. Hence, research studies take dynamic information into account to alleviate this limitation by introducing lagged measurements in constructing the investigated machine learning models. Research findings [67] reveal that significant improvement can be obtained using dynamic machine learning models. In addition, the results highlighted the superior performance of the dynamic GPR compared to the other models (i.e., support vector regression, boosted trees, bagged trees, decision tree, random forest, and XGBoost) by achieving minimal prediction errors and its simplified approach for predicting COVID-19 spread. The inference presented in [68] reveals that among comparing the four methods: ARIMA, ANN, LSTM, and CNN, to predict the COVID-19 spread based on data available from March 12 to October 31, 2020, the CNN model outperformed the other models by achieving a minimal averaged mean absolute percentage error. However, the models are to be trained using small-sized data, making it difficult to get accurate models for forecasting the future trends of COVID-19 spread. As the COVID-19 training and testing data used in the research studies are not the same, it becomes a herculean task to compare the performance of the proposed approach with the various SOTA methods. However, Table 2 summarizes the forecasting performance of the existing methods when applied to small-sized datasets.

The following Table 3 presents a summary of all the models that are considered in this literature study.

5 Conclusion

Based on current situations around the globe, till now, COVID-19 pandemic has not been defeated, but the good news is, there is a huge number of recovered cases after active vaccination and awareness about the disease. COVID-19 disease prediction models are vital at this stage, therefore, accurate forecasting of COVID-19 spread and understanding their dynamics is a key factor in slowing down this pandemic's transmission by providing relevant information to assist healthcare professionals. This work aimed to provide a literature study by reviewing and analyzing the articles related to COVID-19 prediction models. The accuracy of the forecasting models supports hospital management for clinical prognostication and risk stratification. Research studies indicate that the forecasting methods cause a decline in disease progression by providing a word of caution. Many articles that address the statistical models adopted are reviewed, whereas the studies indicating effective models that address disease dynamics and forecast potential contagion scenarios are confined. Hence, the working philosophies of the most cited COVID-19 dynamic prediction model reports are collated in this systematic literature study. The reports of this study are furnished by adopting a search strategy that is refined in terms of geographical purview, time period according to a predefined protocol, and scientific reports presented across the globe. Traditional time-series forecasting methods show promising results in small-sized dataset, whereas these methods also assume a few artificial spikes in the incidence cases during an epoch period so as to increase the prediction accuracy. On the other hand, ML and DL approaches outperform many SOTA models; however, they are inferior while considering the availability of a sufficient amount of data to build an accurate model. Unfortunately, there are no protocols or methods to check this criterion. Essentially, it is inferred that the performance of DL models when applied to a small amount of time-series data shows promising

Table 2 Performance of the existing forecasting methods when applied to small-sized datasets (Ref. [67])

References	Country	Model	Average MAPE %
Ceylan et al.	Italy, Spain, and France	ARIMA	5.59
Balli Serkan	Germany and the USA	Random forest	1.0042
		Linear regression	0.2228
		MLP	0.5153
		SVM	0.1162
Nasution et al.	Jakarta	ARIMA	20.51
		SES	20.435
		HW	47.415
		BATS	33.945
		Prophet	42.27
		PAR	18.435
Istaiteh et al.	China, Eritrea	ARIMA	14.14
		ANN	3.23
		LSTM	4.14
		CNN	3.13
Shaikh et al.	India	Linear regression	27.9
		Polynomial with 2 degrees	13.3
Acosta et al.	Brazil, Chile, Colombia, Mexico, Peru, and the USA	SVM	23.5
		MLP	17
Dairi et al.	Brazil, France, India, Mexico, Russia, Saudi Arabia, and the USA	RBM	18.452
		CNN	20.763
		LSTM	20.394
		GAN-DNN	11.105
		GAN-GRU	5.254
		LSTM-CNN	3.718
Omran et al.	Egypt, Saudi Arabia, and Kuwait	a single-layer GRU	3.0419
		a single-layer LSTM	0.6203
Kafieh et al.	Nine countries, including China, Spain, Italy, and the USA	M-LSTM	0.509
Yasminah Alali et al.	India and Brazil	OGPR	0.1025

Table 3 Summary of all the prediction models considered in this literature study

S. No	Reference list	Year	Model adopted
1	Krishna Regmi	2019	Prisma model
2	ChuanyuHu	2020	Machine learning
3	AlbahriAS	2020	Vikor-method
4	AlbahriOS	2020	Mcmd method
5	Bandyopadhyay	2020	LSTM model
6	Sk ShahidNadim	2020	SEIR, SIR model
7	Ajith Abraham	2020	LSTM model
8	Annwesha Banerjee Majumder	2020	Regression model
9	R Sujath	2020	Regression model
10	TimonRabczuk	2020	SEIR, SIR model
11	Durga Mahesh	2020	ANN model
12	Zahra Taheri	2020	LSTM model
13	Swapan	2020	Prediction
14	Parularora	2020	LSTM & RNN model
15	Ramesh Kumar Mojjada	2020	Regression model
16	Omar Sharif	2020	Holt's method wright modified ULFR
17	Kimberlynloosa	2020	SEIR, SIR model
18	k.c. Santhosh	2020	SEIR, SIR model
19	Linhao Zhong	2020	SEIR, SIR model
20	Matheus Henrique Dal MolinRibeiro	2020	ARIMA model
21	Wie kiang H	2020	Regression model
22	Saleh I Alzahrani	2020	ARIMA model
23	Mrutyunjaya panda	2020	Holt's method wright modified ULFR
24	Shengli cao	2020	SEIR, SIR model
25	Edison	2020	Holt's method wright modified ULFR
26	Kaustuv Chatterjee	2020	SEIR, SIR model
27	Anuradha Tomar	2020	LSTM model
28	Christopher JL Murray	2020	Statistical model
29	Samit Ghosal	2020	Regression model
30	Zirun Zhao	2020	Prediction
31	Andres Hernandez Matamoros	2020	ARIMA model
32	L. J Muhammed	2020	Regression model
33	Farhan Mohammed Khan	2020	ARIMA model
34	Laure Wynants	2020	Prisma model
35	Lishi Wang	2020	Patient information based algorithm
36	Ansari Saleh Ahmar	2020	Sutte ARIMA model
37	Samuel Lalmuanawma	2020	Prediction

(continued)

Table 3 (continued)

S. No	Reference list	Year	Model adopted
38	J Clin Virol	2020	Prediction
39	Anton Civit	2021	Forecasting, statistical model
40	Shaikh S	2021	Regression model
41	Arunkumar K	2021	Arima model
42	Yazeed Zoabi	2021	Gradient boosting machine model
43	Mohamed Marzouk	2021	LSTM model
44	AlbahriAS	2021	Topsis
45	Hongwei Zhao	2021	Prediction
46	Chi Hua Chen	2021	Forecasting
47	Pooja sengupta	2021	SEIQHRf model
48	Andreas Andreou	2021	Regression model
49	Aurelle Tchagnakouanou	2021	Linear regression model
50	Deepak Painuli	2021	Random forest
51	Rohit Salgotra AmirH. Gandomi	2021	Scientific & genetic programming
52	Sk Shahid Nadim	2021	Prediction
53	Vikas Kulshreshtha	2021	ARIMA model
54	Connor Shorten	2021	SEIR, SIR model
55	Darina kapusta	2021	Holt's method wright modified ULFR
56	J. Sultana	2021	Forecasting
57	Zahra Ramezani	2021	LSTM model
58	Upasana	2021	Prediction
59	Soudeh Ghafouri-Fard	2021	LSTM model
60	Sumayh S. Aljameel	2021	Scientific & genetic programming
61	Rubaiya. HossainMondal	2021	Classification
62	Mahdi Mahdavi	2021	Regression model
63	Senthilkumar mohan	2021	Hybrid model
64	M Rohini, K. R Naveena	2021	KNN algorithm
65	Matt J. Keeling	2021	Forecasting
66	Md. Shahriare Satu	2021	Cloud
67	Martí Català,	2021	Prediction
68	Behnam Nikparvar	2021	LSTM model
69	R. Lakshmanan Kumar	2021	Regression model
70	Shuhaida Ismail	2021	Random forest
71	Gaetano Perone	2021	Topsis
72	Mahdi Kalantari	2021	SSA-model
73	Seng Hansun	2021	Holt's method wright modified ULFR
74	Theyazn H. H. Aldhyani	2021	Holt's trend model

(continued)

Table 3 (continued)

S. No	Reference list	Year	Model adopted
75	Erin McNemar	2021	Decision model
76	Bilal Majeed	2022	Curve fitting model
77	Yasminah Alali	2022	Random forest
78	Lieslehto	2022	Logistic regression, SVM
79	Ebubeogu	2022	SEIR, SIR model
80	Bo yang	2022	SEIR, SIR model
81	Roshan Wathore	2022	LSTM model
82	Koushlendra Kumarsingha	2022	Random forest
83	Zeynep Ceylan	2022	Prediction
84	Omar Sherif	2022	Holt's method wright modified ULFR
85	Sheikh Taslim Ali	2022	Meta-analysis
86	Madini O. Alassafi	2022	LSTM & RNN model

results. The scope for improvisation in the future is to incorporate explanatory variables, viz., lockdown indicators, in constructing the deep learning models. Further, conducting comparative studies to investigate the impacts of measures established during this pandemic period is mandatory.

Acknowledgements The contributing authors would like to acknowledge the Indian Council of Medical Research (ICMR)—India. This work was supported by funding from ICMR—Bio-Medical Informatics (BMI) Division, under project id No: 2021-6415.

References

1. Majeed B, Li A, Peng J, Lin Y (2022) A multi-period curve fitting model for short-term prediction of the COVID-19 Spread in the U.S. Metropolitans. Front Public Health 9:809877. https://doi.org/10.3389/fpubh.2021.809877
2. Civit A et al (2021) Fighting COVID-19: emerging techniques and aid systems for prevention, forecasting and diagnosis. https://www.mdpi.com/journal/applsci/special_issues/Fighting_COVID-19
3. Alali Y, Harrou F, Sun Y (2022) A proficient approach to forecast COVID-19 spread via optimized dynamic machine learning models. Sci Rep 12:2467. https://doi.org/10.1038/s41598-022-06218-3
4. Shaikh S et al (2021) Analysis and prediction of covid-19 using regression models and time series forecasting. In: 11th international conference on cloud computing, data science & engineering (confuence). pp 989–995
5. ArunKumar K et al (2021) Forecasting the dynamics of cumulative covid-19 cases (confrmed, recovered and deaths) for top-16 countries using statistical machine learning models: auto-regressive integrated moving average (arima) and seasonal auto-regressive integrated moving average (sarima). Appl. Sof Comput. 103:107161
6. Zoabi Y, Deri-Rozov S, Shomron N (2021) Machine learning-based prediction of COVID-19 diagnosis based on symptoms. npj Digit Med 4:3. https://doi.org/10.1038/s41746-020-00372-6

7. Lieslehto J, Rantanen N, Oksanen LMAH et al (2022) A machine learning approach to predict resilience and sickness absence in the healthcare workforce during the COVID-19 pandemic. Sci Rep 12:8055. https://doi.org/10.1038/s41598-022-12107-6
8. Ebubeogu AF, Ozigbu CE, Maswadi K et al (2022) Predicting the number of COVID-19 infections and deaths in USA. Global Health 18:37. https://doi.org/10.1186/s12992-022-00827-3
9. Hu C, Liu Z, Jiang Y, Shi O, Zhang X, Xu K, Suo C, Wang Q, Song Y, Yu K, Mao X, Wu X, Wu M, Shi T, Jiang W, Mu L, Tully DC, Xu L, Li J, Li S, Tao X, Zhang T, Chen X (2022) Early prediction of mortality risk among patients with severe COVID-19, using machine learning. Int J Epidemiol 9(6):1918–1929. https://doi.org/10.1093/ije/dyaa171
10. Marzouk M, Elshaboury N, Abdel-Latif A, Azab S (2021) Deep learning model for forecasting COVID-19 outbreak in Egypt. Process Saf Environ Prot. 153:363–375. https://doi.org/10.1016/j.psep.2021.07.034
11. Albahri AS, Al-Obaidi JR, Zaidan AA et al (2020) Multi-biological laboratory examination framework for the prioritization of patients with COVID-19 based on integrated AHP and group VIKOR methods. Int J Inf Technol Decis Mak 19:1247–1269
12. Albahri AS, Hamid RA, Albahri OS et al (2021) Detection-based prioritisation: framework of multi-laboratory characteristics for asymptomatic COVID-19 carriers based on integrated Entropy-TOPSIS methods. Artif Intell Med 111:101983
13. Albahri OS, Al-Obaidi JR, Zaidan AA et al (2020) Helping doctors hasten COVID-19 treatment: towards a rescue framework for the transfusion of best convalescent plasma to the most critical patients based on biological requirements via ml and novel MCDM methods. Comput Methods Programs Biomed 196:105617
14. Zeroual F, Harrou A, Dairi YS (2020) Deep learning methods for forecasting COVID-19 time-series data: a comparative study. Chaos, Solitons & Fractals 140
15. Sarkar K, Khajanchi S, Nieto JJ (2020) Modeling and forecasting the covid-19 pandemic in india. Chaos, Solitons & Fractals 139
16. Abbasi Z, Zamani I, Mehra AHA, Shafieirad M, Ibeas A (2020) Optimal control design of impulsive sqeiar epidemic models with application to covid-19. Chaos, Solitons & Fractals 139
17. CH Nkwayep, S Bowong, J Tewa, J Kurths (2020) Short-term forecasts of the covid-19 pandemic: a study case of cameroon. Chaos, Solitons & Fractals 140
18. MHDM Ribeiro, RG da Silva, VC Mariani, L dos Santos Coelho (2020) Short-term forecasting COVID-19 cumulative confirmed cases: perspectives for Brazil. Chaos, Solitons & Fractals
19. Peng Y, Nagata MH (2020) An empirical overview of nonlinearity and overfitting in machine learning using covid-19 data. Chaos, Solitons & Fractals 139
20. Kavadi DP, Patan R, Ramachandran M, Gandomi AH (2020) Partial derivative nonlinear global pandemic machine learning prediction of covid 19. Chaos,Solitons & Fractals 139
21. Nasution BI, Nugraha Y, Kanggrawan JI, Suherman AL (2021) Forecasting of covid-19 cases in jakarta using poisson autoregression. In: 2021 9th international conference on information and communication technology (ICoICT). pp. 594–599
22. Majumder AB, et al (2021) International online conference on engineering response to COVID-19 (IOCER-COVID-19) 2020 8–9 Oct 2020, JIS College of Engineering, Kalyani, West Bengal, India. J Phys: Conf Ser 1797:012011
23. Mahdavi M, Choubdar H, Zabeh E, Rieder M, Safavi-Naeini S, Jobbagy Z et al (2021) A machine learning based exploration of COVID-19 mortality risk. PLoS ONE 16(7):e0252384. https://doi.org/10.1371/journal.pone.0252384
24. Ghosal S, Sengupta S, Majumder M, Sinha B (2020) Linear regression analysis to predict the number of deaths in India due to SARS-CoV-2 at 6 weeks from day 0 (100 cases—March 14th 2020). Diabetes & Metablc Syndr: Clin Res & Rev 14(4):311–315. https://www.sciencedirect.com/science/article/pii/S1871402120300576
25. Taheri Z, et al (2020) Machine learning and deep learning Covid-19 project-final report. https://www.researchgate.net/publication/343685138_Machine_Learning_and_Deep_Learning_Covid-19_Project-Final_Report

26. Deepak P, et al (2021) Forecast and prediction of COVID-19 using machine learning. Data Sci COVID-19. 381–397. https://doi.org/10.1016/B978-0-12-824536-1.00027-7
27. Singh KK, Kumar S, Dixit P et al (2021) Kalman filter based short term prediction model for COVID-19 spread. Appl Intell 51:2714–2726. https://doi.org/10.1007/s10489-020-01948-1
28. Ceylan Z (2021) Short-term prediction of COVID-19 spread using grey rolling model optimized by particle swarm optimization. Appl Soft Comput. 109:107592. https://doi.org/10.1016/j.asoc.2021.107592
29. Sujath R, Chatterjee JM, Hassanien AE (2020) A machine learning forecasting model for COVID-19 pandemic in India. Stoch Environ Res Risk Assess 34:959–972. https://doi.org/10.1007/s00477-020-01827-8
30. Mojjada RK, Yadav A, Prabhu AV, Natarajan Y (2020) Machine learning models for covid-19 future forecasting. Mater Today. https://doi.org/10.1016/j.matpr.2020.10.962
31. Muhammad LJ, Algehyne EA, Usman SS et al (2021) Supervised machine learning models for prediction of COVID-19 infection using epidemiology dataset. SN Comput Sci 2:11. https://doi.org/10.1007/s42979-020-00394-7
32. Shaikh S et al (2021) Analysis and prediction of covid-19 using regression models and time series forecasting. In: 2021 11th international conference on cloud computing, data science & engineering (confuence) 989–995
33. Kouanou AT, Attia TM, Feudjio C, Djeumo AF, Mouelas AN, Nzogang MP, Tchapga CT, Tchiotsop D (2021) An overview of supervised machine learning methods and data analysis for COVID-19 detection. J Healthc Eng 2021:18. https://doi.org/10.1155/2021/4733167
34. Salgotra R, Gandomi AH (2021) Time series analysis of the COVID-19 pandemic in Australia using genetic programming. Data Sci COVID-19 399–411. https://doi.org/10.1016/B978-0-12-824536-1.00036-8
35. Sumayh S, Aljameel et al (2021) Managing big data, visualization and its analytics in healthcare based on scientific programming. https://www.hindawi.com/journals/sp/si/459838/
36. An analysis of COVID-19 clusters in India: two case studies on Nizamuddin and Dharavi. BMC Public Health. 21(1):631. https://doi.org/10.1186/s12889-021-10491-8. PMID: 33789619; PMCID: PMC8011051
37. Perone G (2021) Comparison of ARIMA, ETS, NNAR, TBATS and hybrid models to forecast the second wave of COVID-19 hospitalizations in Italy. Eur J Health Econ 4:1–24. https://doi.org/10.1007/s10198-021-01347-4
38. Harrou F, Dairi A, Sun Y, Kadri F (2018) Detecting abnormal ozone measurements with a deep learning-based strategy. IEEE Sens J 18(17):7222–7232
39. Cheng T, Harrou F, Kadri F, Sun Y, Leiknes T (2020) Forecasting of wastewater treatment plant key features using deep learning-based models: a case study. IEEE Access
40. Harrou F, Sun Y, Hering AS, Madakyaru M et al (2020) Statistical process monitoring using advanced data-driven and deep learning approaches. Theory Pract Appl
41. Dairi F, Harrou YS, Senouci M (2018) Obstacle detection for intelligent transportation systems using deep stacked autoencoder and k-nearest neighbor scheme. IEEE Sens J 18(12):5122–5132
42. Dairi T, Cheng F, Harrou YS, Leiknes T (2019) Deep learning approach for sustainable WWTP operation: a case study on data-driven influent conditions monitoring. Sustain Cities Soc 50(2019):101670
43. Chimmula VKR, Zhang L (2020) Time series forecasting of covid-19 transmission in canada using lstm networks. Chaos, Solitons & Fractals 109864
44. Wang P, Zheng X, Ai G, Liu D, Zhu B (2020)Time series prediction for the epidemic trends of COVID-19 using the improved LSTM deep learning method: case studies in Russia, Peru and Iran. Chaos, Solitons & Fractals
45. Shahid F, Zameer A, Muneeb M (2020) Predictions for covid-19 with deep learning models of lstm, gru and bi-lstm. Chaos, Solitons & Fractals 140
46. Roosa K, Lee Y, Luo R, Kirpich A, Rothenberg R, Hyman JM, Yan P, Chowell G (2020) Short-term forecasts of the COVID-19 epidemic in Guangdong and Zhejiang, China: February 13–23, 2020. J Clin Med 9(2):596. https://doi.org/10.3390/jcm9020596

47. Alassafi MO et al (2022) Time series predicting of COVID-19 based on deep learning. Neurocomputing 468:335–344. https://doi.org/10.1016/j.neucom.2021.10.035
48. Wang L, Li J, Guo S, Xie N, Yao L, Cao Y, Day SW, Howard SC, Graff JC, Gu T, Ji J, Gu W, Sun D (2020) Real-time estimation and prediction of mortality caused by COVID-19 with patient information based algorithm. Sci Total Environ 727
49. Dairi A, Harrou F, Zeroual A, Hittawe MM, Sun Y (2021) Comparative study of machine learning methods for COVID-19 transmission forecasting. J Biomed Inf 118:103791
50. IHME COVID-19 health service utilization forecasting team (2020) Forecasting COVID-19 impact on hospital bed-days, ICU-days, ventilator days and deaths by US state in the next 4 months. MedRxiv. https://doi.org/10.1101/2020.03.27.20043752
51. Kulshreshtha V, Garg NK (2021) Predicting the new cases of coronavirus [COVID-19] in India by using time series analysis as machine learning model in python. J Inst Eng India Ser B 102:1303–1309. https://doi.org/10.1007/s40031-021-00546-0
52. Dal Molin Ribeiro MH, da Silva RG, Mariani VC, dos Santos CoelhoL (2020) Short-term forecasting COVID-19 cumulative confirmed cases: perspectives for Brazil,Chaos. Solitons & Fractals 135:109853. https://doi.org/10.1016/j.chaos.2020.109853
53. Khan FM, Gupta R (2020) ARIMA and NAR based prediction model for time series analysis of COVID-19 cases in India. J Saf Sci Resil 1(1):12–18. https://doi.org/10.1016/j.jnlssr.2020.06.007
54. Hernandez-Matamoros A, Fujita H, Hayashi T, Perez-Meana H (2020) Forecasting of COVID19 per regions using ARIMA models and polynomial functions. Appl Soft Comput 96:106610. https://doi.org/10.1016/j.asoc.2020.106610
55. Alzahrani SI, Aljamaan IA, Al-Fakih EA (2020) Forecasting the spread of the COVID-19 pandemic in Saudi Arabia using ARIMA prediction model under current public health interventions. J Infect Public Health 13(7):914–919. https://doi.org/10.1016/j.jiph.2020.06.001
56. Charles CH (2004) Forecasting seasonals and trends by exponentially weighted moving averages. Int J Forecast 20:5–10
57. Wright DJ (1986) Forecasting data published at irregular time intervals using an extension of Holt's method. Manage Sci 32:499–510
58. Fuller WA (1987) Measurement error models. JohnWiley
59. Ahmar AS, del Val EB (2020) SutteARIMA: short-term forecasting method, a case: Covid-19 and stock market in Spain. Sci Total Environ 729
60. Van den Broeck W, Gioannini C et al (2011) The GLEaMviz computational tool, a publicly available software to explore realistic epidemic spreading scenarios at the global scale. BMC Infect Dis 11:37
61. Mohan S, John A, Abugabah A, Adimoolam M, Singh SK, Bashir Ak, Sanzogni L (2021) An approach to forecast impact of Covid-19 using supervised machine learning model
62. Regmi K, Lwin CM (2020). Impact of social distancing measures for preventing coronavirus disease 2019 [COVID-19]: a systematic review and meta-analysis protocol. https://doi.org/10.1101/2020.06.13.20130294
63. Wynants L, Van Calster B, Collins GS, Riley RD, Heinze G, Schuit E et al (2020) Prediction models for diagnosis and prognosis of covid-19: systematic review and critical appraisal BMJ 2020 369:1328. https://doi.org/10.1136/bmj.m1328
64. Kalantari M (2021) Forecasting COVID-19 pandemic using optimal singular spectrum analysis. Chaos Solitons Fractals 142:110547. https://doi.org/10.1016/j.chaos.2020.110547
65. Albahri AS, Al-Obaidi JR, Zaidan AA (2020) Multi-biological laboratory examination framework for the prioritization of patients with COVID-19 based on integrated AHP and group VIKOR methods. Intl J Inf Technol Decis Mak. 19:1247–1269
66. Satu M et al (2021) Short-term prediction of covid-19 cases using machine learning models. Appl Sci 11:4266
67. Alali Y, Harrou F, Sun Y (2022) A proficient approach to forecast COVID-19 spread via optimized dynamic machine learning models. Nature 12:2467. https://doi.org/10.1038/s41598-022-06218-3

68. Istaiteh O, Owais T, Al-Madi N, Abu-Soud S (2020) Machine learning approaches for covid-19 forecasting. In: 2020 international conference on intelligent data science technologies and applications (IDSTA). pp 50–57

Cough Sound Disease Detection with Artificial Intelligence

Sarah Jane Kho, Brian Loh Chung Shiong, Vong Wan Tze, and Patrick Then Hang Hui

Abstract According to the World Health Organization (WHO), since the discovery of the severe acute respiratory syndrome coronavirus 2 (SARS-CoV-2), there have been a total of 530,896,347 confirmed cases of COVID-19 of which 6,301,020 deaths have been reported as of June 2022. The virus (SARS-CoV-2) primarily affects the respiratory system causing dry cough to be prominent among other symptoms. Sound and the type of cough are said to contain useful features that can contribute to the diagnosis of a disease. Artificial Intelligence (AI) and signal processing show promising potential in the prediction of pulmonary diseases. This provides a platform for a non-invasive method of screening for COVID-19. In turn, this allows for a faster way to screen for the presence of the virus promoting early detection. This paper studies the usage of AI and signal processing methods for the detection of COVID-19 using cough sounds collected from crowdsourced applications. This paper focuses on three different models trained using a Support Vector Machine (SVM), ResNet, and a traditional Convolutional Neural Network (CNN). The datasets used in this study were obtained from the University of Cambridge and the Coswara project. Preliminary investigations show that for models trained and tested using the Cambridge Dataset, CNN performed the best with an AUC of 0.816 followed by the ResNet model and the SVM model with an AUC of 0.738 and 0.671, respectively. Further investigation shows that the models perform poorly in classifying COVID-19-positive and -negative classes when tested on the Coswara dataset. Testing results show that the SVM model performed the best, with an AUC of 0.564, followed by the ResNet and the CNN models (AUCs = 0.522 and 0.443, respectively). Our future work will focus on improving the models' performance by applying various data pre-processing and feature extraction methods on the datasets.

Keywords COVID-19 · Artificial intelligence · Machine learning · Deep learning · Signal processing

S. J. Kho (✉) · B. L. C. Shiong · V. W. Tze · P. T. H. Hui
Swinburne University of Technology, Sarawak, Jalan Simpang Tiga, 93350 Kuching, Sarawak, Malaysia
e-mail: sjkho@swinburne.edu.my

C. S. Choo et al. (eds.), *Proceedings of ASEAN-Australian Engineering Congress (AAEC2022)*, Lecture Notes in Electrical Engineering 1072,
https://doi.org/10.1007/978-981-99-5547-3_2

1 Introduction

The global spread of the severe acute respiratory syndrome coronavirus 2 (SARS-CoV-2) virus has resulted in a global pandemic lasting for more than 2 years. As of June 2022, this has resulted in a total of 530,896,347 confirmed cases of COVID-19 of which 6,301,020 deaths have been reported to the World Health Organization. Among the common symptoms of COVID-19 is the presence of cough [13].

The sound and type of the cough are said to contain useful features that can contribute to the diagnosis of a disease [1]. Recent studies have utilized feature-based engineering with artificial intelligence (AI) for the classification of different respiratory diseases using cough sounds. This lays out the groundwork for the detection of COVID-19 through cough sounds. This in turn can help in mitigating the spread in the community through a quick and convenient way to pre-screen for the presence of the virus. In the long run, these results can aid healthcare professionals in their decision-making and promote early intervention.

Common models implemented for cough sound detection include Linear Regression (LR), Convolutional Neural Network (CNN), and Support Vector Machine (SVM). These models make up for 25 peer-reviewed articles in the year 2021 [8]. The present study features the training of a Support Vector Machine (SVM), a ResNet, and traditional CNN models based on the works of [2, 5, 11] models, respectively. The aim is to investigate the generalizability of the models across datasets from different sources. Two datasets were obtained from different crowdsourced data collection applications, which include the COVID-19 sound dataset from the University of Cambridge [2] and the audio data from the Coswara project by the Indian Institute of Science (IISc) Bangalore [15]. The models were trained and tested on different datasets to analyze the model's performance.

2 Related Work

Typically, a cough sound can be divided into three phases being the explosive sound that is first heard during the explosive phase which mainly consists of a noise-type waveform. This is followed by an intermediate phase that corresponds to a decrease in airflow and a decrease in sound amplitude. The final voiced or glottal phase corresponds to a partial closure of the glottis that vibrates to produce a regular periodic noise [4]. In patients affected with the disease, prolonged series of explosive coughs have the tendency to occur after a single breath or several breaths interspersed. This is known as cough "epoch", "peals", "bouts", or "attacks" [9]. These phases are reflected in cough recording waveforms as seen in Fig. 1.

Features extracted from cough sound are commonly denoted by Mel-frequency cepstral coefficients (MFCC), Short-Time Fourier Transform (STFT), Log-Mel, and Mel Spectrogram. In a study conducted by Cai and Xu [3], these features were experimented as input features into a CNN model on MATLAB to determine the

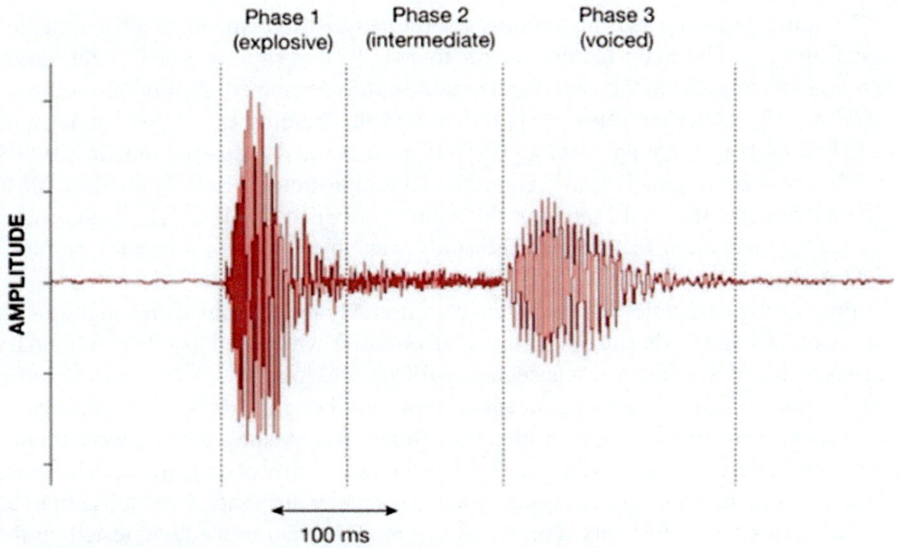

Fig. 1 Cough phases reflected on waveform [10]

best-performing features for cough detection. From the study, it is found that the best features were obtained from the Mel spectrum followed by STFT. From the study, it is seen that the features that were selected were able to obtain accuracies ranging from 89.17 to 92.67% in the detection of cough sounds.

Several studies regarding the classification of pulmonary diseases using AI trained on cough sounds have been conducted in the past. Models such as SVM as well as CNN have been reported to have promising performances in the classification of pulmonary diseases using cough sounds. Sharan et al. [14] conducted a study on the classification of croup cough sounds using an SVM model. The study was able to obtain 91.21% accuracy, 88.37% sensitivity, and 91.59% specificity. Looking into deep learning, [16] studied the use of a CNN model for the classification of pneumonia, bronchial asthma, and chronic pulmonary obstructive disease (COPD). The authors report a 98.59% accuracy and 98.17% sensitivity. In addition, [6] utilized SpecAugment in their study with an attention-based hybrid CNN-LSTM model for COVID-19 diagnosis using cough sound. The study was able to obtain 91.13% AUC.

Since the beginning of the pandemic, several studies have been conducted using cough sounds for the detection of COVID-19. Machine learning models such as SVM have also been utilized in the classification of COVID-19-positive and -negative coughs. Brown et al. [2] studied the classification of COVID-19-positive individuals and healthy individuals (COVID-19-negative with no symptoms, no comorbidities, and non-smokers) using VGGish-based features with onset, duration, and tempo from cough and breath samples combined. The VGGish model extracts log Mel-spectrogram features from the raw audio files. Their method resulted in a training AUC of 0.80.

Looking into deep learning models, [5] trained a ResNet model using augmented spectrograms. The main feature set for this study was the log-STFT spectrogram that was obtained from the cough and breath sounds combined resulting in a training AUC of 0.83. Another study by [11] featured the training of a CNN model with cough sounds originating from COVID-19-positive and -negative individuals. By employing data augmentation techniques including time stretch, pitch shift, shift, trim, and gain to the audio samples, the authors obtained 15 MFCCs from the cough sounds as their main feature set. The study was able to obtain a training result of 0.87 AUC.

From previous studies, it is seen that common AI models trained on cough sounds such as SVM and CNN can obtain high performance when used to detect pulmonary diseases. Moreover, feature sets such as MFCCs and log Mel-spectrograms trained on AI models show promising means to represent cough sounds through previous works. Furthermore, the current literature shows that existing models were trained and tested solely on one dataset source. The generalizability of existing models across datasets from different sources is uncertain. This paper discusses the results obtained from a series of experiments. Three existing models were trained and tested on the Cambridge dataset and the Coswara dataset.

3 Methodology

The overall methodology used in this study can be broken down into four main stages being the data collection stage, the data-preprocessing and feature extraction stage, the model training stage, and the model testing stage. In the first stage, data containing cough and breath audio recordings were collected from various sources. This includes the cough dataset from the University of Cambridge, (Cambridge dataset) as seen in the works of [2], and the audio dataset from the Coswara project (Coswara dataset), as featured in the works of [15]. Table 1 shows a comparison of the two datasets used in this study.

During the data collection stage, the data were prepared for model training and testing. The audio data were first classified based on the symptoms that have been declared by the participants. In this study, training data from the Cambridge dataset

Table 1 A comparison of the dataset used

Features	Cambridge dataset	Coswara dataset
Data collection application	Mobile application & web-based application	Web-based application
Type of respiratory audio	Cough, breathing	Cough heavy
Total number of files	439 files	877 files
Original sampling frequency (Hz)	22,050	48,000

consist of both the cough sound data that was obtained from the web-based application and the mobile application source. There are two main classes, namely, covid_positive and healthy_nosymp. For the covid_positive class, data from this class comprises of users who have declared COVID-19-positive and are from both symptomatic and asymptomatic classes. Conversely, for healthy_nosymp, data from this class are from users who are COVID-19-negative. In addition, the participants in this category have declared no known comorbidities, are non-smokers, and also show no symptoms. The same steps were repeated for the Coswara dataset.

The second stage consists of the data pre-processing and feature extraction stage The data were pre-processed based on the steps taken by the respective authors [2, 5, 11] for each of the models being the SVM Model, the ResNet model, and the CNN model. This method was done for both the Cambridge dataset and the Coswara dataset. The data preprocessing and feature extraction stage is summarized in Fig. 2.

The third stage is the model training stage. For audio management, the Python package *librosa* [12] was used. The audio files were resampled at a uniform sampling rate of 22050 Hz. At this stage, the models were trained and tested on the Cambridge dataset. The models that were trained comprised of an SVM model [2], a ResNet model [5], and a CNN model [11]. Different features were extracted from the audio files making up the feature set used for the model training and testing. Features that were obtained from the models are summarized in Table 2.

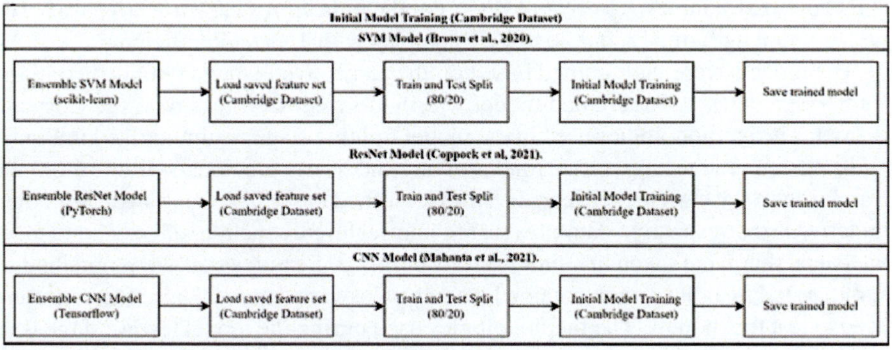

Fig. 2 Data pre-processing steps used in the models

Table 2 Features used in model training

Model	Feature set
SVM model [2]	Transfer learning (VGGish)
ResNet model [5]	Log-STFT spectrogram
CNN model [11]	15 MFCC coefficients

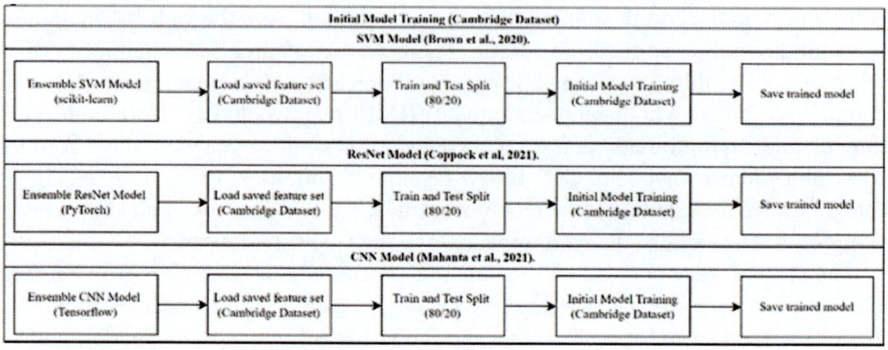

Fig. 3 The methodologies used in the model training stage

Transfer learning features were obtained from a pre-trained model, VGGish. The model extracts features from the raw audio files sampled at 16 kHz. The features obtained from the model are the Log-Mel spectrogram with 64 Mel bins covering the range of 125–7500 Hz. The model divides the data samples into 0.96 s of non-overlapping sub-samples [7]. For each 0.96 s, the model returns a 128-feature dimension. Log-STFT spectrograms were extracted from the audio files by using *librosa.stft*. The hop length was set to 512 samples with a Fast Fourier Transform (FFT) of 1024 samples. To obtain MFCC coefficients, *librosa.feature.mfcc* was used with a hop length of 512 and 2048 FFT samples with 15 Mel-coefficients.

The models were then trained based on the respective features obtained from the Cambridge dataset. The trained model was then saved to be tested on the Coswara dataset. The methodologies used in the model training stage is summarized in Fig. 3.

In the model testing stage, the models that were previously trained on the Cambridge dataset were then tested on the Coswara dataset. The Coswara dataset then underwent the respective data processing and feature extraction stage. The respective steps that were taken are summarized in Fig. 2. Features that were obtained as summarized in Table 2 were obtained from the Coswara dataset. These features were used to test the models. The methodologies used during the model testing stage is as summarized in Fig. 4.

4 Results

The model's generalizability was verified by a series of experiments. The experiments conducted can be divided into two parts, model training (Cambridge dataset) and model testing (Coswara dataset). The models were first trained and tested on the Cambridge dataset. The models were evaluated based on the Area Under the Curve (AUC) obtained from the training process. The results of the training process show

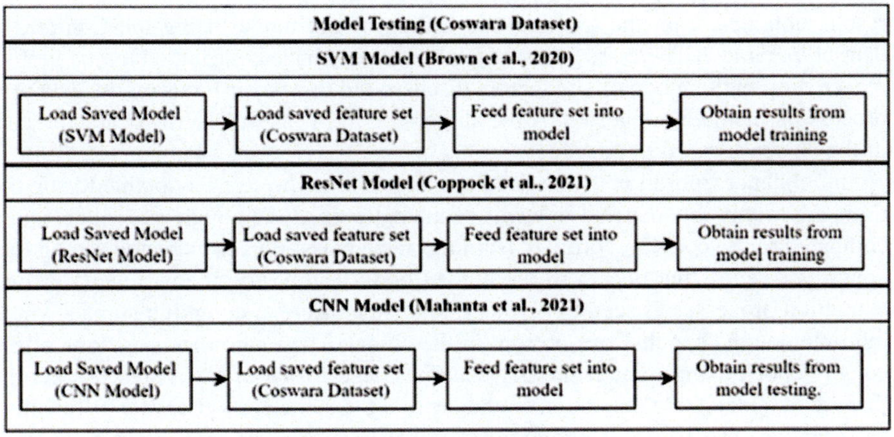

Fig. 4 The methodologies applied during the model testing stage

Table 3 Results obtained from the model training and testing stage

Models	Training dataset (80%)	Testing dataset (20%)	Area Under the Curve (AUC)
SVM model [2]	Cambridge dataset	Cambridge dataset	0.671
	Cambridge dataset	Coswara dataset	0.564
ResNet model [5]	Cambridge dataset	Cambridge dataset	0.738
	Cambridge dataset	Coswara dataset	0.522
CNN model [11]	Cambridge dataset	Cambridge dataset	0.816
	Cambridge dataset	Coswara dataset	0.443

that the CNN model performed the best among all the models with the highest AUC, followed by the ResNet and SVM models.

Further investigations include the testing of the models trained using the Cambridge datasets with the Coswara dataset. Testing results show that the model performs poorly in classifying the covid_positive and healthy_nosymp classes when using the Coswara dataset. This reflects the model's ability to generalize across datasets from different sources. Table 3 summarizes the results from the model training and testing stage.

5 Discussion

The results conclude that the models that were initially trained on the Cambridge dataset are unable to generalize when tested on the Coswara dataset as seen in Table 3. These results show a performance drop in the models when compared to the initial results obtained from the training and testing of the model using the Cambridge dataset in Table 3. There are several factors that may have contributed to this performance drop in the models. The factors discussed in this paper focus on the data preprocessing methods suggested by the authors [2, 5, 11] as well as the characteristics of the databases used in this study.

Initial data exploration of the databases shows that the signals obtained from the users are highly inconsistent in terms of the patterns of the cough recordings when compared to the cough recordings from the same class. This is expected seeing that cough is a reflex that occurs in the human body leading to inconsistency. Feature extraction processes as seen in [2, 5, 11] show the usage of coughing signals where multiple cough "epochs" are present in the signal. The randomize positions of the cough "epochs" throughout the datasets may lead to varying features or patterns that are obtained from the cough sounds within the same class (COVID-19-positive or -negative). This impact is significant when comparing cough recordings from different datasets. A comparison of waveform among datasets shows that the number of "epochs" as well as the duration in between phases are different. Waveforms of COVID-19-positive cough sounds from the Cambridge dataset and from Coswara dataset are shown in Fig. 5a, b, respectively. Figure 5c, show the waveforms obtained from COVID-19-negative users from the Cambridge dataset and from the Coswara dataset.

The data pre-processing methods suggested by [2] deal with the trimming of leading and trailing silences using *librosa.trim*. However, these methods do not normalize the patterns of coughing that may occur within the recording time. By trimming the leading and trailing silences, silences that occur within the coughing signal are not dealt with, leading to an inconsistency in the patterns of the coughing waveform. This in turn causes the feature extraction process to show an inconsistent representation of the cough signals. The effect of *librosa.trim* on the waveforms is illustrated in Fig. 6.

Coppock et al. [5] and Mahanta et al. [11] proposed to normalize the duration of a recording by either zero-padding the audio or trimming it. This was accomplished by setting a fixed duration of the audio samples throughout the dataset. Trimming was done to the audio when it was longer than the value set and zero-padding the audio when it's shorter than the value. The effects of zero-padding allow for silences to exist during the feature extraction process.

The effect of zero-padding audio signals can be seen in the waveforms, as shown in Fig. 7a. On the other hand, trimming cough signals which are longer than the set duration may lead to an abnormal signal pattern. Figure 7b shows the effects of trimming the signal that occurs in the middle of a cough "epoch".

When investigating data augmentations methods suggested by [5], the authors utilized SpecAugment to augment the original spectrograms. This involves the masking of the time and frequency axes of the spectrograms randomly. By applying SpecAugment to the cough spectrogram, it is unclear how much information is lost through the masking of the spectrograms. While investigating the implementation of SpecAugment to coughing spectrograms, it is found that in worst-case scenarios

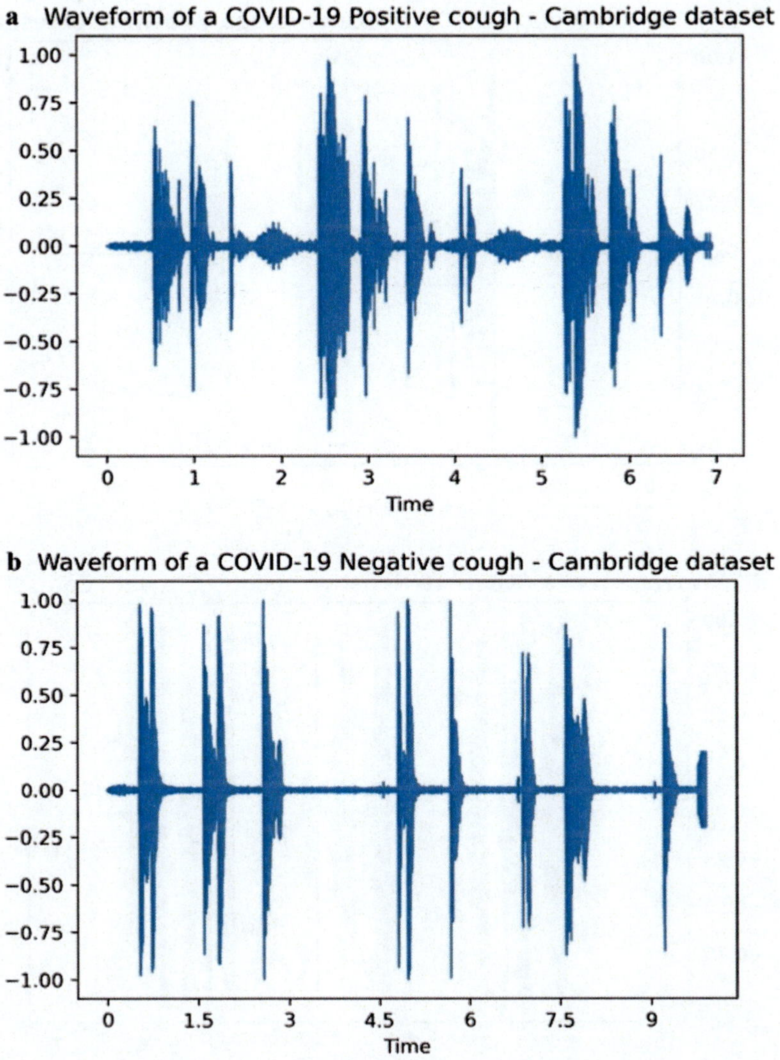

Fig. 5 **a** Waveform of COVID-19-positive user from Cambridge dataset, **b** Waveform of COVID-19-negative user from Cambridge dataset, **c** Waveform of COVID-19-positive user from Coswara dataset, **d** Waveform of COVID-19-negative user from Coswara dataset

where users only have one cough "epoch" in the recordings, the signal is completely masked causing total information loss.

Figure 8a shows the spectrogram after SpecAugment in an ideal situation. Figure 8b shows the impact of SpecAugment on the audio spectrograms masking the coughing signal completely leading to loss of information.

Furthermore, as reported by [5], a fixed window size of eight seconds was used. This meant that for every sample that was longer than 8 seconds, the signal was

c
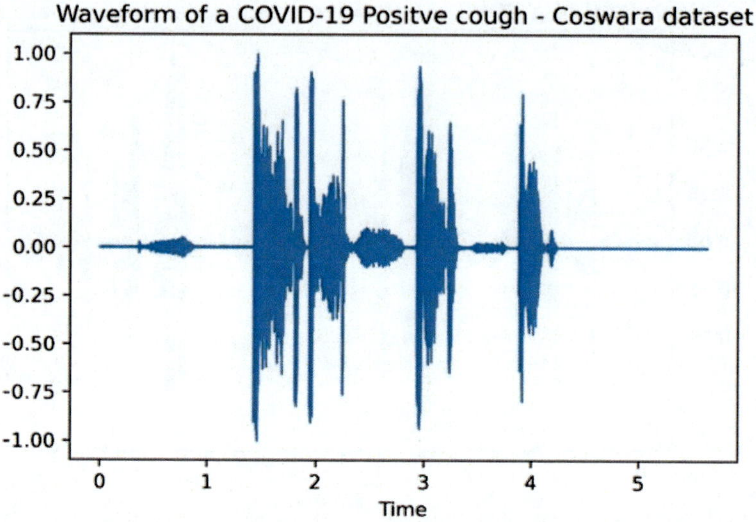

d
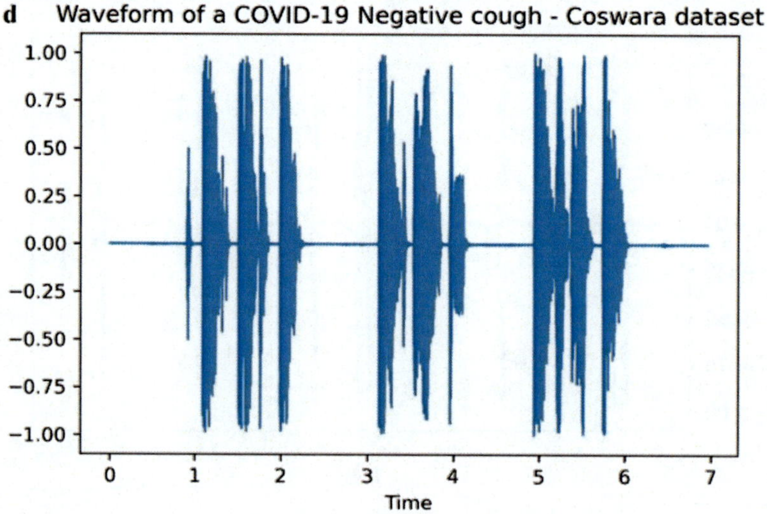

Fig. 5 (continued)

trimmed. The trimmed signal was then augmented with SpecAugment, leading to the masking of the time and frequency axis of the spectrogram. This poses issues with the pattern of the cough as well as information loss due to the masking of the spectrogram. Taking an example audio signal with the original duration of 11.28 s, the effect of trimming causes the removal of the last cough "epoch" as seen in Fig. 9. Moreover, the effect of masking causes more loss of information. The results show that SpecAugment may not be suitable as a method of augmentation for cough audio signals.

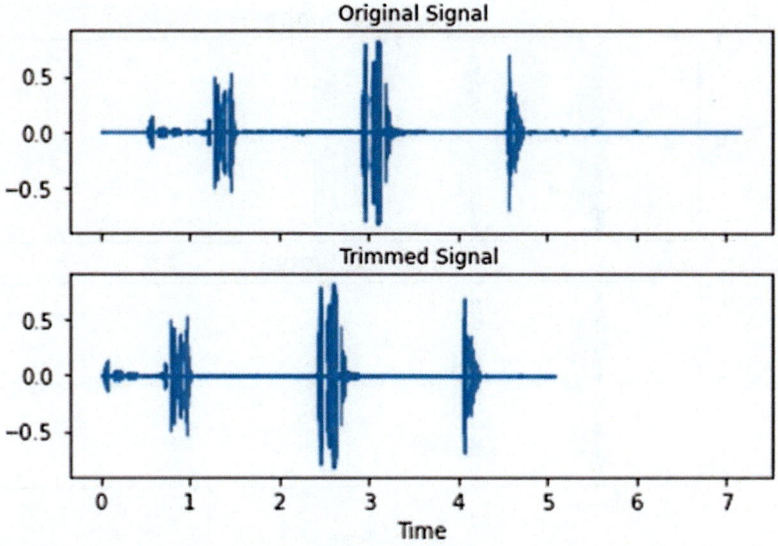

Fig. 6 Effect of trimming the leading and trailing silences

A limitation of this study is the usage of crowdsourced data. The methods by which data is collected to form the Cambridge dataset and the Coswara dataset are through crowdsourcing applications. The statuses of the users (COVID-19-positive or -negative) are not clinically verified. This may cause asymptomatic users to be misclassified as negative. In addition, the use of crowdsourcing applications meant that the cough sound recordings were collected in an uncontrolled environment. The quality of the cough sound recording can be affected by the presence of background noise and the variation of the distance between the microphone and the user during the time of recording.

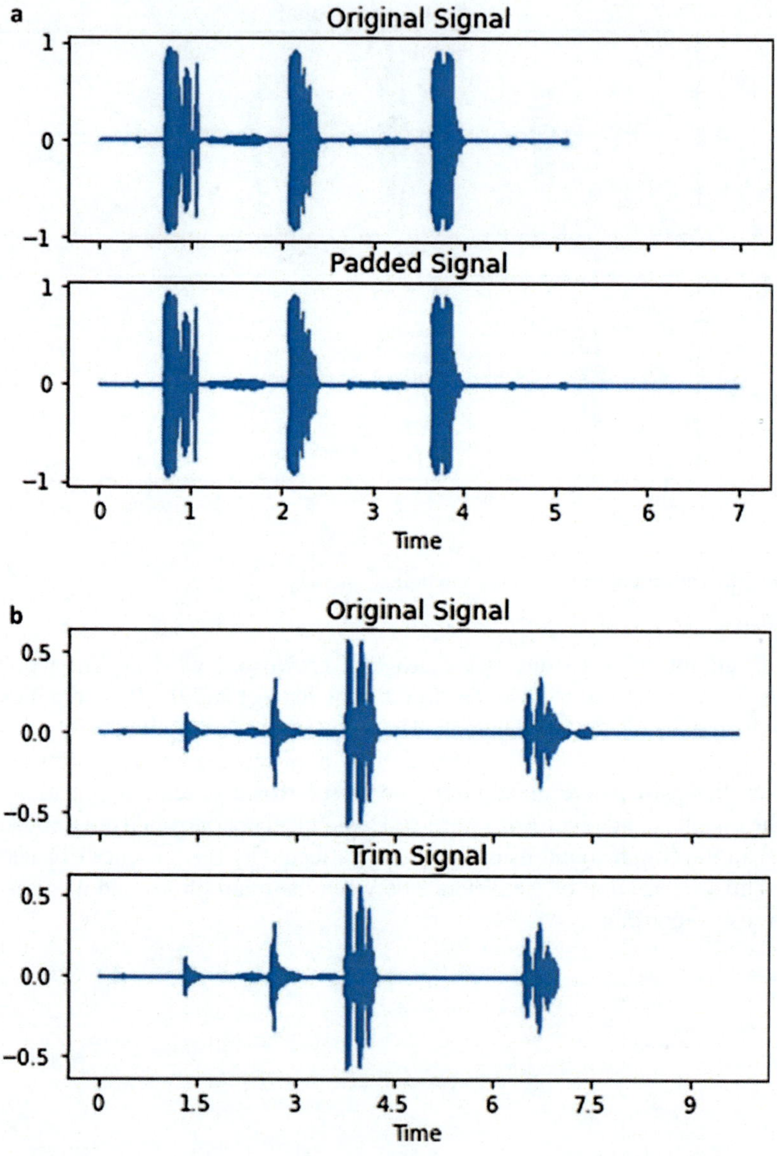

Fig. 7 **a** Effect of zero-padding of audio when fixing duration of audio, **b** Effect of trimming audio due to fixing duration of audio

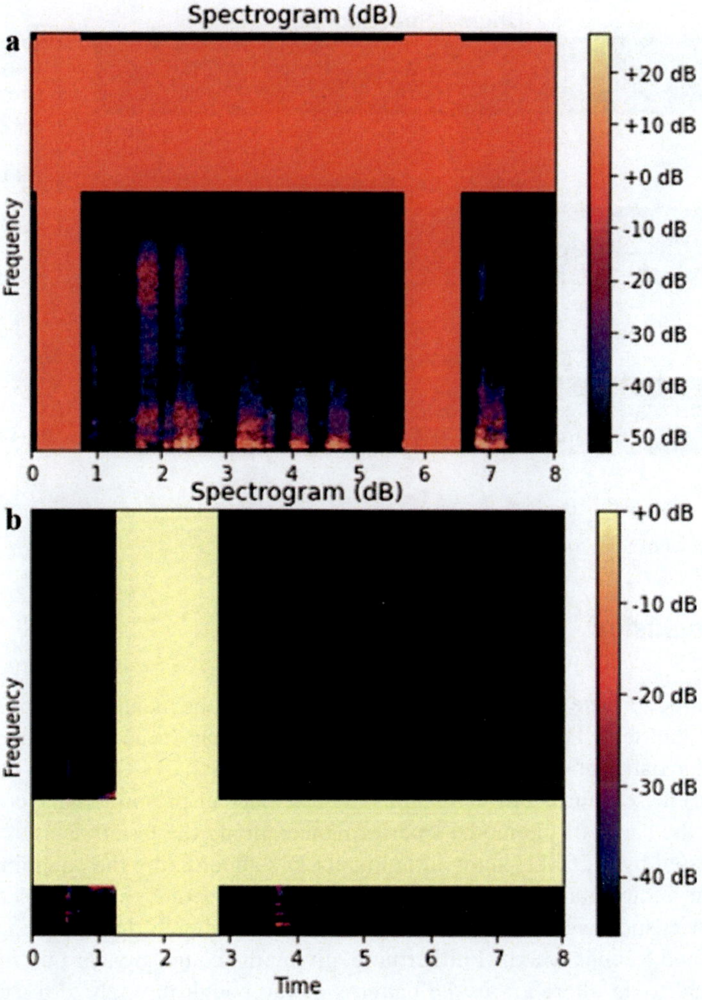

Fig. 8 **a** Effect of SpecAugment in ideal situation, **b** Effect of SpecAugment masking cough signals

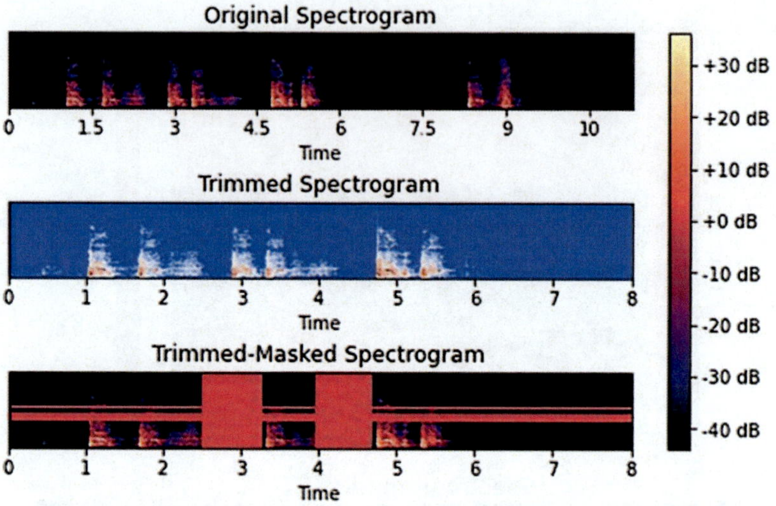

Fig. 9 Effect of SpecAugment and trimming

6 Conclusion

Current models pose a limitation where it suffers from the inability to generalize with different datasets. The model testing results show that the models tested with the Cambridge dataset performed poorly in the classification of COVID-19-positive and -negative classes. Further investigations suggest that data pre-processing could potentially be the cause of the model's performance among the tested datasets. Methods as suggested by [2, 5, 11] show little impact in standardizing the coughing patterns within the audio data. In addition, data augmentation methods such as SpecAugment poses issues such as information loss from the cough signals through the masking of the time and frequency axis. Furthermore, giving the cough signals a uniform length might lead to an alteration in the patterns of the cough through zero-padding and trimming of the audio signal.

Our future work involves the standardization of the feature extraction process that can be done by improving the data pre-processing methods. When considering a cough signal, the focus should lie on the capturing phases of the cough. It is proposed that, the feature extraction should be taken in smaller windows (shorter durations) in an attempt to standardize the coughing signals from person to person. It is hoped that by extracting features this way, the signatures that lie within the phases of the cough can be identified by the AI model when classifying COVID-19-positive and -negative classes.

Acknowledgements The authors would like to thank Swinburne University of Technology, Sarawak Campus, for providing the necessary resources to carry out this study.

References

1. Alqudaihi KS, Aslam N, Khan IU, Almuhaideb AM, Alsunaidi SJ, Ibrahim NMAR, Alhaidari FA, Shaikh FS, Alsenbel YM, Alalharith DM, Alharthi HM, Alghamdi WM, Alshahrani MS (2021) Cough sound detection and diagnosis using artificial intelligence techniques: challenges and opportunities. IEEE Access: Pract Innov, Open Solut 9:102327–102344. https://doi.org/10.1109/ACCESS.2021.3097559
2. Brown C, Chauhan J, Grammenos A, Han J, Hasthanasombat A, Spathis D, Xia T, Cicuta P, Mascolo C (2020) Exploring automatic diagnosis of COVID-19 from crowdsourced respiratory sound data. In: Proceedings of the 26th ACM SIGKDD international conference on knowledge discovery & data mining. Association for Computing Machinery, pp 3474–3484. https://doi.org/10.1145/3394486.3412865
3. Cai Y, Xu W (2021) The best input feature when using convolutional neural network for cough recognition. J Phys: Conf Ser 1865(4):042111. https://doi.org/10.1088/1742-6596/1865/4/042111
4. Chung KF, Mazzone SB (2016) 30—cough. In: Broaddus VC, Mason RJ, Ernst JD et al (eds) Murray and Nadel's textbook of respiratory medicine (Sixth Edition). W.B. Saunders, Philadelphia, pp 497–514.e495. https://doi.org/10.1016/B978-1-4557-3383-5.00030-0
5. Coppock H, Gaskell A, Tzirakis P, Baird A, Jones L, Schuller B (2021) End-to-end convolutional neural network enables COVID-19 detection from breath and cough audio: a pilot study. BMJ Innovations 7(2):356–362. https://doi.org/10.1136/bmjinnov-2021-000668
6. Hamdi S, Oussalah M, Moussaoui A, Saidi M (2022) Attention-based hybrid CNN-LSTM and spectral data augmentation for COVID-19 diagnosis from cough sound. J Intell Inf Syst. https://doi.org/10.1007/s10844-022-00707-7
7. Hershey S, Chaudhuri S, Ellis DP, Gemmeke JF, Jansen A, Moore RC, Plakal M, Platt D, Saurous RA, Seybold B (2017) CNN architectures for large-scale audio classification. In: 2017 IEEE international conference on acoustics, speech and signal processing (icassp). IEEE, pp 131–135
8. Ijaz A, Nabeel M, Masood U, Mahmood T, Hashmi MS, Posokhova I, Rizwan A, Imran A (2022) Towards using cough for respiratory disease diagnosis by leveraging artificial intelligence: a survey. Inform Med Unlocked 29:100832. https://doi.org/10.1016/j.imu.2021.100832
9. Kelsall A, Decalmer S, Webster D, Brown N, McGuinness K, Woodcock A, Smith J (2008) How to quantify coughing: correlations with quality of life in chronic cough. Eur Respir J 32(1):175–179. https://doi.org/10.1183/09031936.00101307
10. Lee KK, Davenport PW, Smith JA, Irwin RS, McGarvey L, Mazzone SB, Birring SS, Abu Dabrh A, Altman KW, Barker AF, Birring SS, Blackhall F, Bolser DC, Brightling C, Chang AB, Davenport P, El Solh AA, Escalante P, Field SK, Fisher D, French CT, Grant C, Harding SM, Harnden A, Hill A, Irwin RS, Iyer V, Kahrilas PJ, Kavanagh J, Keogh KA, Lai K, Lane A, Lim K, Madison JM, Malesker M, McGarvey L, Murad MH, Narasimhan M, Newcombe P, Oppenheimer J, Rubin B, Russell RJ, Ryu JH, Singh S, Smith MP, Tarlo SM, Vertigan AE (2021) Global physiology and pathophysiology of cough: part 1: cough phenomenology—CHEST guideline and expert panel report. Chest 159(1):282–293. https://doi.org/10.1016/j.chest.2020.08.2086
11. Mahanta SK, Kaushiky D, Jainz S, Van Truongx H, Guha K (2021) COVID-19 diagnosis from cough acoustics using ConvNets and data augmentation
12. McFee B, Raffel C, Liang D, Ellis DP, McVicar M, Battenberg E, Nieto O (2015) librosa: audio and music signal analysis in python. In Proceedings of the 14th python in science conference, vol 8. pp 18–25
13. Nitin P, Nandhakumar R, Vidhya B, Rajesh S, Sakunthala A (2022) COVID-19: invasion, pathogenesis and possible cure—a review. J Virol Methods 300. https://doi.org/10.1016/j.jviromet.2021.114434

14. Sharan RV, Abeyratne UR, Swarnkar VR, Porter P (2017) Cough sound analysis for diagnosing croup in pediatric patients using biologically inspired features. In: 2017 39th annual international conference of the IEEE engineering in medicine and biology society (EMBC). IEEE, pp 4578–4581
15. Sharma N, Krishnan P, Kumar R, Ramoji S, Chetupalli S, Nirmala R, Ghosh P, Ganapathy S (2020) Coswara—a database of breathing, cough, and voice sounds for COVID-19 diagnosis. https://doi.org/10.21437/Interspeech.2020-2768
16. Wang H-H, Liu J-M, You M, Li G-Z (2015) Audio signals encoding for cough classification using convolutional neural networks: a comparative study. In: 2015 IEEE international conference on bioinformatics and biomedicine (BIBM). IEEE, pp 442–445
17. WHO Coronavirus (COVID-19) Dashboard. https://covid19.who.int/

Data Analysis of Fly Ash Geopolymer Compressive Strength Using Machine Learning Method

Hissyam Bin Hazmi, Idawati Ismail, Annisa Jamali, and Mohamad Nazim Bin Jamli

Abstract Geopolymer is an alternative material that is suitable to substitute Ordinary Portland Cement (OPC) to produce concrete. A mixture of geopolymer paste that binds coarse and fine aggregate and other unreacted materials together is called geopolymer concrete. Previous studies stated that alkaline activator molarity, water–binder ratio, and type of activator played a significant role in the compressive strength of geopolymer concrete. Machine learning or artificial neural networks are particularly appropriate for modeling nonlinear relationships, and they are characteristically used to accomplish pattern recognition and categorize objects or signals in vision, speech, and control systems. This research is to analyze compressive strength datasets of geopolymer concrete by using the machine learning method. The result comparison of compressive strength is divided into three parameters which are based on molarity, water–binder ratio, and the type of activators in the ratio between sodium hydroxide (NaOH) and sodium silicate (Na_2SiO_3). The materials used for the preparation of geopolymer concrete in this study are fly ash as a binder, fine and coarse aggregates, water, sodium silicate sodium hydroxide, and (NaOH) (Na_2SiO_3) as activators. A total of 240 samples were cast and cured at 80 °C for 24 h with 28 days age of maturity before they have been tested for the compressive strength. This study confirms that the molarity, water-binding ratio, and the type of activator pointedly affect the compressive strength of geopolymer concrete. The compressive strength was further analyzed by MATLAB to observe the neural network and clustering of the compressive strength data. It is found that there are nine clusters discovered. The

H. B. Hazmi · I. Ismail (✉)
Department of Civil Engineering, University Malaysia Sarawak, Kota Samarahan, Sarawak, Malaysia
e-mail: iidawati@unimas.my

A. Jamali
Department of Mechanical Engineering, Universiti Malaysia Sarawak, Kota Samarahan, Sarawak, Malaysia

M. N. B. Jamli
Faculty of Computer Science and Information Technology, Universiti Malaysia Sarawak, Kota Samarahan, Sarawak, Malaysia

C. S. Choo et al. (eds.), *Proceedings of ASEAN-Australian Engineering Congress (AAEC2022)*, Lecture Notes in Electrical Engineering 1072,
https://doi.org/10.1007/978-981-99-5547-3_3

clustering of the compressive strength shows that there is a likeness of material usage in the creation of geopolymer concrete.

Keywords Geopolymer concrete · Fly ash · Sodium hydroxide (NaOH) · Sodium silicate (Na_2SiO_3) · Artificial neural network · Machine learning

1 Introduction

The demand for concrete is very high due to the increased development all around the world. When the concrete becomes the main element of construction, it indirectly raises the higher demand for OPC. However, the production of OPC is responsible for carbon dioxide release that comes from the burning process of the limestone at a high temperature [7]. Geopolymer is used as a binder as an alternative to cement paste to produce concrete as it is believed to have a higher compression strength of 28 days when compared to conventional concrete with the same volumetric material proportions [5]. The microstructure of GPC is denser than normal cement concrete [4]. Besides, geopolymer concrete is gaining momentum as an innovation due to its potential in solving issues related to the environment [2]. Sodium hydroxide (NaOH) and potassium silicate or sodium silicate are the greatest common alkaline liquid used in geopolymerization. The high concentration of NaOH solution surges the geopolymerization reaction. The workability of GPC is subject to the ratio of Na_2SiO_3 to NaOH and the concentration of NaOH.

The molarity of alkaline binder is believed to play a significant role in affecting the compressive strength of the geopolymer concrete mixture. Also, the different types of activators as a binder to the geopolymer concrete contribute to the compressive strengths of the geopolymer concrete. Other than that, the water–binder ratio also influences the characteristics of geopolymer concrete in terms of workability and strength of the concrete. These days, artificial intelligence (AI)-based algorithms were broadly used to solve very complex problems in engineering [1]. An artificial neural network is a system designed similar to a brain of a human that learns by using interrelated neurons or nodes in a layered structure. ANN was accomplished to recognize shapes, categorize data, and estimate future events. As artificial neural networks concede modeling of nonlinear processes, it has turned into a very prevalent and convenient tool for solving many problems such as classification, clustering, pattern recognition, regression, dimension reduction, anomaly detection, structured prediction, decision making, and visualization and machine translation [8]. In the situation of classification and data breakdown, the Self-Organizing Map (SOM) method focuses on the neighborhood structure between clusters. Topology preservation is the correspondence between this clustering and the input nearness [6]. The samples will be evaluated by the machine learning method to evaluate their clustering.

2 Methodology

2.1 Sample Preparation

In the mixing of the geopolymer concrete, the selection of materials is important to make sure the mixture fully complies with the standards and guidelines. The quality of the material is according to ASTM C618 class F classification to ensure the cube specimens give better results in structural response, strengths, and workability. Figure 1 shows the materials needed for the mixing of the geopolymer concrete.

For the purpose of this research, several parameters have been used in order to perform the strength of geopolymer concrete. Firstly, the molarity of the NaOH is 8M and 10M with a water-binding ratio of 0.3 and 0.35. Second is the ratio between

Fig. 1 Cube specimens **a** Geopolymer concrete inside the oven, **b** Out from the oven after 24 h, **c** Demolding of samples, **d** Sample for curing and testing

Table 1 Mix proportion for molarity and w/b ratio

Molarity	8M	8M	10M	10M
Fly ash content (kg)	10	10	10	10
Fine aggregates (kg)	20	20	20	20
Coarse aggregates (kg)	40	40	40	40
Alkaline activator (NaOH) (g)	960	1120	1200	1400
Water-binding ratio	0.3	0.35	0.3	0.35
Water consumption (ml)	3000	3500	3000	3500
Maximum size of aggregates (mm)	20	20	20	20

Table 2 Mix proportion for the type of activator

Ratio between NaOH to Na_2SiO_3		1:0.5	1:1	1:0.5	1:1
Fly ash content (kg)		10	10	10	10
Fine aggregates (kg)		20	20	20	20
Coarse aggregates (kg)		40	40	40	40
Alkaline activator (g)	8M/10M of NaOH	960	960	1200	1200
	Na_2SiO_3	480	960	600	1200
Water-binding ratio		0.3	0.3	0.3	0.3
Water consumption (ml)		3000	3000	3000	3000
Maximum size of aggregates (mm)		20	20	20	20

NaOH and Na_2SiO_3 where the ratio takes 1:0.5 and 1:1 with a 0.3 water-binding ratio. The mix proportion of the geopolymer concretes prepared can be shown in Table 1 as a function of molarity and water/binder (w/b) ratio while Table 2 shows the mix proportion as a function of activator ratios. The ratio of the mixture is 1:2:4 (fly ash: fine aggregates: coarse aggregates). There are 30 samples of cube specimens with sizes of 100 mm × 100 mm × 100 mm for each parameter.

2.2 Compressive Strength Test

After mixing geopolymer concrete, it is then compacted inside the cube mold and dried in the oven for 24 h with a temperature of 80 °C. Then, the mold cooled before demolding and cured at room temperature. The curing process for the geopolymer concrete is based on the normal temperature after drying in the oven. The cube specimens rest for 28 days before being compressed with a 0.5 MPa/s loading rate to get their strengths according to ASTM C618, 2005.

2.3 Neural Network

MATLAB software was used to obtain the second objective of this study, which is to analyze the compressive strength of geopolymer concrete. The process of clustering is related to grouping data by similarity. In order to obtain that, the first step is to create the network with a single click on the software. Then, by using the Self-Organizing Feature Map (SOM), the neuron is organized in a grid form. The second step is to specify the map size, this corresponds to the number of rows and columns in the grid. For this study, the map size values to 30 $\times$ 8. The total number of neurons is equal to the number of points in the grid which in the map shows nine neurons. Next in line is to analyze the result. The weight vector allied with each neuron moves to develop the center of input vectors. The neuron that is adjacent to each other should also move close to each other in the input space, consequently, it is visualized as a high-dimensional input space in the two dimensions of the topology network. The default topology of the SOM is in a hexagonal shape.

3 Result and Discussion

3.1 Compressive Strength

The average compressive strength of eight different mixtures of GPC and its clustering are shown in Fig. 2. From the chart, it shows that the average strength of 1:1 10M NaOH to Na_2SiO_3 with 0.3w/b (24.516 MPa) is higher compared to the other samples which are 1:05 10M NaOH to Na_2SiO_3 with 0.3w/b (16.614 MPa), followed by 10M NaOH with 0.35 w/b (12.375 MPa), next is 8M NaOH with 0.35w/b (12.047 MPa), 10M NaOH with 0.35 w/b (9.920 MPa), 8M with 0.3w/b (7.908 MPa), 1:1 8M NaOH to Na_2Si0_3 with 0.3w/b (4.67 MPa), and lastly 1:0.5 8M NaOH to Na_2Si0_3 with 0.3w/b (3.123 MPa). The strength of the GPC mixture is mostly influenced by the molarity of the alkaline binder in it. It can be proved from the result as the highest strength is the mixture with the highest molarity. Even with the differences between NaOH and Na_2SiO_3, a mixture with higher molarity is always on the top of the list. This is also meaning that the type of alkaline binder is influencing a little strength but it is more to the molarity of the alkaline binder in it. After all, the addition of Na_2SiO_3 did enhance the strength of geopolymer concrete. Consequently, it can be said that the higher the molarity, the highest its compressive strength. It can also be concluded that the strengths of the GPC mixture depend on the amount, type, and combination of the alkaline binder in it. Also, as the assumption proved by the data, the higher the water-binding ratio in geopolymer concrete, the lower its strength.

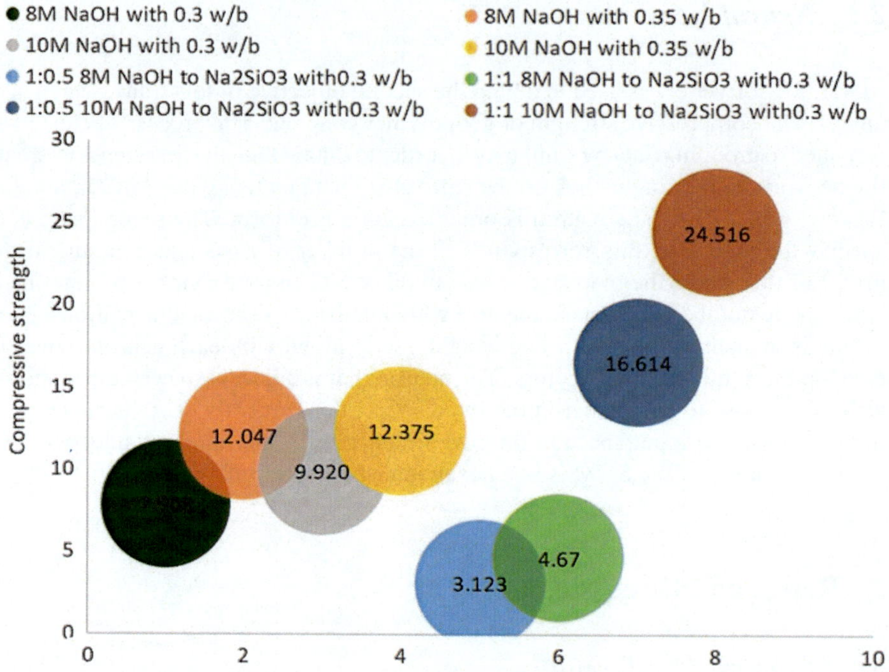

Fig. 2 A cluster of compressive strength range as a function of molarity, water–binder ratio, and type of activator

3.2 Neural Network

Figure 3 shows that most of the compressive strength is located in the same place proving that it is connected to each other. This is due to the mixture of the geopolymer concrete consisting of likely identical materials which contain NaOH and Na_2SiO_3 as the alkaline binder of GPC. This is the initial neuron network for the compressive strength to be clustered. While Fig. 4 shows the whole intensity of similarities for every network connection in this study. There are eight (8) inputs which have been formed based on the eight (8) types of parameters. The lighter color of the structure connection shows the very good similarity of the neuron which means the compressive strength of each parameter of geopolymer concrete has similarity production that contributes to the value of compressive strength. This could be the indication of likely similar material used in the mixture even with different amounts of the material itself. The darker color of the neuron structure shows dissimilarity of the neuron.

From Table 3, it can be presumed that most of the neurons have similarities to each other which can contribute to a cluster. The colors represent the input parameters of the geopolymer mixes. The same color means similarities to each other in terms of compressive strength. It also shows the relationship between input and similarities between each mix. Based on the result of the compressive strength, the neuron is

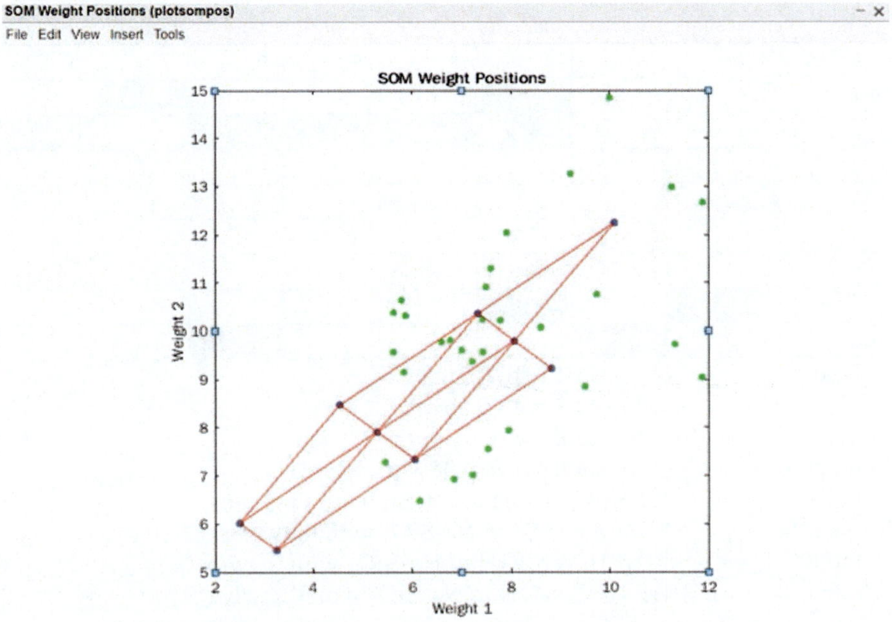

Fig. 3 Neuron network of geopolymer concrete compressive strength

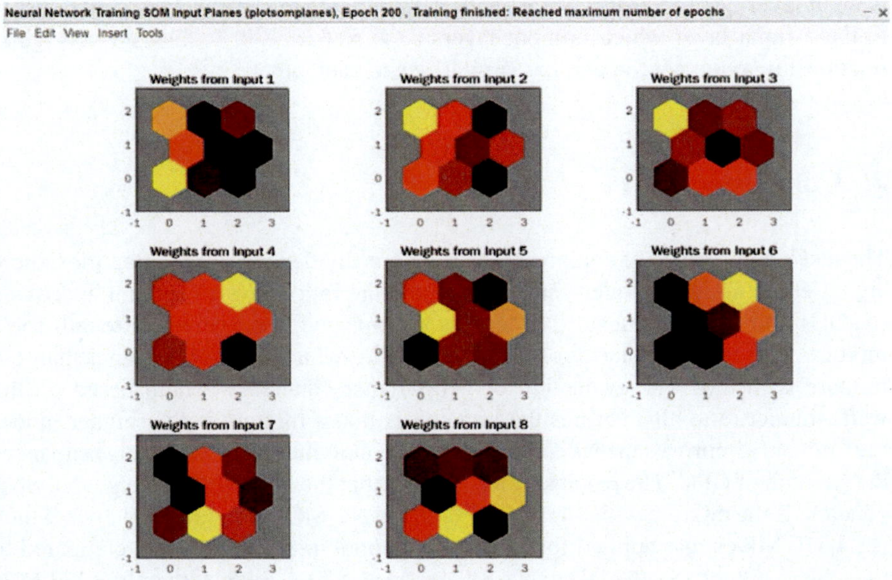

Fig. 4 Neural network input

Table 3 Neuron similarity and dissimilarity

Input	1	2	3	4	5	6	7	8
1								
2								
3								
4								
5								
6								
7								
8								

Input 1	8M NaOH with 0.3w/b
Input 2	8M NaOH with 0.35w/b
Input 3	10M NaOH with 0.3w/b
Input 4	10M NaOH with 0.35w/b
Input 5	1:0.5 8M NaOH to Na_2SiO_3 with 0.3w/b
Input 6	1:1 8M NaOH to Na_2SiO_3 with 0.3w/b
Input 7	1:0.5 10M NaOH to Na_2SiO_3 with 0.3w/b
Input 8	1:1 10M NaOH to Na_2SiO_3 with 0.3w/b

classified and connected to each other. It clarified that the most similar input layer is input layer 2 which is 8M NaOH with 0.35 w/b where it has a strong relationship to the 5-input layer which is input layers 3, 4, 5, 6, and 8. It shows a very good relationship to the neuron and has similarities to each other.

4 Conclusion

The result comparison of compressive strength is divided into three categories, there are based on molarity, water–binder ratio, and the ratio between sodium hydroxide (NaOH) and sodium silicate (Na_2SiO_3). As expected, the concrete strength for a mixture with higher molarity of alkaline mixture which is 10M is higher than the mixture with 8M. The evaluation of geopolymer concrete strength based on the water–binder ratio also verifies the hypothesis that a mixture with a higher binder ratio not only improves the workability of the geopolymer concrete but also enhances the strength of GPC. The result similarly shows that the highest strength geopolymer concrete is the mixture with a 1:1:10M NaOH to Na_2SiO_3 ratio (1:05), 0.3w/b which is 24.516 MPa stress applied for 28 days of geopolymer concrete age compared to the lowest strength 8M NaOH to Na_2SiO_3 ratio (1:0.5), 0.3w/b with only 3.123 MPa stress applied. It can be said that the analysis shows that the ratio of NaOH/Na_2SiO_3 significantly affects the compressive strength of geopolymer concrete. This study also basically applied the result of geopolymer concrete strength on MATLAB to produce neural network. The aim is to analyze the data portrayed in self-organizing map to evaluate its cluster. There are nine clusters discovered. The clustering of the compressive strength shows the likeness of material usage in the production of geopolymer concrete.

5 Recommendations

It is recommended to further study this analysis as it could be the innovation of concrete technology to be in line with the Sustainable Development Goals (SDGs). It is advisable to prepare and test the geopolymer concrete strength for 3, 7, 14, 28, and 60 days to consider the strength and durability of geopolymer concrete over time to determine the types of structure which is the most suitable to use geopolymer concrete to prove the hypothesis that the strength of geopolymer concrete is higher at the early stage. The high early compressive strength can be suitable for emergency construction and repair works. It is also recommended to further study the application of Artificial Neural Networks (ANN) to predict the compressive strength of geopolymer concrete.

Acknowledgements The authors wish to thank Universiti Malaysia Sarawak for research funding through Cross Disciplinary Research Grant (No.: F02/CDRG/1841/2019).

References

1. Chou JH, Ghaboussi J (2001) Genetic algorithm in structural damage detection. Comput Struct 79(14):1335–1353
2. Davidovits J (1994) High-alkali cements for 21st century concretes in concrete technology, past, present and future. Mater Sci 1004
3. Guo X, Shi HS, Dick WA (2010) Compressive strength and microstructural characteristics of class C fly ash geopolymer. Cement Concr Compos 142–147
4. Mehta P, Paulo JMM (2016) Concrete: microstructure, properties, and materials. McGraw Hill Professional, New York City
5. Purwanto, Han AL, Nuroji, Ekaputri JJ (2018) The influence of molarity variations to the mechanical behavior of geopolymer concrete. In: MATEC web of conferences
6. Rousset P, Guinot C, Maillet B (2006) Understanding and reducing variability of SOM neighbourhood structure. Neural Netw 838–846
7. Wang WC, Wang HY (2016) Study on engineering properties of alkali-activated ladl furnace slag geopolymer. Constr Build Mater
8. Zhang YS, Wei S, Li Z (2006) Impact behavior and microstructural chara teristics of PVA fiber reinforced fly ash-geopolymer boards prepared by extrusion technique. J Mater Sci 2787–2794

Feature Reduction of Relational Oil Drilling Data Before Propositionalization and Harmonization by Measuring Relational Data Missingness

Clement Pek Wen Ting and Patrick Hang Hui Then

Abstract Knowledge Discovery in Database (KDD) is a process undertaken by many data scientists who aim to extract knowledge from previously collected, secondary data. KDD is difficult when the scientist must analyze big data without prior due to the amount of knowledge required to understand and select relevant features for propositionalization, which is the conversion of raw relational data (stored in several tables) to propositional data (stored as one table) which is more compatible with established feature selection and data mining techniques. In this paper, we describe the Database Attribute Health Feature Reduction process (DAHFR) which is used as a filter-method feature selection metric for Feature Selection before Propositionalization (FSbP) to reduce the scope of features for observation, analysis, harmonization, and propositionalization before data mining. We employ FSbP using DAHFR during the KDD of multi-sourced oil drilling data which is contained in about 20,000 features in about 700 tables in 1 database per source. Each database follows an identical schema but has varying feature-wide missingness and semantic patterns which makes effective propositionalization difficult. DAHFR is a semi-automated process which allows the user to measure relational missingness of each feature for feature reduction and compare missingness patterns across databases. DAHFR is an improvement to existing feature selection processes because it requires minimum initial knowledge of the database's business rules and could be used with relational data without the need for direct data observation nor prior propositionalization.

Keywords KDD · Feature selection before propositionalization · Missing data · oil drilling data · Data harmonization

C. P. W. Ting (✉) · P. H. H. Then
Centre for Digital Futures, School of Information and Communications Technologies, Faculty of Engineering, Computing and Science, Swinburne University of Technology Sarawak Campus, Kuching, Sarawak, Malaysia
e-mail: cpting@swinburne.edu.my

C. S. Choo et al. (eds.), *Proceedings of ASEAN-Australian Engineering Congress (AAEC2022)*, Lecture Notes in Electrical Engineering 1072,
https://doi.org/10.1007/978-981-99-5547-3_4

1 Introduction

Knowledge Discovery in Database (KDD) is the process for extracting useful knowledge in a database. Its process is broken down to the phases of Selection, Preprocessing, Transformation, Data Mining, and Interpretation [4].

Under regular procedures, the selection stage of KDD is performed by domain experts who know which data and features are relevant to the KDD goals. However, domain experts are not always available. In these cases, the data scientist must gather their own priors to perform the KDD Selection. This research process is difficult, time-consuming, and does not scale well into the domain of big data where the number of features for analysis can reach thousands and the amount of data recorded surpasses one gigabyte.

The KDD Selection process is important for KDD involving big data as the preprocessing and transformation of each individual feature can be resource-intensive; resource would be wasted on underperforming or unusable features otherwise.

KDD Selection of data for oil drilling data is especially difficult without domain and database knowledge. The data's characteristics, simply described as multi-sourced, big data, and extreme sparsity at feature and tabular levels, makes data exploration, observation, and compilation, which are needed to priors during KDD Selection unfeasible.

FSbP refers to the automated feature selection at the KDD Selection phase. A FSbP methodology has previously only been documented once, by Motl [11]. His methodology automates the data compilation and aggregation of relational data for random feature subsets and using the performance of predictive models trained from those data to determine the best features to be used for predictive modeling. Motl's methodology allows the user to perform feature selection on relational data without priors and without the need for the user to directly observe the relational data.

FSbP for oil drilling data (see Fig. 1) must be performed differently due to the scalability and missing data issue which prevents successful, automated joining of cross-table data and direct observation of data. We create two filter feature selection methods: DAHFR and Reports Feature Count Matrix (RFCM) for FSbP which can be performed without needing to join or directly observe the raw relational data. These metrics serve to measure two dimensions of data quality: completeness and relevancy, respectively [13].

It must be reiterated that we differ FSbP from the common definition Feature Selection (FS) which is a preprocessing step during the data mining stage [2, 14].

Feature Selection is a crucial step in the Machine Learning process which selects a subset of relevant features for model construction while improving learning performance and the final model's simplicity and ability to generalize [7].

This ability is especially important with high-dimensionality data which are considered sparse, in which only a small subset of features out of hundreds to tens of thousands are relevant to the prediction label [8, 15].

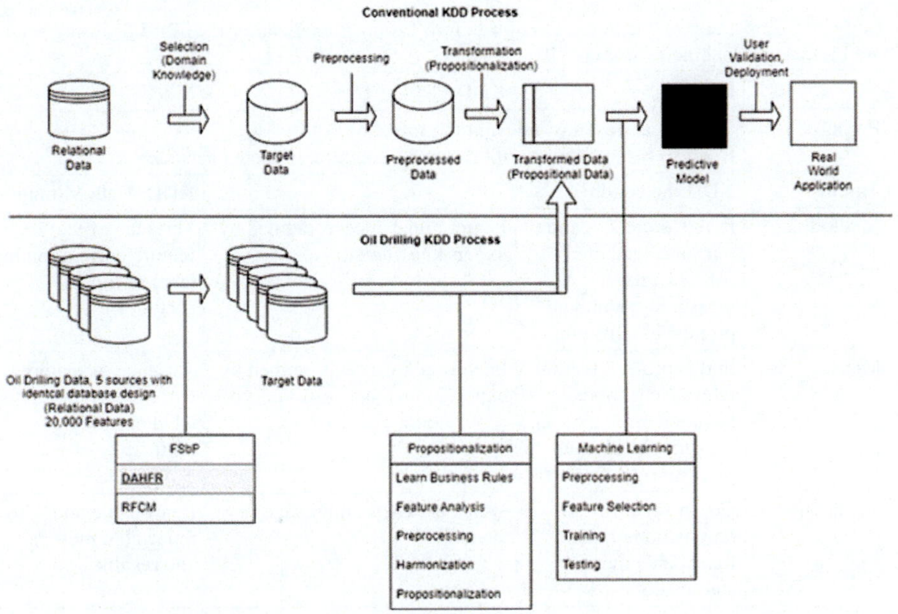

Fig. 1 Comparison of conventional KDD process versus oil drilling KDD

However, FS during ML is reliant on the data being in a propositionalized state, which is not possible during KDD of oil drilling data due to the abovementioned scalability and missing data issue. Hence, the need for a specialized process of FSbP.

A brief comparison of FSbP, Feature Selection during KDD: Selection and Feature Selection during Machine Learning can be seen in Table 1.

This research paper aims to document DAHFR as a FSbP methodology, which is a filter feature selection method to measure, visualize, compare, and select relevant features from a large relational feature set based on their relational missingness.

2 FSbP Via DAHFR

Missing data is often treated as a problem to be solved at the preprocessing stage of Machine Learning. Missing data are usually assumed to be a problem to be solved, a hole that must be filled in (via imputation) or otherwise case-deleted because it reduces the quality of the dataset and because conventional machine learning and statistical analysis not compatible with data with missing value [3].

Oil drilling data is unique in which it is both sparse and heavily missing to the point that relational data in database table forms can't be simply joined just by tracing the database foreign keys to map the table relationships. This problem is compounded by the multi-sourced nature of oil drilling data, where each oil drilling company has

Table 1 Comparison of Feature Selection before Propositionalization (FSbP), FS during KDD, and FS during Machine Learning (ML)

	FSbP	KDD FS	ML FS
Purpose	Select Features most relevant to project objective Reduces project scope with minimal reduction to model accuracy		
Timing	KDD: Selection		KDD: Data Mining
Methodology	In this project, feature relevance gathered from factors measurable without propositionalization	Feature relevance as defined by expert knowledge	Variance of feature, feature correlation to label or model accuracy
Metrics	In this project, feature relevance gathered from factors measurable without propositionalization	Feature relevance as defined by expert knowledge	Variance of feature, feature correlation to label or model accuracy
Advantage	Doesn't require propositionalization Repeatable metrics and results	Quick if expert knowledge is available	Reasonable and repeatable metrics and results
Limitation	Data have not been or cannot be easily propositionalized Efficacy is difficult to measure (Long project duration between KDD: Selection and completion of KDD: Data Mining)	Requires access to expert knowledge of database and business rules. Reliance on existing business rules and may miss new knowledge. Less scalable when multiple experts (one or more for each source) must be consulted Reliance on expert knowledge which may change from expert to expert	Requires standardized, propositionalized dataset

differing missing data patterns due to data entry patterns, business practices, and semantics.

Overall, missing data acts as a sparsity factor in oil drilling data which prevents progress in the Preprocessing and Transformation stage of KDD.

Conventional wisdom from the ML preprocessing stage of dealing with missing data may cause users to ignore features with high missing rates. Yet these data may yet be highly relevant to the final model [10, 14]. Some form of KDD Selection must be performed to reduce the scope for KDD Preprocessing and Transformation, and yet missing data is interfering with Selection.

Features with high missing rates may turn out to be useful for data mining only after further analysis later down the KDD life cycle to determine the cause of missingness, which is more important to imputation than % missingness [1, 9].

And how should a pre-compiled, relational feature's missing rate be measured anyway?

DAHFR is used to solve this issue as a semi-automatic filter-based feature reduction method. As part of its processes, Data Attribute Health (DAH) is measured as a relative missing rate, which gives the data scientist the ability to observe, detect, document, and compare missing data patterns early in KDD. These capabilities are crucial to inform feature selection [10, 14] and imputation methodologies throughout the KDD life cycle which may have to be catered to each specific feature [6, 10, 12].

3 Data

In this paper, we employ DAHFR as one of the FSbP steps during the KDD of multi-sourced, high-dimensionality oil drilling data. The data was collected by a data solutions provider from 5 oil drilling companies over the duration of up to 10 years and was stored in databases of identical base design. Each oil drilling company's data however differ in missing data patterns, data entry patterns, business practices, and semantics.

As such, the ability to observe and compare data between the databases is highly valued because they will be harmonized as part of the KDD process to ensure comparability for the Machine Learning process.

The database pattern employed contains about 20,000 features stored in about 700 tables, with some tables containing up to 200,000 instances of data. Many of the table or table groups store data in irregular-interval time series, as they contain schedular data and therefore they can't be easily joined together.

The data on per-database basis is considered sparse mainly because many tables and features may contain little data, no data, or only test data which could not be identified without DAHFR.

One aspect which makes DAHFR easier for these databases is that the tables are hierarchical in nature, containing foreign keys of all databases above it. For example, the daily table would contain foreign keys for Wellbore_uid, operation_uid, well_uid, etc.simultaneously.

4 Methodology

This study aims identify relevant features for machine learning in a relational high-dimensionality database without preprocessing or compiling the data into a dataset. We used a quantitative research design, in which we analyzed the dataset using a novel filter feature selection method called DAHFR to features' missing rates relative to a baseline feature with the above limitations. The identification of the relative missing rate is significant because the data missingness is preventing data preprocessing and compilation.

4.1 Determining a Baseline Feature

A baseline feature is selected based on four criteria. It must contain unique data. It must be relatable to as many tables in the database as possible (Relatability). It must contain a high number of unique values (Granularity) to ensure a better resolution of DAH score. It must be meaningful such that the DAH result measured relative to it can easily be understood as significant and representative of the whole.

4.2 Relative Missing Rate Measurement

The Database Attribute Health (DAH) formula used to quantify the relative missing rate is as follows:

$$\text{Table Health } H_t = \frac{o_t \cap o_v}{o_v}.$$

$$\text{Column Health } H_C = \frac{c_o}{o_v} / H_t$$

where o_t = count of distinct, valid baseline uids represented by the table.

o_v = count of valid baseline uids in the baseline table.

c_o = count of valid baseline uids represented by the column. A baseline uid is represented when at least one row holding the baseline UID contains data in the column.

Relational missing % = 1 – Health.

Features are scored based on the number of baseline features' unique values which are represented by the feature. Each unique baseline feature value is given identical weight and is considered fully represented (1) if at least 1 occurrence happens, and 0 if none occurs.

Although more complex formulas could be made to represent missing data, e.g., giving a fraction of the weight based on how much missing data each weight has or increasing weights based on how representative the weight is on the whole database, it is decided to keep the formula simple for ease of understanding. A more complex DAH result may cause the user to look past the nuances during feature reduction phase by focusing purely on the DAH % result, instead of handling these nuances further down the KDD process where more information is available to make a more informed decision for feature selection (e.g., Feature Selection during Machine Learning).

Table 2 Filtering threshold for feature selection. Harmonization threshold is used when comparing features shared by several databases

	Column health H_C	Table health H_t
One database threshold	≥60%	>5%
Harmonization threshold	≥60% of databases passes one database threshold	

4.3 Feature Relevance Determination

To identify relevant features, we set a threshold for the following relative missing rate rates (Table 2).

One database threshold is used when only one database is being measured, while harmonization threshold is used when multiple databases which are going to be harmonized will be measured.

4.4 Validation

This methodology is performed early in the KDD lifecycle. Due to the complexity of the future stages, KDD preprocessing, propositionalization, and harmonization, immediate validation could not be performed and would be done using the predictive models created from the dataset, whereby it is expected that features with low missing rates would perform better, but features with higher missing rates are not expected to perform the worst due to missingness mechanism factors.

Validation has not been performed and its results would be published in a future paper.

5 Results and Discussion

During experimentation, it is found over several iterations and analysis of DAH and DAHFR that the combination of Attribute Health and Table Health is a better method to observe missing data for each feature (100% health = 0% relational missing data) (Fig. 2).

If the general relational missing data definition is used (Health = Attribute Health * Table Health), features that are less often used would fail the threshold test used for feature reduction.

But in the sample (Fig. 2), it could be observed that in Database 5, the pairing reveals that the entire table is less often used, and therefore may be caused by Missing at Random (MAR) instead of Missing Completely at Random (MCAR) missing data and therefore the missing data mechanism for the table may require analysis. For this table, it was later determined that these data were validly missing because the

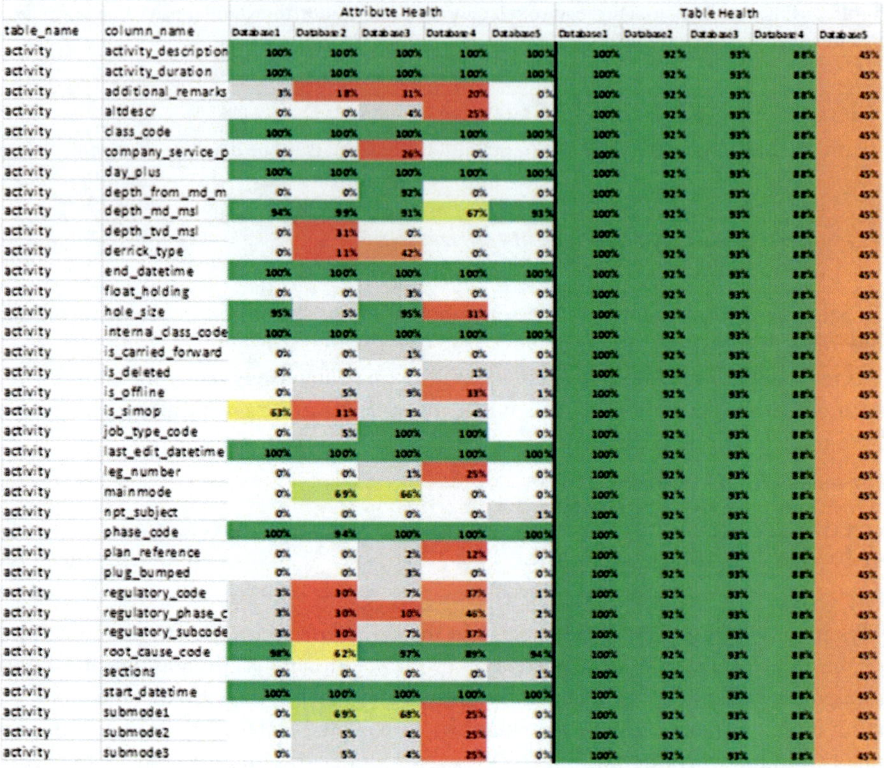

Fig. 2 Visualization of DAH score for some features across five databases

data collected in this table are only available for drilling data (for drilling operations) while other types of operations' data are stored in the database.

Table 3 Number of features in various stages of DAHFR (harmonizing five databases or just Database4)

	Raw	Not including unique identifier features: Harmonize DAHFR $H_C < 5\%$ or $H_t < 5\%$ (unused features)	DB4 DAHFR $H_C < 5\%$ or $H_t < 5\%$ (unused features)	Harmonize DAHFR (Harmonization threshold)	DB4 DAHFR (One database threshold)	Harmonize DAHFR (Harmonization threshold, no foreign keys)
# Tables	~700	121	75	58	51	37
# Features	~20,000	2967	1422	964	1188	707

By using the Attribute Health and Table Health pair, the observer could focus observation the Attribute Health while being aware of the context which is the table-wide MAR data, which shows the quality of data (in terms of missingness) for the features are comparable to other databases.

Although DAH can capture 1 version of MAR, other MARs may be found within the Attribute Health which will require analysis further down the KDD process.

The definition used where DAH score is calculated by the existence of data (true/false) instead of % of data available both obfuscates and simplifies the information but is acceptable for the purpose feature reduction of KDD when no data has been observed and only the worst features are to be removed, which are those revealed and marked as missing.

This purpose of DAHFR of removing only the worst features is also the reason for the generous amount of missingness allowed through, as seen in Table 3 [5]. The 60% column health threshold used for this project assumes that actual missing rate (as defined by regular literature) during propositionalization would be much higher after considering the table health and the method which health is measured (it checks only for 1 row containing data per operation). There are further factors to be considered:

(1) Missing rates are less important to imputation than missing data mechanisms [1, 9].
(2) Literature has not agreed on the best threshold for missing rates (of which, DAH =/= missing rate) and therefore it should be on a project-by-project basis.

Table 3 shows the results of the DAHFR in terms of the number of features reduced. Taking the database at its raw form shows about 20,000 features to be processed. The use of DAHFR to remove the worst features ($H_C < 5\%$ or $H_t < 5\%$) instantly reduces the scope of features to be observed to below 3,000 features.

Despite the generous threshold used in FSbP via DAHFR, we successfully reduce the scope of features for analysis and propositionalization from about 700 tables and 20,000 features to 37 tables and 707 unique features.

Among these tables are a group of tables which are not related to Operation, which are later found to be system tables for supporting the database and accompanying software instead data related to oil drilling.

This scope reduction of features has improved the feasibility of manual, feature-specific analysis, and studying as well as propositionalization attempts by guiding the process to target features that may be more viable for the data mining stage of KDD.

6 Conclusion

The DAHFR process is a great asset to observe and perform feature reduction on big data datasets during FSbP where missing data is of great concern and business and database knowledge is limited while being scalable for multi-sourced data harmonization attempts. After the initial feature reduction process, the DAH score provides the user with knowledge of features which would require more attention in terms of analysis to determine the type and source of data missingness and the best method to process the missing data during the data mining stage. Furthermore, this process scales well with multiple sources in KDD projects where data harmonization of data from multiple sources with similar database designs must be performed because less business knowledge is required to understand the rules and semantics each source in advance. For the DAHFR, an area that is still being explored is the choice of threshold for feature selection. With this paper, the threshold of 60% column health and above is used. Experimentation is actively being performed to determine the effects of various threshold values on the data mining models' prediction accuracy.

Acknowledgements This research was supported by Independent Data Services (IDS) who provided the research grant and monthly stipend for the project as well as Yayasan Sarawak whom through the Sarawak Foundation Tun Taib Scholarship (BYSTT) provided support for the tuition fee and monthly stipend.

References

1. Dong Y, Peng CYJ (2013) Principled missing data methods for researchers. Springerplus 2(1):1–17
2. Doquire G, Verleysen M (2012) Feature selection with missing data using mutual information estimators. Neurocomputing 90:3–11
3. Emmanuel T et al (2021) A survey on missing data in machine learning. J Big Data 8:1–37
4. Fayyad U, Piatetsky-Shapiro G, Smyth P (1996) The KDD process for extracting useful knowledge from volumes of data. Commun ACM 39(11):27–34
5. Lee JH, Huber JC Jr (2021) Evaluation of multiple imputation with large proportions of missing data: how much is too much? Iran J Pub Health 50(7):1372
6. Leyrat C, Carpenter JR, Bailly S, Williamson EJ (2021) Common methods for handling missing data in marginal structural models: what works and why. Am J Epidemiol 190(4):663–672
7. Li J, Cheng K, Wang S, Morstatter F, Trevino RP, Tang J, Liu H (2017) Feature selection: a data perspective. ACM Comput Surv (CSUR) 50(6):1–45
8. Li X, Wang Y, Ruiz R (2020) A survey on sparse learning models for feature selection. IEEE Trans Cybern 52(3):1642–1660
9. Madley-Dowd P, Hughes R, Tilling K, Heron J (2019) The proportion of missing data should not be used to guide decisions on multiple imputation. J Clin Epidemiol 110:63–73
10. Mostafa SM, Eladimy AS, Hamad S, Amano H (2020) CBRG: a novel algorithm for handling missing data using Bayesian ridge regression and feature selection based on gain ratio. IEEE Access. 8:216969–216985
11. Motl J, Kordík P (2018) Do we need to observe features to perform feature selection? ITAT 44–51

12. Perkins NJ, Cole SR, Harel O, Tchetgen Tchetgen EJ, Sun B, Mitchell EM, Schisterman EF (2018) Principled approaches to missing data in epidemiologic studies. Am J Epidemiol 187(3):568–575
13. Pipino LL, Lee YW, Wang RY (2002) Data quality assessment. Commun ACM 45(4):211–218
14. Seijo-Pardo B, Alonso-Betanzos A, Bennett KP, Bolón-Canedo V, Josse J, Saeed M, Guyon I (2019) Biases in feature selection with missing data. Neurocomputing 342:97–112
15. Tan M, Tsang IW, Wang L (2014) Towards ultrahigh dimensional feature selection for big data. J Mach Learn Res

Automation and Sensors

Investigation of the Effect of Polyvinyl Acetate Coating on Multiple U-shaped Fibre Optic Sensors in Different Solution Conditions

Alan Wen Zheng Yek, Alfred Jia Yee Tan, Sing Muk Ng, and Hong Siang Chua

Abstract U-shaped fibre optic sensors (FOSs) have been developed by many researchers in view of the low-cost and miniaturised size. The sensitivity of the FOS towards refractive index (RI) changes is affected by the fibre core diameter, bending radius and numerical aperture of the fibre. However, there are limited resources available to prove that the sensitivity of the sensor can be improved by a coating layer and can perform well in different types of solutions. Therefore, this project proposes the development of three U-shaped probes that can perform well in solutions with normal and viscous conditions. Glycerol and honey solutions were used to represent the normal and viscous conditions, respectively. To increase the sensitivity of the proposed sensor, fibre cladding removal and chemical material coating on the fibres were implemented. From the literature, the polyvinyl acetate (PVAc) coating material was proposed as the fabrication process of PVAc coating was simpler compared to other coating materials such as polyimide and Norland Optical Adhesive 61 (NOA61). By conducting some coating experiments, the coating thicknesses proposed in this project are 10 and 20 μm for 2 different fibre probes through the dripping method while another probe remained uncoated with the cladding removed. All the results are repeatable, and the accuracy of the FOS is 99.49%.

Keywords Coating · Fibre optic sensors · Polyvinyl acetate · Refractive index · U-bent · U-shaped

A. W. Z. Yek · A. J. Y. Tan · H. S. Chua (✉)
Faculty of Engineering, Computing and Science, Swinburne University of Technology, Kuching, Sarawak, Malaysia
e-mail: hschua@swinburne.edu.my

S. M. Ng
Research and Development Department, Sarawak Energy Berhad, 93050 Kuching, Sarawak, Malaysia

C. S. Choo et al. (eds.), *Proceedings of ASEAN-Australian Engineering Congress (AAEC2022)*, Lecture Notes in Electrical Engineering 1072,
https://doi.org/10.1007/978-981-99-5547-3_5

1 Introduction

The introduction of low-loss fibre optics in the 1970s has piqued the interest of academics all over the world [1]. FOS is a non-invasive sensor that does not change the characteristic of the solution. FOSs have unique characteristics compared to conventional sensors such as biological, electrochemical and semiconductor sensors. For example, FOS is a sensor that is miniaturised in size; has electromagnetic interference resistance, multiplexing and remote sensing capability; is elastic, lightweight, the ability to sense various physical disturbances such as acoustic, temperature, strain and pressure; and as an insulator [2]. Several research papers reported that the modification of the fibre optic geometry can improve the sensitivity of the FOS in different solutions or medium conditions [3, 4]. Moreover, there is another strategy to improve the sensitivity of FOS which is through the coating of a certain chemical material on the fibre [5, 6].

However, there are limited resources available to prove that the sensitivity of the sensor can be improved by a coating layer and can perform well in different types of solutions. For example, in a research paper [7], the U-shaped FOS was developed and exposed to different refractive index (RI) and temperature. The sensor did not apply the chemical coating layer on the fibre. Another example is that researchers have only reported that the coating layer of polyvinyl acetate (PVAc) on the fibre can perform well in transformer oil [5]. Therefore, this project proposes the development of three U-shaped probes with PVAc coating that can perform well in solutions with normal and viscous conditions. Glycerol and honey solutions were used to represent the normal and viscous conditions, respectively. Also, this project involves wider coverage for RI compared to the normal FOSs, i.e. RI range from 1.34 to 1.49.

2 Background

2.1 U-shaped FOS

U-shaped FOS is a type of sensor with a U-shaped structure with or without the cladding at the fibre probe. It is highly sensitive, simple to fabricate, sturdy, and highly mobile in the field, which is commonly used with chemical components [8]. Indeed, multiple recent articles have been published on the development of U-shaped FOSs theoretically [9, 10] and experimentally [11–14]. According to the research [4], a simple and low-cost U-shaped FOS was developed to sense a very small concentration change of ethanol. The range of concentration changes is from 0.00633%v/v to 0.0633%v/v per litre based on the initial bio-ethanol production rate. The reason for using the U-shaped FOS is that it has high sensitivity in the low concentration range, which enables the measurement of minute changes in the solution RI. Besides, the sensor can operate in a visibly dense solution due to its sturdiness and is capable of online measurement applications in the industry. The sensitivity of the sensor is 0.6448 O.D/%v/v with 97.14% linearity. Moreover, there are different sensitivities of U-shaped FOSs in different applications [15, 16]. The sensitivities of the FOSs are 0.4%/%RH and 154 dB/RIU, in monitoring relative humidity and RI, respectively.

2.2 *Polyvinyl Acetate (PVAc)*

In the research paper [5], the authors proved that the coating of PVAc can improve the sensing performance of the U-shaped fibre optic in an oily solution. The PVAc coating can also be applied to the straight fibre and tapered fibre. The coating layer can change the RI of fibre optic cladding to a suitable RI range for humidity sensing. After coating, the light penetrating into the coating layer is scattered out when the medium RI is changed. This is because the porous material PVAc with incoherent structure can scatter light [17]. According to the research paper [5], the PVAc that can be dissolved in ethanol alcohol is coated layer by layer on the fibre probe. After that, the thickness of the layer is observed and measured by three dipping cycles. The coating layer is approximately 50 μm. A drying process is required for each dipping cycle. The coated fibre is dried by putting it in the hot air oven. To ensure the coating is fully cured and dried, the coated fibre needs to let set for at least 1 day for each layer.

3 Methodology

3.1 *Equipment and Material Used*

Plastic fibre optics used in the research were produced by Zhongshan Haijia Lighting Technology Co., Ltd. with a core diameter of 980 ± 60 μm and cladding thickness of 10 ± 10 μm. The RI of the core layer and the cladding are 1.49 and 1.41, respectively, and the NA is 0.5. A 3.6 V green LED (WL-TMRW 151053GS03000, Wurth Elektronik) with a dominant wavelength of 522 nm which was used as the light source while an 8 MPs Raspberry Pi Camera Board v2 was used as a detector. The Raspberry Pi 4 Model B (4 GB RAM) was used to interface with the Raspberry PI Camera to extract and analyse the data from the fibre optics. Besides, pure glycerol and honey were used to prepare the solutions with different RI and their RIs were measured using an Atago pocket refractometer. Lastly, PVAc beads with an average Mw of ~140,000 were purchased from Robert Scientific Company Pte Ltd. Ethyl acetate solution was used to remove the cladding of the fibre optics. Next, the ethanol was used to dissolve the PVAc completely and the 10 μL to 100 μL micropipette (Orange Scientific Micropipette) was used to drip the PVAc solution onto the fibre probe. Figure 1a shows the schematic of the experimental setup and

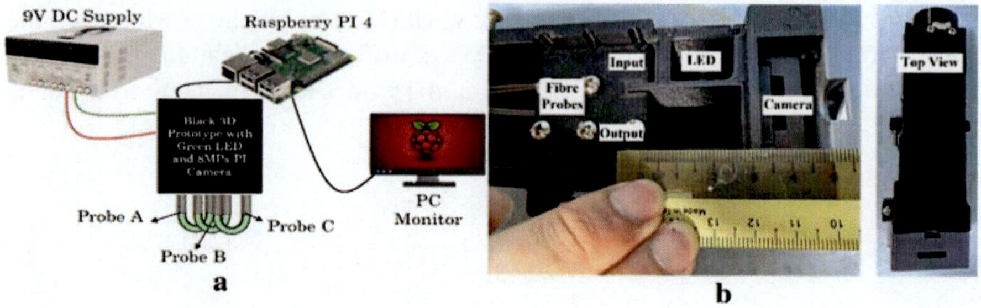

Fig. 1 **a** Schematic of the experimental setup for the three U-shaped fibre probes and **b** photos of the actual sensor prototype

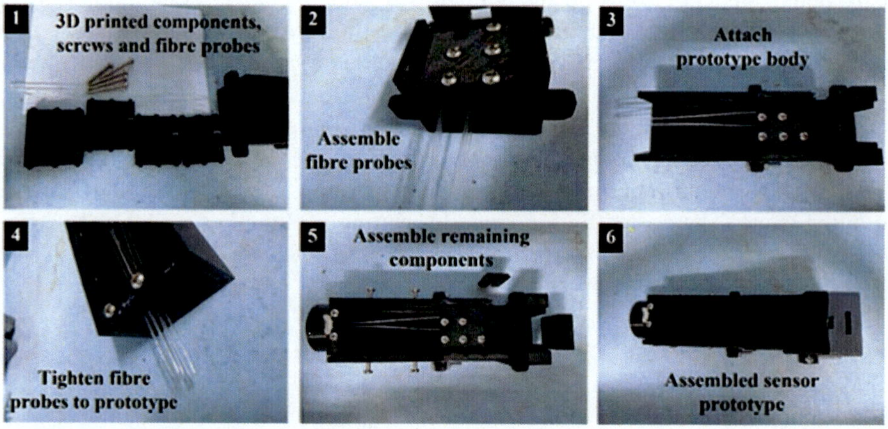

Fig. 2 Sensor prototype assembly phases

the photos of the sensor prototype. Figure 2 shows the sensor prototype assembly phases.

3.2 *Fabrication of Fibre Probe*

The plastic fibres require heat energy to change the structure permanently with the presence of forces [4, 18]. Therefore, constant temperature and bending force were applied to the fibre for the U-shaped fibre probe fabrication. Firstly, a plastic fibre was inserted into the 3D U-shaped mould. The hot air oven temperature was set to 90 °C and waited until the temperature was stable at 90 °C. Next, the U-shaped mould with the fibre was put in the oven for 15 min. After 15 min, the mould with the fibre was removed and placed on the table for cooling purposes. After 5 min, the fibre was cleaned by dipping it into clean water and dried using a tissue. The fibre was then dipped in the ethyl acetate solution for 2 min to remove the cladding. The fibre cannot be exposed to the chemical for a longer period to prevent damage [19]. Lastly, the fibre was cleaned by dipping it into clean water and dried with a tissue. All procedures were repeated to obtain 10 units of fibres. All the optical fibre end was cut by a fibre cutter and polished with emery papers.

3.3 Fibre Coating with Dripping Method Experiment

Before the coating process, the PVAc solution was prepared by pouring 60 mL of 95% ethanol into a beaker. PVAc of 2 g was measured and put into the beaker. The ethanol was stirred for almost 1.5 h until the PVAc dissolved completely. After that, 10 μL PVAc solution was dropped onto the fibre probe by using the micropipette. Lastly, the fibre was dried with cover by putting it in the original position for 10 min (fully dried out).

3.4 Glycerol Solution (RI) Experiment

Distilled water with a volume of 30 mL was poured into a glass vial covered with black tape. The U-shaped fibre probes were submerged in the distilled water, which acts as the reference. Next, 30 mL glycerol solution with RI of 1.34 was poured into another glass vial covered with black tape. The Python program was executed to obtain 20 sets of data for distilled water. The vial with distilled water was then replaced by the vial with the glycerol solution. After that, the Python program was executed continuously to obtain 40 sets of data for the glycerol solution. The procedures were repeated by using the glycerol solutions with RI from 1.35 to 1.39 with steps of 0.01.

3.5 Honey Solution (RI) Experiment

The experimental procedures for collecting the data for honey samples are similar to that of the glycerol solutions. Here, the honey sample RI ranges from 1.40 to 1.49 with steps of 0.01. Therefore, this range of RI represents the solution with higher viscosity or level of stickiness. With 20 sets of data of distilled water collected as the reference solution, another 40 sets of data were collected for the honey samples, which were then normalised for comparison purposes.

3.6 Data Analysis

Different solution conditions have different RI. Therefore, different solution conditions should provide different light intensity levels at the output as the RI affects the light intensity level. In other words, the image captured at the output can be used to monitor any changes in the solution RI.

Figure 3 shows the method of analysing the image captured at the output for three U-shaped fibre probes. The region of interest (ROI) is determined by cropping the image of the fibre outputs. The cropped 8-bit image consists of a total of 40,000 pixels ranging from 0 to 255 values. The pixel values are integrated for each fibre output and the result represents the intensity of the respective fibre. As described earlier, the experimental results were analysed by normalising the data obtained for both the glycerol and honey solutions. With the distilled water acting as the reference, the following simple equation is used for the normalisation:

$$P_{normalised,x} = \frac{P_x}{P_{x(distilledwater)}} \times 100\% \tag{1}$$

where x represents probe A, B and C; P represents the intensity power of the probe and P(x(distilled water)) is the reference intensity power of probe x when submerged in distilled water. After that, the average normalised intensity for each fibre probe

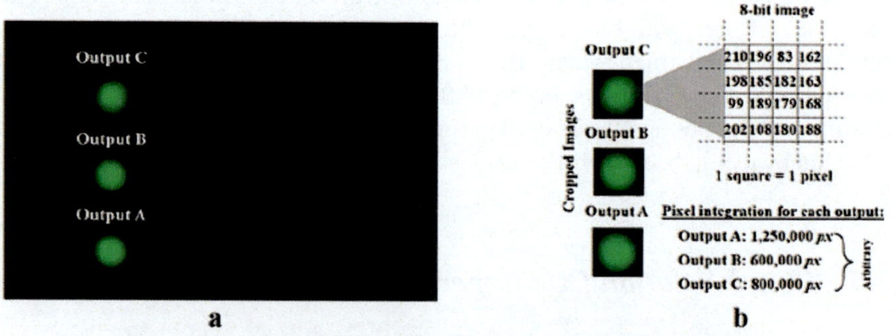

	Output A (px)	Output B (px)	Output C (px)	Solution
Unique Patterns	1,750,000	1,400,000	1,020,000	I
	1,500,000	700,000	800,000	II
	1,490,000	1600,000	700,000	III
	1,200,000	340,000	350,000	IV

c

Fig. 3 **a** Visualisation of the fibre output image, **b** cropped images with the pixel integration for every single output, **c** unique patterns for each solution (example)

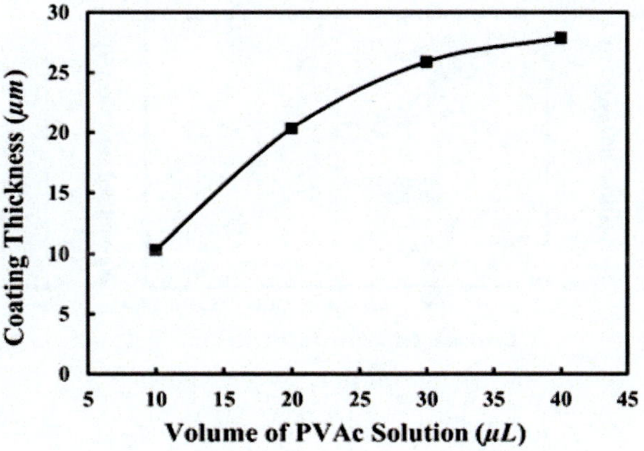

Fig. 4 Depiction of the coating thickness of PVAc with respect to the volume of the PVAc solution dripped onto the U-shaped fibre probe

from different RI values of different solution conditions was plotted as the calibration curves.

4 Result and Discussion

4.1 Fibre Coating with Dripping Method Experiment

The U-shaped fibre was coated with PVAc through the dripping method, where each coating cycle involves a drop of 10 μL of the PVAc solution. For each drip of the solution onto the U-shaped fibre, the thickness of the coating was measured and portrayed in Fig. 4. A slightly linear relationship is observed for the first three drops of the PVAc solution, following which the coating thickness starts to become stagnant.

In practice, it is not necessary to achieve a greater coating thickness as greater thickness leads to a slower sensor response [20]. Hence, the desired range of the coating thickness is within 10 to 30 μm or to be more specific, 15 μm as advised elsewhere [20]. This is because lower coating thickness would expedite the deterioration of the coated PVAc on the probes. In this paper, 10 and 20 μm coating thicknesses are chosen for the two U-shaped fibre probes.

4.2 Glycerol Solution (RI) Experiment

With the two U-shaped fibre probes coated with PVAc and another probe without any coating, the three probes are exposed to the glycerol solutions of different RIs. Figure 5a portrays the RI calibration curve of the three probes in the RI range of 1.34

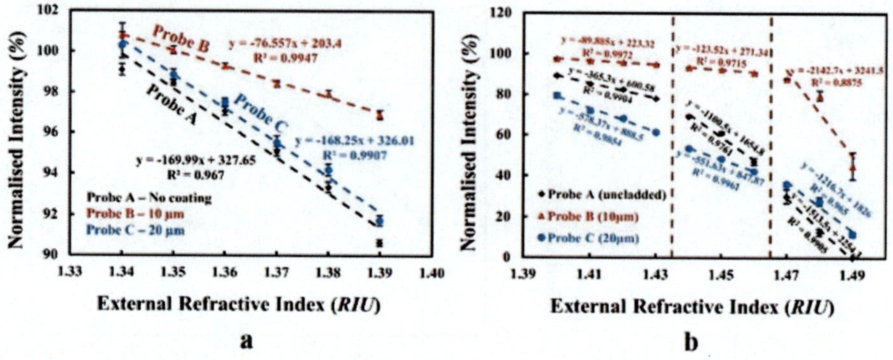

Fig. 5 RI calibration curves of the three U-shaped fibre probes for RI range of **a** 1.34 to 1.39 (glycerol solution) and **b** 1.40 to 1.49 (honey solution)

to 1.39. It is impractical to compare the RI sensitivity of different probes here. This is because different probes are expected to yield different sensitivity as demonstrated elsewhere [21]. Regardless, the two coated fibres demonstrated good linearity in sensing the RI variation for the non-viscous solution. Three fibre probes are used for the purpose of providing a diverse sensing response when subjected to the same sample as portrayed in Fig. 5a. With this setup, the sensing functionality is ensured because as long as one of the probe is functioning, the sensor will still work as usual.

4.3 Honey Solution (RI) Experiment

To investigate the RI ranging from 1.40 to 1.49, honey solution was chosen and prepared with different concentrations accordingly. Figure 5b demonstrates the RI calibration curve in this RI range for the honey solution.

The sensor response, as a whole, was found to be non-linear. This is understandable as the sensor loss at higher RI is too significant to the extent that negligible light intensity is received at the detector side. In this case, the calibration curve is segregated into three regions, which will be used later to verify the sensor's accuracy in determining RI changes.

In Fig. 5b, there are specific instances in which the calibration curves do not cover. The vertical dotted lines are also denoted as the unknown regions, which segregate the three regions, are void of any trendlines that are necessary to estimate the experimental RI given the normalised intensity. To cater for this issue, the trendlines from the two adjacent regions are extrapolated to this unknown region. Since there are two trendlines at this unknown region for each fibre probe, the recorded normalised intensity is substituted into both trendlines, following which an average RI that represents the estimated RI is computed. As such, samples with unknown RI from the range of 1.34 to 1.49 can be used to test the sensor's functionality.

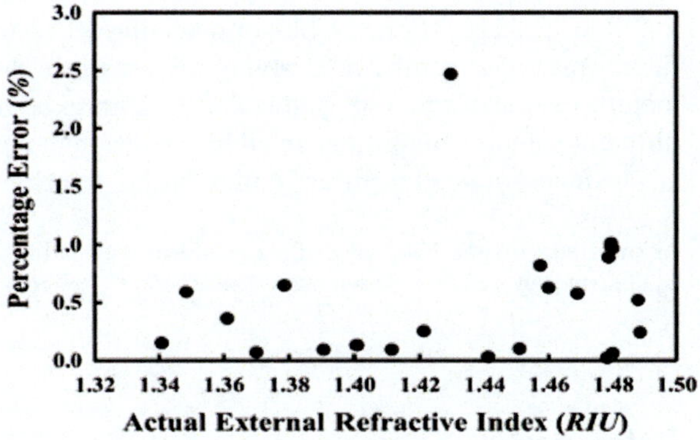

Fig. 6 Percentage error graph to depict the difference between the actual and experimental RI for the 23 samples

4.4 Sensor Prototype Testing Equipment

The sensor prototype was exposed to a total of 23 samples with unknown RI in the range of 1.34 to 1.49. The obtained normalised intensity is substituted into the trendline equations of Fig. 5 to obtain the respective experimental RI. Since there are three fibre probes, there will be three estimated RIs, and these will be averaged and compared with the actual RI of the sample measured by the Atago refractometer. Consequently, the percentage errors between the experimental and actual RI are computed and plotted in Fig. 6 for each sample.

For each sample, the sensor is able to estimate the RI with errors that are within the acceptable range. The highest error around 2.47% belongs to the viscous solution with an actual RI of 1.4298. The increased error may be caused by some errors pertaining to the involvement of human work while conducting the experiment. Regardless, when the errors of the 23 samples are averaged and a value of 0.51% is obtained, which gives an averaged sensor accuracy of 99.49% for detecting RI changes in the range of 1.34 to 1.49.

5 Conclusion

A small fibre optic sensor with multiple U-shaped fibre probes was developed successfully to sense refractive index (RI) variation in the range of 1.34 to 1.49 with two different solution conditions—normal (1.34–1.39) and viscous (1.40–1.49). Polyvinyl acetate (PVAc) was selected as the coating material due to simplicity in terms of the coating procedure. Two U-shaped probes were coated, and another probe remained uncoated. These three probes were submerged in glycerol and honey solutions with different RIs, following which the calibration curves were plotted.

The sensor prototype was tested with 23 samples of unknown RI within the desired RI range and the error bars are within the acceptable range. An averaged sensor accuracy of 99.49% is observed. This paper has demonstrated the feasibility of the multiple U-shaped fibre probes in detecting RI changes from 1.34 to 1.49. In the future, the focus can be in applying the sensor in different applications such as in monitoring transformer oil quality since the fibres demonstrated good performance in different solution conditions. In addition, different coatings or sensor configurations can be incorporated for different fibre probes to achieve multiple parameter sensing.

Acknowledgements Yek, AWZ offers a sincere appreciation for the learning opportunities and lab equipment provided by the Swinburne University of Technology Sarawak Campus.

References

1. Horiguchi M, Osanai H (1976) Spectral losses of low-OH-content optical fibers. Electron Lett 12(12):310–312
2. Tan AJY, Ng SM, Stoddart PR, Chua HS (2020) Theoretical model and design considerations of U-shaped fiber optic sensors: a review. IEEE Sens J 20(24):14578–14589
3. Chandani SM, Jaeger NAF (2005) Fiber-optic temperature sensor using evanescent fields in D fibers. IEEE Photonics Technol Lett 17(12):2706–2708
4. Memon SF, Lewis E, Ali MM, Pembroke JT, Chowdhry BS (2017) U-bend evanescent wave plastic optical fibre sensor for minute level concentration detection of ethanol corresponding to biofuel production rate
5. Rodriguez-Rodriguez JH, Martinez-Pinon F, Alvarez-Chavez JA, Jaramillo-Vigueras D (2008) Polymer optical fiber moisture sensor based on evanescent-wave scattering to measure humidity in oil-paper insulation in electrical apparatus. In Proceedings of IEEE Sensors, pp 1052–1055
6. Jiang J, Zhong L, Duan X, Song M, Li L, Wu Z, Li Z, Huang Z, Hong Z, Jin Z, Yao L (2019) Bend limitation of a polyimide-coated optical fiber at cryogenic temperature of 77 K. IEEE Trans Appl Supercond 29(2)
7. Jing N, Teng C, Yu F, Wang G, Zheng J (2016) Temperature dependence of a refractive index sensor based on side-polished macrobending plastic optical fiber. IEEE Sens J 16(2):355–358
8. Tan AJY, Ng SM, Stoddart PR, Chua HS (2021) Trends and applications of U-shaped fiber optic sensors: a review. IEEE Sens J 21(1):120–131
9. Danny CG, Danny Raj M, Sai VVR (2020) Investigating the refractive index sensitivity of U-bent fiber optic sensors using ray optics. J Light Technol 38(6):1580–1588
10. Tan AJY, Ng SM, Stoddart PR, Chua HS (2022) Dual U-shaped fibers refractometer with enhanced sensitivity based on the coupling effect. Opt Fiber Technol 71:102935
11. Wang Q, Du N, Zhao W, Wang L, Cong X, Zhu A, Qiu F, Zhang K (2022) Highly sensitive U-shaped optical fiber surface Plasmon resonance. IEEE Trans Instrum Meas 71
12. Filoteo-Razo J, Vera-Duarte O, Estudillo-Ayala J, Hernandez-Garcia J, Jauregui-Vazquez D, Martinez-Angulo J, Elizondo-Leal J, Rojas-Laguna R (2021) U-shaped plastic fiber optic sensor for measuring adulteration in liquids via RGB color changes. IEEE Sensors Lett 5(12)
13. Feng J, Yang L, Wang L, Chen J, Xue P, Liu Z, Liu J, Jiang Q, Shang H, Ti G (2022) Sensing performance of U-shaped fiber in chaotic correlation fiber loop ring down system. Opt Commun 511
14. Mohd Arif NAA, Ehsan AA (2021) Sensitivity optimization of U-shaped fiber optics based on the Taguchi method. OSA Contin 4(7):2024–2035
15. Chen GY, Wu X, Kang YQ, Yu L, Monro TM, Lancaster DG, Liu X, Xu H (2017) Ultra-fast hygrometer based on U-shaped optical microfiber with nanoporous polyelectrolyte coating. Sci Rep 7(1):1–7
16. Jing N, Zheng J, Zhao X, Teng C (2015) Refractive index sensing based on a side-polished macrobending plastic optical fiber. IEEE Sens J 15(5):2898–2901

17. Xu L, Fanguy JC, Soni K, Tao S (2004) Optical fiber humidity sensor based on evanescent-wave scattering. Opt Lett 29(11):1191–1193
18. He Y, Li Y, Li N (2017) Temperature-independent evanescent wave sensor made of a stress-released silica optical fiber taper. Opt Fiber Technol 36:237–244
19. Gowri A, Sai VVR (2016) Development of LSPR based U-bent plastic optical fiber sensors. Sensors Actuators B Chem 230:536–543
20. Pavan Kumar GLG, Ahmad MH, Piah MAM, Mohd Noor MY, Nawawi Z, Sidik MAB, Jambak MI (2019) Modelling of polymer coated-based optical fiber sensor for moisture measurement in transformer oil. In ICECOS 2019 - 3rd International Conference on Electrical Engineering and Computer Science, Proceeding, pp 45–50
21. Tan AJY, Ung KS, Ng SM, Stoddart PR, Chua HS (2021) Refractive index, temperature, and heat source origin sensing with dual U-shaped fiber probes. In 2021 IEEE International Conference on Sensors and Nanotechnology (SENNANO), pp 29–32

Temperature Modulation of Metal Oxide Semiconductor Gas Sensors for Black Pepper Geo-Tracing

Yen Jin Pui, Hui En Lee, Zehnder J. A. Mercer, Sing Muk Ng, and Hong Siang Chua

Abstract It is essential to determine whether the food produced for commercialization is genuine and of good quality to protect consumer rights. Black pepper is one of the most common spices and its quality and price differ based on geographical origin (geo-origin). This paper reports the employment of temperature modulation in an array of metal oxide semiconductor (MOS) gas sensors to improve their accuracy and robustness in profiling black pepper volatile organic compounds (VOCs) by geographical origin. The sensor temperature was adjusted by varying the sensor heater voltage (3, 4 and 5 V). It was observed that the same black pepper sample showed varying MOS gas sensor responses magnitude and ratio for different sensor heater voltages. Employing principal component analysis (PCA) demonstrated different sample PCA scores distribution when sensor heater voltage was varied, which was further verified by varying the classification accuracy from classification models training. The result in this preliminary study had proven that variation of MOS gas sensor responses pattern through temperature modulation could lead to increase in black pepper sample data dimensionality, hence better geo-tracing result.

Keywords Black pepper · Electronic nose · Food analysis · Gas sensors · Headspace sampling · Machine learning · Metal oxide semiconductor · Principal component analysis · Volatile organic compound

---

Y. J. Pui · H. E. Lee · H. S. Chua (✉)
Faculty of Engineering, Computing and Science, Swinburne University of Technology, Kuching, Sarawak, Malaysia
e-mail: hschua@swinburne.edu.my

Z. J. A. Mercer
Malaysian Pepper Board, Tanah Putih, Lot 1115, Jalan Utama, , 93916 Kuching, Sarawak, Malaysia

S. M. Ng
Research and Development Department, Sarawak Energy Berhad, 93050 Kuching, Sarawak, Malaysia

C. S. Choo et al. (eds.), *Proceedings of ASEAN-Australian Engineering Congress (AAEC2022)*, Lecture Notes in Electrical Engineering 1072,
https://doi.org/10.1007/978-981-99-5547-3_6

1 Introduction

The importance of food in sustaining humans' lives emphasizes the importance of analyzing food quality. In this case, human's sense of smell lacks consistency and provides very little information regarding food quality compared to metal oxide semiconductor (MOS) gas sensors. A study proved that using a metal oxide sensor to determine compound thresholds in wine is superior to a human sensory panel [1]. On the other hand, a study used MOS gas sensors to detect odor markers in olive fruits infested with olive scale, indicating whether olive oil originated from healthy olive fruits or infested olive fruits [2]. Another notable use of MOS gas sensors was their use for detecting fish degradation [3]. The roles of MOS gas sensors in various applications of food analysis inspired us to conduct research on the capability of MOS gas sensors in geo-tracing black pepper as the quality and price of black pepper are heavily determined by their country of origin [4].

2 Literature Review

Prior to this research, a background study was made to develop understanding related on the background of black peppercorns and existing research related to the use of MOS gas sensors for food analysis.

Piper Nigrum, or black pepper, is a well-known spice used globally in various cuisines. According to the International Pepper Community (IPC), as of May 26th, 2022, Malaysia Black Pepper—Kuching ASTA—fetched the highest price of USD 5900 per metric ton compared to other producing countries—Brazil, India, Vietnam, and Indonesia—that fetched a price range of USD 3800 to 7000 per metric ton. In addition, Malaysia is one of the three countries which produces black pepper that is recognized as high quality by the American Spice Trader's Association (ASTA) [4].

However, the Covid-19 pandemic has caused black pepper prices to skyrocket due to substantial decline in production [5]. The rise in black pepper prices magnifies the importance of accurately geo-tracing black pepper products on the market to prevent skepticism of their quality, more importantly to prevent adulteration of Malaysia's superior black pepper by those of inferior quality. Previous research work by [6] demonstrated the success of geo-tracing Malaysia, India and Vietnam black pepper using an array of MOS gas sensors.

Trirongjitmoah et al. [7] reported an attempt of classifying garlic cultivars using an electronic nose consisting of an array of MOS gas sensors. A two-dimensional principal component analysis (PCA) was used to classify the sensor response patterns by garlic cultivar. The classification result obtained from the electronic nose was verified through cluster analysis on the genetic links between garlic cultivars using amplified fragment length polymorphism (AFLP) markers and quantification of sulfur-containing chemicals in each garlic cultivar using gas chromatography-mass spectrometry (GC–MS). Four Thai garlic cultivars were characterized and divided into three groups using the electronic nose, AFLP and GC–MS methods. This suggested that electronic nose can classify garlic cultivars accurately and rapidly.

Kiani et al. [8] conducted a study to classify eleven types of saffron samples based on odor profile. A portable electric nose system, consisting of an array of MOS gas sensors was developed to detect the volatile organic compounds (VOCs) of saffron. By employing principal component analysis (PCA), hierarchical clustering analysis (HCA), multilayer perceptron (MLP) neural networks and partial least squares (PLS) on the sensor responses, the saffron samples were successfully classified by eleven types. The study had demonstrated the capabilities of MOS gas sensors in classifying saffron of different qualities.

As MOS gas sensors' operation requires heating the sensing element to excite the sensing material, studies were conducted to examine the effect of sensor temperature on their performance. Varying sensor temperature, also known as temperature modulation, had been proven to modify the sensitivity and selectivity of MOS gas sensors, therefore exhibit varying sensor responses towards the same sample. Variation in sensor responses led to increased dimensionality of MOS gas sensors response data, hence more accurate classification of samples [9–11].

Wen et al. [12] proposed temperature modulation by applying low frequency triangular heater voltage to improve the gas sensing robustness of MOS gas sensors. In their study, three types of gas samples—methanol, ethanol and methanol-ethanol mixture—were prepared to compare the gas sensing accuracy between a two-sensor array utilizing temperature modulation and a six-sensor array operating at constant temperature. The result demonstrated that temperature modulation on the two-sensor array (97.7% sensing accuracy) outperformed the six-sensor array operating at a constant temperature (94.4% sensing accuracy).

Therefore, temperature modulation can be employed to improve sensor data dimensionality, hence detection accuracy and robustness of MOS gas sensors. Temperature modulation can also minimize the number of sensors required in an array, leading to smaller hardware footprint and lower cost.

The work presented in this paper reports an early attempt to determine whether temperature modulation of MOS gas sensors can result in varying sensor responses towards India, Indonesia, Malaysia and Vietnam black pepper.

3 Methodology

Each black pepper sample used in this study was prepared as a batch of fifty whole black pepper berries. No sample pre-processing was done on the black pepper samples. Black pepper samples from four producing countries—India, Indonesia, Malaysia and Vietnam—were investigated. Five black pepper samples were prepared for each producing country, therefore 20 sets of sample data in total were analyzed and compared.

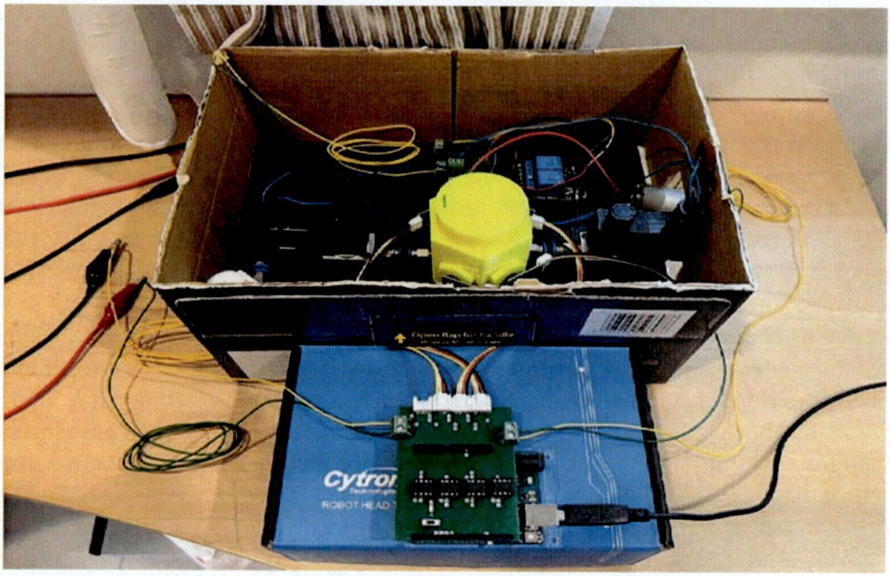

Fig. 1 Physical appearance of assembled e-nose, consisting of an array of three MOS gas sensors—TGS2600, TGS2602 and TGS2620

The black pepper sample data was generated through headspace VOC sampling of the fifty-berry sample using an assembled electronic nose (e-nose), comprising of an array of MOS gas sensors as shown in Fig. 1.

The e-nose has the same architecture and sampling procedure as the system assembled by [6] in their studies. To demonstrate that temperature modulation could improve MOS gas sensor responses and minimize sensor count, the number of MOS gas sensors used in the e-nose was minimized to three only, consisting of the Figaro TGS2600, TGS2602 and TGS2620 MOS gas sensors.

Using the assembled e-nose, headspace VOC sampling was conducted for 300 s for each sample so that the sensor responses could reach a steady state. The sample data consisted of three variables, which were the steady state sensor responses from the three MOS gas sensors used. Steady state sensor response of each gas sensor was presented as the percentage change in sensor output voltage at steady state compared to its baseline [13]. Temperature modulation of MOS gas sensors was employed by varying the sensor heater voltage to represent different temperature levels. As a preliminary work, black pepper sampling was repeated for three sensor heater voltage levels –3, 4 and 5 V respectively.

The sensor responses data obtained for each sensor heater voltage level was processed by employing PCA for dimension reduction, followed by classification models training, which is a supervised machine learning method. 22 classification models were used in this study, which consisted of models from decision trees, discriminant analysis, naïve bayes, support vector machines (SVM), k-nearest neighbor (KNN) and ensemble classifiers category. The distribution of sample data on PCA score scatter plots and samples classification accuracy for each sensor heater voltage level was compared to demonstrate how changing temperature affects MOS gas sensor responses on the same black pepper samples. In addition to that, it was evaluated whether sample data obtained at different heater voltage level could be concatenated in future work to increase sample data dimensionality for more robust and accurate black pepper geo-tracing.

4 Results and Discussion

Table 1 shows the mean and standard deviation of MOS gas sensor responses on each geo-origin's (India, Indonesia, Malaysia and Vietnam) black pepper samples for sensor heater voltage of 3, 4 and 5 V respectively.

In general, the magnitude of MOS gas sensor responses on black pepper samples increased as the sensor heater voltage increased. This showed that the responsiveness of MOS gas sensors increased as temperature increased. The only exception was observed for Vietnam black pepper samples, where TGS2600 and TGS2620 showed decrease in sensor response magnitude when sensor heater voltage was increased from 3 to 5 V. Table 1 also shows that the sensor responses ratio of TGS2600, TGS2602 and TGS2620 of each geo-origin's black pepper samples changed when a different sensor heater voltage was applied. This demonstrated that temperature modulation altered the MOS gas sensors' selectivity, resulting in different sensor responses pattern for the same black pepper samples, and such pattern variation

Table 1 MOS gas sensor responses on black pepper samples for different sensor heater voltages

Sensor heater voltage	Black pepper sample geo-origin	MOS gas sensor responses (%ΔV) Mean $\pm$ Standard deviation		
		TGS2600	TGS2602	TGS2620
3V	India	1.125 $\pm$ 0.445	2.634 $\pm$ 0.552	1.321 $\pm$ 0.479
	Indonesia	1.809 $\pm$ 0.446	3.799 $\pm$ 2.442	2.444 $\pm$ 0.245
	Malaysia	2.302 $\pm$ 1.377	4.583 $\pm$ 1.595	3.184 $\pm$ 1.437
	Vietnam	1.568 $\pm$ 0.616	1.660 $\pm$ 0.637	1.220 $\pm$ 0.734
4 V	India	2.776 $\pm$ 1.158	4.310 $\pm$ 0.824	3.634 $\pm$ 0.924
	Indonesia	3.860 $\pm$ 0.692	5.236 $\pm$ 3.780	4.136 $\pm$ 0.344
	Malaysia	4.191 $\pm$ 1.420	6.473 $\pm$ 0.888	4.632 $\pm$ 1.357
	Vietnam	3.130 $\pm$ 0.976	2.590 $\pm$ 0.770	2.874 $\pm$ 1.736
5 V	India	3.629 $\pm$ 1.131	15.375 $\pm$ 1.850	4.492 $\pm$ 0.635
	Indonesia	4.548 $\pm$ 0.759	16.962 $\pm$ 3.438	6.942 $\pm$ 0.579
	Malaysia	6.497 $\pm$ 3.005	21.205 $\pm$ 3.739	9.506 $\pm$ 4.637
	Vietnam	1.954 $\pm$ 0.863	10.732 $\pm$ 3.231	2.845 $\pm$ 1.235

could be used to improve distinguishability between black pepper samples of different geo-origins.

After employing PCA on the sensor responses data, the variation in sensor responses pattern due to temperature modulation was visualized by the change in the samples' PCA score distribution on the principal component (PC) scatter plots—PC2 versus PC1 and PC3 versus PC1 graphs—as shown in Figs. 2–5.

Three-dimensional training of 22 classification models on PC1, PC2 and PC3 scores for 3, 4 and 5 V sensor heater voltage respectively yielded the following black pepper samples classification accuracy as shown in Table 2.

While Fine KNN, Weighted KNN, Bagged Trees and Subspace KNN yielded 100% classification accuracy for all sensor heater voltage, the classification accuracy of other models showed different trends, whether increasing, decreasing or random, when sensor heater voltage was increased from 3 to 4 V and 5 V. Every classification model, which has its own unique classification algorithm, has varying performance for different data distribution patterns. Therefore, the observation of varying classification accuracy of each classification model further verified that temperature modulation altered data distribution pattern of the black pepper samples. An analysis that

Fig. 2 PC scatter plots legend

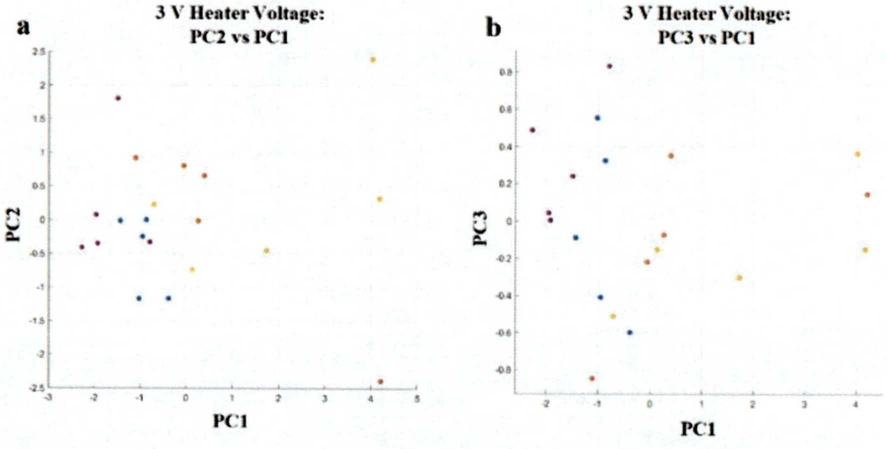

Fig. 3 PC scatter plots for 3 V sensor heater voltage **a** PC2 versus PC1 **b** PC3 versus PC1

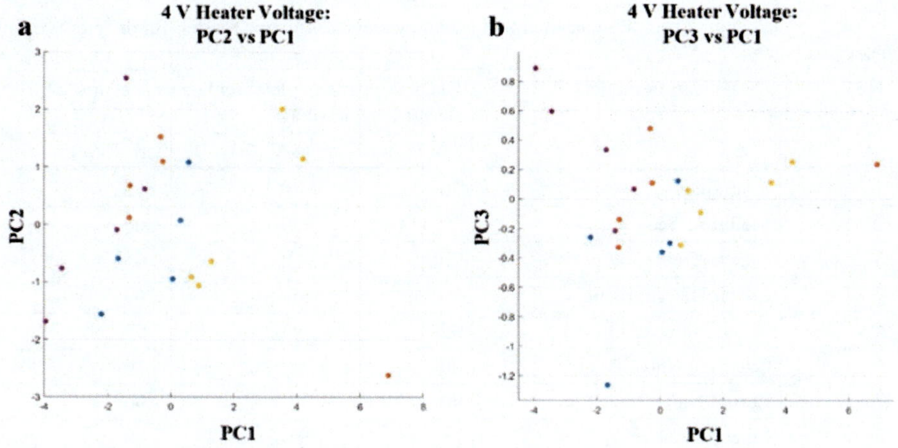

Fig. 4 PC scatter plots for 4 V sensor heater voltage **a** PC2 versus PC1 **b** PC3 versus PC1

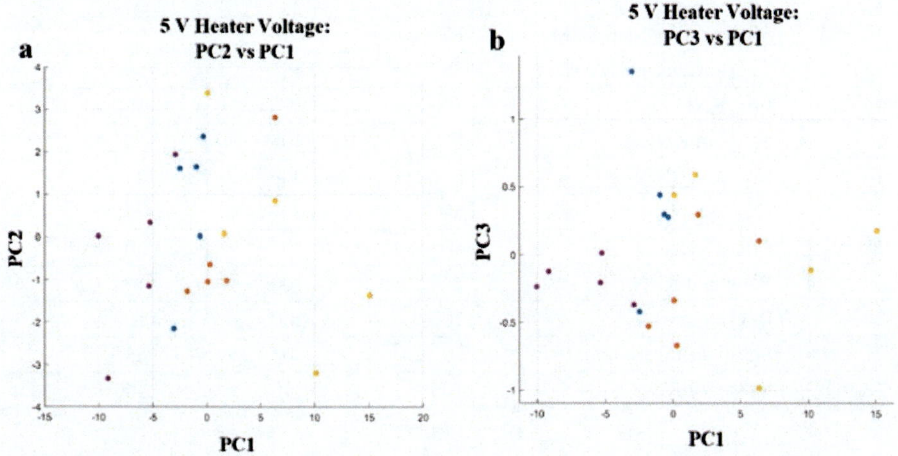

Fig. 5 PC scatter plots for 5 V sensor heater voltage **a** PC2 versus PC1 **b** PC3 versus PC1

considers all the variations of data distribution could potentially enhance the MOS gas sensors array capability in black pepper geo-tracing.

Table 2 Black pepper samples classification accuracy by classification models for different sensor heater voltages

No	Classification model	Black pepper sample classification accuracy for sensor heater voltage		
		3 V (%)	4 V (%)	5 V (%)
1	Fine Tree	70	70	65
2	Medium Tree	70	70	65
3	Coarse Tree	70	70	65
4	Linear Discriminant	70	60	65
5	Quadratic Discriminant	95	95	95
6	Gaussian Naïve Bayes	80	60	70
7	Kernel Naïve Bayes	90	80	90
8	Linear SVM	75	80	60
9	Quadratic SVM	60	75	60
10	Cubic SVM	60	75	45
11	Fine Gaussian SVM	70	5	55
12	Medium Gaussian SVM	85	85	95
13	Coarse Gaussian SVM	75	45	50
14	Fine KNN	100	100	100
15	Medium KNN	50	40	50
16	Coarse KNN	0	0	0
17	Cosine KNN	50	60	45
18	Cubic KNN	40	40	50
19	Weighted KNN	100	100	100
20	Bagged trees	100	100	100
21	Subspace discriminant	70	60	65
22	Subspace KNN	100	100	100

5 Conclusion

This preliminary study demonstrated that temperature modulation could alter the magnitude and ratio of sensor responses in an array of MOS gas sensors, resulting in a varying sample data distribution pattern. This further verifies the fact that temperature modulation could modify MOS gas sensors' sensitivity and selectivity, hence improve the dimensionality of sensor data for more accurate sample classification. At the current stage of study, data distribution pattern for each sensor heater voltage was analyzed separately. In the future, the sensor responses obtained for different sensor heater voltages—hence different temperature—should be concatenated to increase the dimensionality of sample data for potentially better black pepper geo-tracing result. Backed up by a promising preliminary result, further expansion of this research work could lead to better distinguishability of sensor data by black pepper country of origin. Better distribution of data to work could potentially reduce the quantity of sensors needed in the current sensors' array and reduce the computational load to analyze the sensor data.

Acknowledgements The authors would like to express their deepest gratitude to the Malaysian Pepper Board for providing the black pepper samples.

References

1. Santos JP, Lozano J, Aleixandre M, Arroyo T, Cabellos JM, Gil M, Horrillo MdC (2004) Comparison between an electronic nose and a human sensory panel for wine compound detection. In: SENSORS, 2004 IEEE, pp 341–344, vol 341. https://doi.org/10.1109/ICSENS.2004.1426172
2. Kishimoto N, Kashiwagi A (2019) Prediction of specific odor markers in oil from olive fruit infested with olive scale using an electronic nose. In: 2019 IEEE International Symposium on Olfaction and Electronic Nose (ISOEN), pp 26–29 , pp 1–3. https://doi.org/10.1109/ISOEN.2019.8823146
3. Zambotti G, Soprani M, Gobbi E, Capuano R, Pasqualetti V, Natale CD, Ponzoni A (2019) Early detection of fish degradation by electronic nose. In: 2019 IEEE International Symposium on Olfaction and Electronic Nose (ISOEN), pp 1–3. https://doi.org/10.1109/ISOEN.2019.8823461
4. International Pepper Community (2022) FOB Daily Prices. https://www.ipcnet.org/daily-price/. Accessed 27 May 2022
5. Wong J (2021) Tight supply pushing pepper prices up. The Star
6. Lee HE, Mercer ZJA, Ng SM, Shafiei M, Chua HS (2022) Metal oxide semiconductor gas sensors-based e-nose and two-stage classification: authentication of Malaysia and Vietnam black pepper samples. In: 2022 IEEE International Symposium on Olfaction and Electronic Nose (ISOEN), pp 1–4. https://doi.org/10.1109/ISOEN54820.2022.9789618
7. Trirongjitmoah S, Juengmunkong Z, Srikulnath K, Somboon P (2015) Classification of garlic cultivars using an electronic nose. Comput Electron Agric 113:148–153. https://doi.org/10.1016/j.compag.2015.02.007
8. Kiani S, Minaei S, Ghasemi-Varnamkhasti M (2016) A portable electronic nose as an expert system for aroma-based classification of saffron. Chemom Intell Lab Syst 156:148–156. https://doi.org/10.1016/j.chemolab.2016.05.013
9. Martinelli E, Catini A, Natale CD (2013) An active temperature modulation of gas sensor based on a self-adaptive strategy. In: 2013 Transducers & Eurosensors XXVII: The 17th International Conference on Solid-State Sensors, Actuators and Microsystems (TRANSDUCERS & EUROSENSORS XXVII), pp 2045–2048. https://doi.org/10.1109/Transducers.2013.6627200
10. Saeki M, Okura Y, Yoshikawa T, Iwata T (2021) Single-sensor gas discrimination and quantification based on novel temperature modulation method. In: 2021 IEEE Sensors, pp 1–4. https://doi.org/10.1109/SENSORS47087.2021.9639516
11. Zhao WJ, Ding KL, Chen YS, Xie FY, Xu D (2020) Optimized low frequency temperature modulation for improving the selectivity and linearity of SnO2 gas sensor. IEEE Sens J 20(18):10433–10443. https://doi.org/10.1109/JSEN.2020.2993055
12. Wen WC, Liu YT, Tang KT (2019) Gas mixture analysis based on metal-oxide semiconductor gas sensors with temperature modulated method. In: 2019 IEEE International Symposium on Olfaction and Electronic Nose (ISOEN), pp 1–4. https://doi.org/10.1109/ISOEN.2019.8823144

13. Lee HE, Mercer ZJA, Ng SM, Shafiei M, Chua HS (2020) Geo-tracing of black pepper using metal oxide semiconductor (MOS) gas sensors array. IEEE Sens J 20(14):8039–8045. https://doi.org/10.1109/JSEN.2020.2981602

Education 4.0

Online Learning: Students' Barriers and Feedback on the Instructional Delivery Methods

Nor Azalina Rosli, Rosmina Ahmad Bustami, Inawati Othman, and Jethro Henry Adam

Abstract Due to the current pandemic of COVID-19, the educational system is required to convert conventional face-to-face instruction to online learning. This study assessed the effect of learning devices availability and internet connection capability towards online learning and the students' view of current practice for online learning. A total of 108 Year 1 students from Civil Engineering Program, UNIMAS participated in surveys and answered quantitative and qualitative questions to highlight the challenges they experienced during their online learning. The findings showed that device availability and poor internet signal are the barriers among students who are from low-income families which give the greatest impact on the quality of their online learning. The study also found that asynchronous learning is preferred by students to reduce the gap in online learning thus benefitting the group of students affected by internet connectivity issues.

Keywords Online learning · Synchronous learning · Asynchronous learning · Breakout room

1 Introduction

The COVID-19 pandemic has not only impacted human life and the economic sector, but the unprecedented pandemic in this modern era has also greatly impacted the educational system. COVID-19 has forced schools and educational institutions to close due to mandatory lockdowns and made it impossible to conduct physical learning [1]. However, education cannot be withheld nor postponed for too long. Education has also changed its canvas dramatically with the distinctive rise of online learning supported by advancement in technologies [2]. Schools, higher educational

N. A. Rosli (✉) · R. A. Bustami · I. Othman · J. H. Adam
Department of Civil Engineering, Universiti Malaysia Sarawak, 94300 Kota Samarahan, Malaysia
e-mail: rnazalina@unimas.my

C. S. Choo et al. (eds.), *Proceedings of ASEAN-Australian Engineering Congress (AAEC2022)*, Lecture Notes in Electrical Engineering 1072,
https://doi.org/10.1007/978-981-99-5547-3_7

institutions, students and teachers were demanded to adopt e-learning technologies which indirectly tested their readiness to achieve effective online learning [3]. Online learning and e-learning refers to a learning environment that uses the Internet and other technological tools for synchronous and asynchronous instructional delivery, respectively, and management of academic programs [4].

Synchronous online learning involves real-time interactions between the teacher and students, while asynchronous e-learning occurs without a strict schedule for different students [5]. Most of the students prefer to have asynchronous online learning as it provides a solution for those with internet connectivity issues. The students are allowed to work self-paced and independently of time and place, with more time to retain and explore the provided materials [6]. Nevertheless, asynchronous online learning required students to have self-study skills to stay on track, including motivation to avoid procrastination [7]. For that, the biggest advantage of synchronous online learning would be its real-time interpersonal communication and students' commitment and motivation [8].

The availability of learning tools and stable internet connections are also important factors supporting online and e-learning. Although we are in an era where technological instruments and apparatus are within reach, there are students who are deprived of such convenience [9]. Some students do not possess laptops or desktop computers and have very limited access to the internet [10]. These challenges may disrupt students' learning performances and experiences as well inducing stress, anxiety and depression [11].

This study aims to assess the effect of learning devices availability and internet connection capability for online learning among Civil Engineering students. This study explores the students' feedback from their point of view of the current practice for online learning conducted at their program level. The findings can become a basis for improving online learning instructional design in the future.

2 Materials and Methods

2.1 Target Respondent

The study was conducted in Universiti Malaysia Sarawak (UNIMAS), where the respondents are Year 1 students from Civil Engineering Program. The data for this study were obtained using a questionnaire survey, distributed while online learning were still conducted, from 8 to 16 December 2021.

2.2 Questionnaire Distribution

A survey link on Google Forms was created for the survey and was posted on e-Learning Enrichment and Advancement Platform (eLEAP) and WhatsApp groups. eLEAP is an online educational platform built by UNIMAS to enable the students to interact with their course instructors and course mates. All the materials and resources related to the course can be found in the eLEAP. Besides eLEAP, each batch of students are gathered in a WhatsApp group as a medium of communication. Updates and announcements are also posted in the WhatsApp group.

2.3 Questionnaire Design

The online survey consisted of three components (as in Appendix A): (1) Students' profile that included their age, gender, region of residence and family's economic background. (2) Students' internet accessibility to identify the connection issues, device availability and other supporting accessories which might affect the teaching and learning process. (3) Mode of teaching and delivery used; to get students' feedback on the suitability to have an online live class or pre-recorded lecture. The survey also included the students' views on the effectiveness of having peer study using breakout room sessions. All sections applied multiple-choice questions to obtain the data with a few questions requiring the students to provide a short answer.

2.4 Statistical Analysis

Statistical analysis was performed using Minitab 17 software. The mean values, standard deviations and standard errors were reported for each of the measurements. Regression analysis was used to study the relationship and correlation between the parameter of interest. The fitted line plot with R squared more than 99% and P value less than 0.05 indicate a strong correlation between parameters.

3 Results and Discussion

3.1 Students' Background

There were 120 registered first year students, and only 108 students responded to the survey. The background of the respondents are illustrated in Fig. 1a–d. Although the targeted students were first-years, their age ranged from 18 to 25 years, with majority being in the 19-year old category, as shown in Fig. 1a. From Fig. 1b, 55%

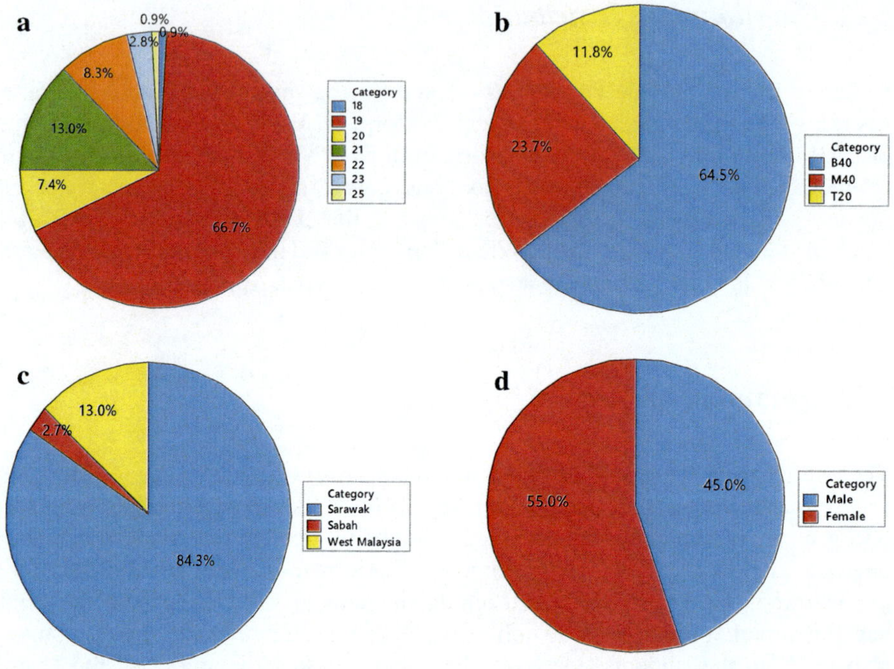

Fig. 1 Students' background **a** Age groups, **b** Gender, **c** Regions of residence **d** Level of economy

respondents were female with majority of the students originating from Sarawak (84%). In terms of their family economic level, 64.5% of the students came from B40 families with only 23.7% and 11.8% from M40 and T20, respectively. It is very relevant to assess the students' economic level as it indirectly affects the students learning due to the lack of facilities and necessity required during the online learning process. Previous studies on student academic performance prior to the COVID-19 pandemic revealed that students exposed to economic instability showed lower levels of academic achievement than students in more stable households [12].

3.2 Students' Accessibility

Despite the numerous benefits of online learning, a few limitations need to be addressed especially on the students' accessibility to the online platforms, which included the lack of sufficient learning devices and internet access [13]. A study reported that not all students are capable to acquire personal laptops and smartphones which were the main tools used in online learning. These devices are considered luxury items for students from economically disadvantaged circles [14]. Malaysian government together with bigger industries have taken initiatives in helping students in online learning by providing free laptops and tablets to those in need and a laptop discount for students. This initiative was taken to ensure no one was left behind during online teaching and learning. From Fig. 2a, majority of the respondents (98.1%) own personal desktops or laptops. However, each household might need more than one laptop as online learning occurred simultaneously for all age groups [15] within the same household. Figure 2b shows that from the 98.1% students with own desktop/laptop, 19.4% of them still had to share the devices with their siblings. Hence, about 8.3% of the students were still attending classes from their smartphones as shown in Fig. 3.

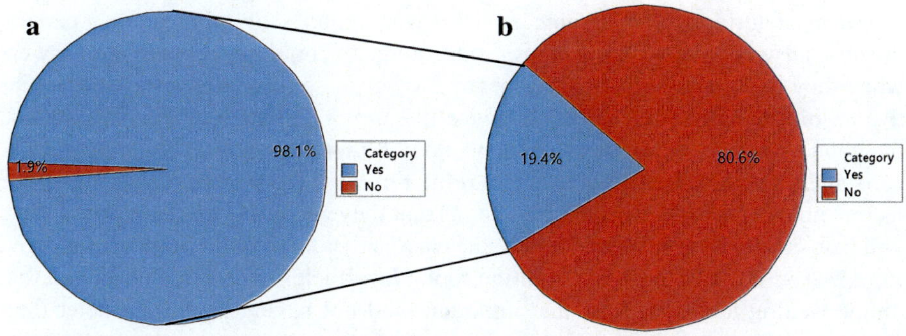

Fig. 2 **a** Number of students that have desktop/laptop, **b** Number of students that need to share desktop/laptop with siblings

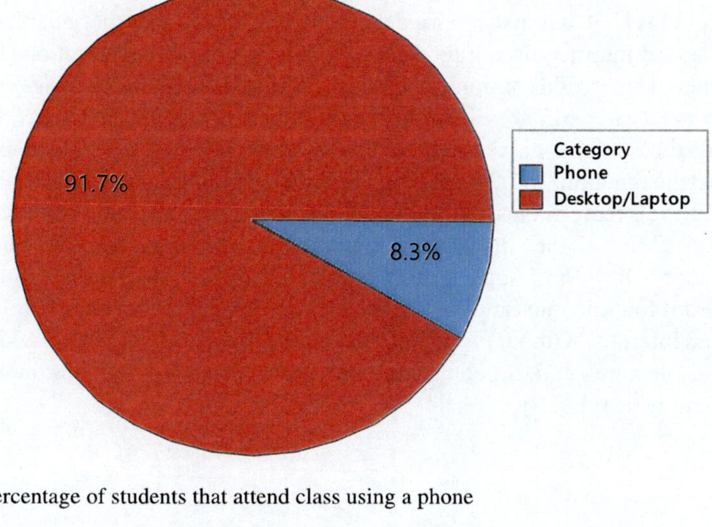

Fig. 3 Percentage of students that attend class using a phone

Having a laptop or desktop alone was not sufficient. Those devices must be equipped with accessories like microphones, headsets, scanners, printers and ink cartridges as well as online writing tools such as tablets and digital pens. These accessories are key to a smooth and engaging online learning. Insufficient tools and accessories proved that online learning losses are likely to be particularly acute for students from economically disadvantaged families [14]. Based on Fig. 4, 28% of the students lack access or had difficulty with microphone and headset operations. As a result, it was difficult to communicate with the lecturers, either to ask or answer questions to/from the lecturers. Additionally, group work (breakout room) with peers may also pose as a difficulty activity to these group of students as they were unable to effectively and actively participate because of their malfunctioning microphones and/or headsets. Although they could respond through chat boxes, however, it will take some time for them to write a question, respond to questions, or wait for responses from their peers. 31.5% of the students had challenges with their webcam or camera. Sometimes, during online learning, students were required to switch on their camera to show answers and working steps. Some students could not provide the answers when they used laptops without webcam. However, some students took the initiative to join the class with a second device like a mobile phone where they can use the phone camera. 43.5% of the students do not have access to a printer and/or ink cartridges. It is very important to have a printer because the students can print out the lecture notes and write down the key points that they get during an online class. This will help ensure their readiness before the class starts. Most of the students surveyed (63.9%) did not have an online writing tool. Nevertheless, the students agreed that online writing tools were not that important and not having it will not deter their online learning experience. Only 18.5% students had access to scanners. Students preferred using their phones to scan assignments before submitting them online.

Connectivity or internet connection is important be it synchronous or asynchronous as internet is required for the learning process to happen. Previous research showed that online learning was constrained by unstable internet signals and expensive data quotas [14]. It is also reported that limited access to internet-enabled devices and high-speed Internet discouraged engagement with remote instruction [12]. Even though there were various options of telcos that offered numerous packages on online data, network issues still arise especially for students living in rural areas. Based on survey, all the students have access to the internet of which 52.8% are using Wifi access and the remaining 47.2% are using mobile data for online learning and internet usage (Fig. 5a). Nevertheless, not all students get to have fast and reliable internet connections. Most of the students have internet connectivity issues, and only 4.6% of the students have no issues at all (Fig. 5b). Studies show that students that are coming from low-income families are the ones that are facing a lack of reliable and high-speed internet. Without proper internet connections, the students could not even attend class lectures and get connected with teachers and classmates, and even do independent research [10].

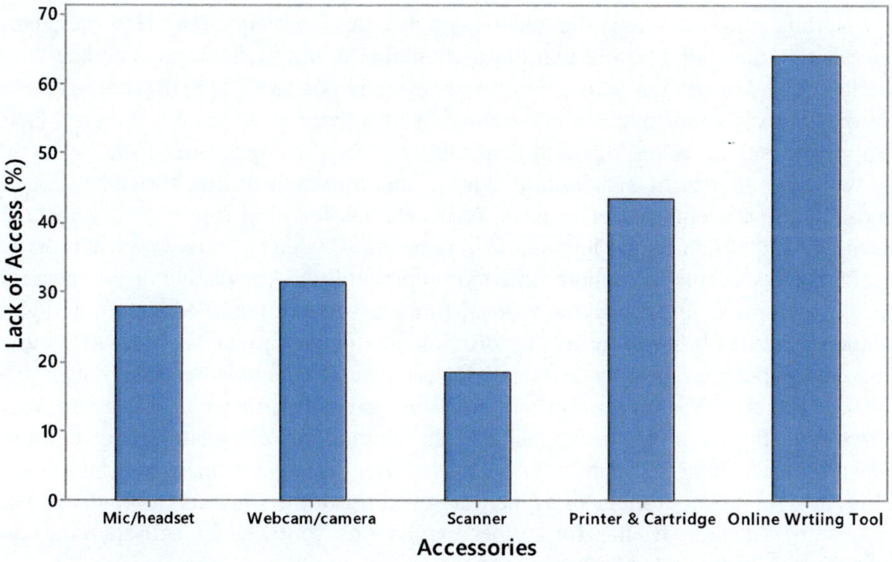

Fig. 4 Percentage of students that have challenges with desktop or laptop accessories

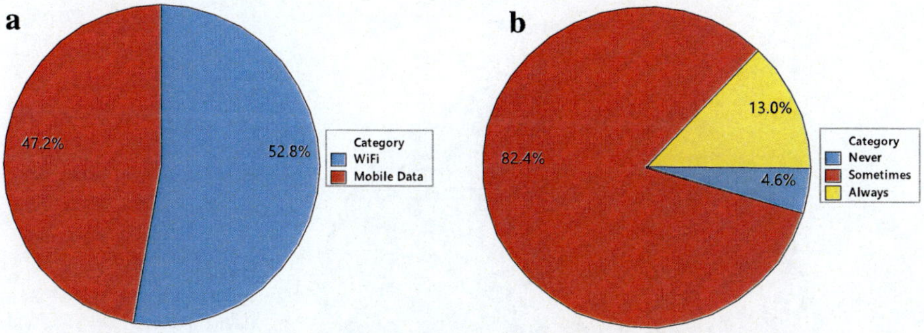

Fig. 5 **a** Internet connectivity of percentage of students using Wi-Fi and mobile data **b** frequency of the connection issues

3.3 Mode of Teaching

3.3.1 Synchronous Versus Asynchronous

The pandemic of COVID-19 has caused all educational institutions to close to curb the spread of the disease. Due to that, all physical classes are cancelled and shifted to the virtual classroom through online learning. There is a concern about the learning outcomes for e-learners. Previous research reported that the quality and quantity of students' participation in synchronous interaction had a higher statistical significance on overall course grade as compared to asynchronous interaction [16]. However, there are also studies that showed that the participants in the asynchronous course were delighted and rated their learning experiences more positively [17]. Recently, review studies revealed that there is no significant difference in learning outcomes [18]. Improvements in technology and increasing bandwidth capabilities have led to the growing popularity of synchronous and asynchronous learning. Based on Fig. 6, most of the students prefer to have asynchronous learning (pre-recorded lecture) compared to synchronous lecture. This is because, when having an asynchronous lecture, the students have more time to comprehend the knowledge or the message as they can pause and repeat the video. Moreover, the pre-recorded lecture will give students the flexibility in terms of more time to engage with and explore the course material, especially for slow learner students and those affected by connectivity issues [19]. Another 35.9% prefer synchronous lectures because they can ask the question directly to the lecturer during class and the method can eliminate procrastination. Nevertheless, only 15.7% prefer to ask the lecturer directly during synchronous class (Fig. 7a). Be it synchronous or asynchronous, most of the students (more than 45%) choose to ask their friends for further explanation followed by self-studying and communicating with the lecturers (Fig. 7a and b).

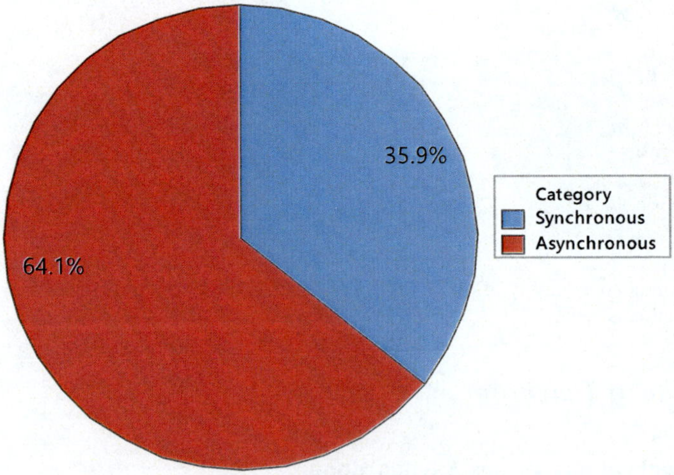

Fig. 6 Majority of students prefer to have asynchronous learning

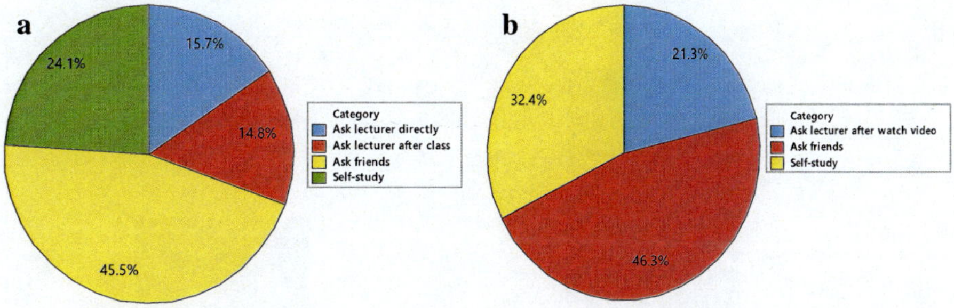

Fig. 7 Students' approach to solving a problem **a** synchronous learning **b** asynchronous learning

3.3.2 Breakout Room in Online Platform

Breakout rooms are sessions that split off from the main meeting room in any online platform like Webex, Zoom, Googe Meet, and Microsoft Teams. Through these breakout rooms, the students are allowed to meet in smaller groups for collaboration, discussion, and exchange points of view in a more comfortable and intimate space [20]. It is recommended to have 2 to 5 students in each breakout room to ensure everyone has the opportunity to contribute and to avoid sleeping partners [21]. For group members selection, 58.3% of the students prefer to choose their members over those assigned by the lecturers (Fig. 8a). This is because some of the students already know their classmates (i.e., from Asasi UNIMAS) so they feel comfortable working with someone they already know. Nevertheless, some of the students are still new and they think that the group assigned by the lecturer can help them to make new friends and not to get distracted by old friends. Based on the feedback, breakout rooms session can be either effective or not effective (Fig. 8b). The effectiveness (57%) is due to active discussion among group members, and they get input and knowledge from each other. For that, they can complete most of the tasks on time and they can also see improvement in their mark for assessments like quizzes and tests. Another 43% found it not effective because their group members are not active, probably because of shyness and they end up doing their own work until the session ends. This group claimed that they do not see any improvement in their marks and performance.

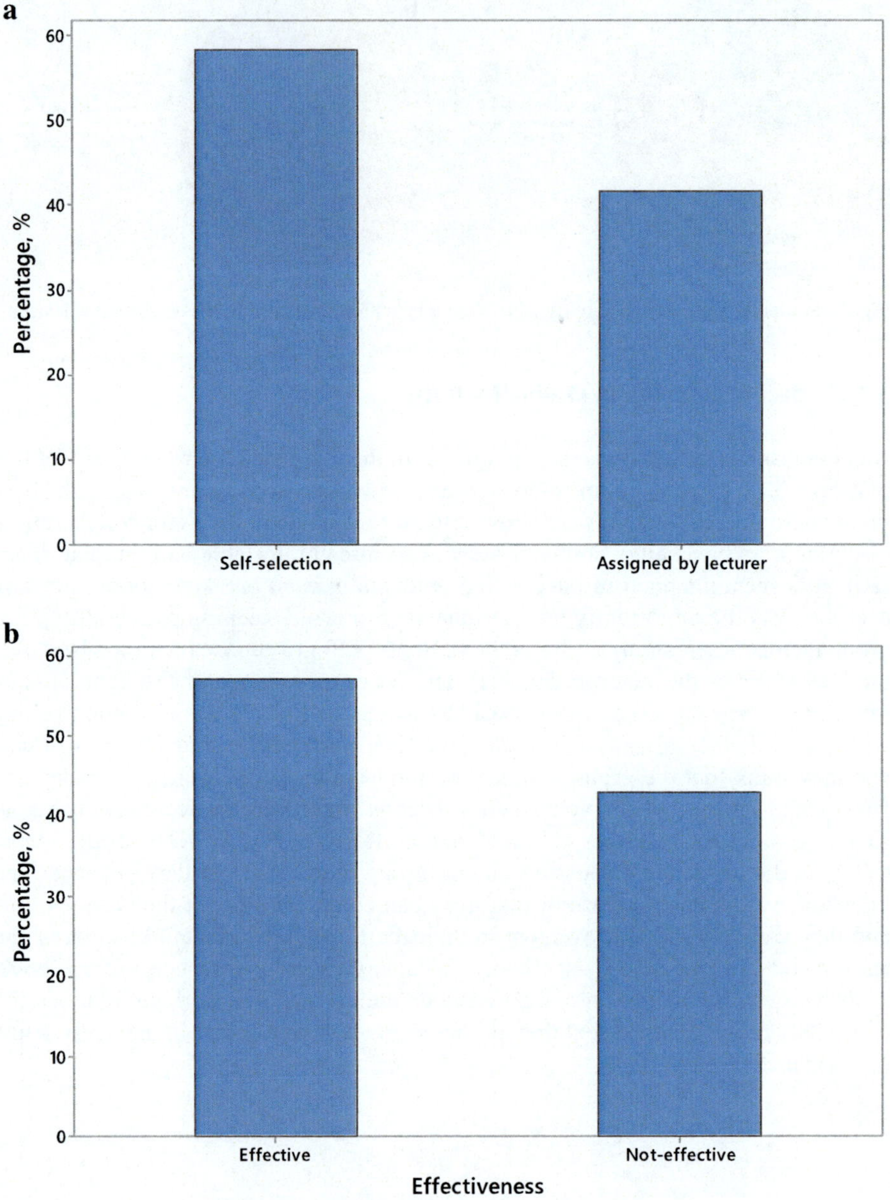

Fig. 8 **a** Group members selection and **b** the effectiveness of breakout room sessions

4 Conclusion

This study shows that device availability and issue with internet connectivity might become a barrier among students that are coming from low-income families in their online learning. Due to that, most of the students preferred to have asynchronous learning compared to synchronous learning. Besides that, they opt to ask friends and do self-study rather than communicating with the lecturers be it synchronous or asynchronous class. For that, breakout room sessions can be very effective especially when they get to choose their group member as it is part of peer-study and monitored by the lecturers. It is recommended to do further study on the student's stress level during online learning and the impact of online learning on the assessment tools.

Acknowledgements We would like to thank Universiti Malaysia Sarawak (UNIMAS) for the facilities and financial support.

Appendix A

Questionnaire.

Section A: Background

Q1. Gender

Male

Female

Q2. Age

19

20

21

22

Other: (please state)

Q3. Origin

Johor

Kedah

Kelantan

Kuala Lumpur

Labuan

Melaka

Negeri Sembilan

Perak

Perlis

Pulau Pinang

Sabah

Sarawak

Selangor

Terengganu

Q4. Parents Salary

<RM 2,500

RM 2,500 – RM 3,999

RM 4,000 – RM 4,849

RM 4,850 – RM 7,099

RM 7,110 – RM 10,959

RM 10,960 – RM 15,039

>RM 15,040

Q5. No of Siblings: __________

Section B: Accessibility

Q1. I own a Desktop / laptop (Yes / No)

Q2. I need to share desktop / laptop with my siblings (Yes / No)

Q3. I have Internet / Network problem (Yes / No)

Q4. I used (Wifi / Mobile data) for connections

Q5. Check all that you had challenges with (lack of access or difficulty of operations)

Mic /headset

Webcam/ camera

Scanner

Online writing tool (tablet and digital pen)

Printer and cartridge

Section C: Mode of Teaching

Q1. I prefer ________

Live Lecture

(State the reason)________________

Pre-recorded lecture

(State the reason)________________

Q2. During live lecture, if I do not understand,

I will ask the lecturer directly

I will ask the lecturer after the class

I will ask my friends

I try to find the answer by doing self-study

I am not doing anything to find the answer

Q3. During pre-recorded lecture, if I do not understand,

I will ask the lecturer

I will ask my friends

I try to find the answer by doing self-study

I am not doing anything to find the answer

Q4. I communicate with my lecturer via

Email

Whatsapp

Eleap Messenger

Q5. The best duration for live lecture is

45 minutes

60 minutes

90 minutes with 10 minutes break

120 minutes with 20 minutes break

Q6. The best duration for pre-recorded lecture is

30 minutes

45 minutes

60 minutes

Q7. The use of breakout rooms are effective or inefficient tools?

Effective

(State the reason)________________--

Inefficient

(State the reason)________________--

Q8. The appropriate number of students per breakout rooms session is _____

Q9. Selection of member in breakout rooms

Assigned by lecturers

(State the reason)________________--

Self-selection

(State the reason)________________--

Q10. Based on your experience, do you think breakout rooms is a success or not?

Yes.

My group is very active

I gained a lot of knowledge from my friends

I get good marks in assessments (quiz, test, etc..)

I managed to complete most of the tutorials by having a discussion with friends

No.

My group is not active

We are not really discussing during the breakout room session

My marks are not improving

I am doing the tutorial all by myself

References

1. AlAteeq DA, Aljhani S, AlEesa D (2020) Perceived stress among students in virtual classrooms during the COVID-19 outbreak in KSA. J Taibah Univ Med Sci 15(5):398–403
2. Li C, Lalani F (2020) The Covid-19 pandemic has changed education forever. https://www.weforum.org/agenda/2020/04/coronavirus-education-global-covid19-online-digital-learning/
3. Mukhtar K, Javed K, Arooj M, Sethi A (2020) Advantages, limitations and recommendations for online learning during COVID-19 pandemic era. Pak J Med Sci, 36(COVID19-S4), S27
4. Usher M, Barak M (2020) Team diversity as a predictor of innovation in team projects of face-to-face and online learners. Comput Educ 144:103702
5. Singh V, Thurman A (2019) How many ways can we define online learning? A systematic literature review of definitions of online learning (1988–2018). Am J Distance Educ 33(4):289–306
6. Fabriz S, Mendzheritskaya J, Stehle S (2021) Impact of synchronous and asynchronous settings of online teaching and learning in higher education on students' learning experience during COVID-19. Front Psychol 12
7. Hartnett M, St George A, Dron J (2011) Examining motivation in online distance learning environments: complex, multifaceted, and situation-dependent. Int Rev Res Open Dist Learn 12:20–38. https://doi.org/10.19173/irrodl.v12i6.1030
8. Blau I, Weiser O, Eshet-Alkalai Y (2017) How do medium naturalness and personality traits shape academic achievement and perceived learning? An experimental study of face-to-face and synchronous e-learning. Res Learn Technol 25. https://doi.org/10.25304/rlt.v25.1974
9. Asio JMR, Gadia E, Abarintos E, Paguio D, Balce M (2021) Internet connection and learning device availability of college students: Basis for institutionalizing flexible learning in the new normal. Stud Humlties Educ 2(1):56–69
10. Cleofas JV, Rocha ICN (2021) Demographic, gadget and internet profiles as determinants of disease and consequence related COVID-19 anxiety among Filipino college students. Educ Inf Technol 1–16
11. Barrot JS, Llenares II, Del Rosario LS (2021) Students' online learning challenges during the pandemic and how they cope with them: The case of the Philippines. Educ Inf Technol 1–18
12. Domina T, Renzulli L, Murray B, Garza AN, Perez L (2021) Remote or removed: Predicting successful engagement with online learning during COVID-19. Socius 7:2378023120988200
13. Aboagye E, Yawson JA, Appiah KN (2021) COVID-19 and E-learning: The challenges of students in tertiary institutions. Soc Educ Res 1–8
14. Sumadi CD, Hidayat A, Agustina I (2022) Literature study: analysis of learning facilities in the pandemic era on the effectiveness of online learning in elementary school. Widyagogik: J Pendidik Dan Pembelajaran Sekh Dasar 9(2):183–190
15. Siddiquah A, Salim Z (2017) The ICT facilities, skills, usage, and the problems faced by the students of higher education. EURASIA J Math, Sci Technol Educ 13(8):4987–4994
16. Duncan K, Kenworthy A, McNamara R (2012) The effects of synchronous and asynchronous participation on students' performance in online accounting courses. Acc Educ 21(4):431–449
17. Buxton EC (2014) Pharmacists' perception of synchronous versus asynchronous distance learning for continuing education programs. Am J Pharm Educ 78(1):8
18. Nieuwoudt JE (2020) Investigating synchronous and asynchronous class attendance as predictors of academic success in online education. Australas J Educ Technol 36(3):15–25
19. Alqurashi E (2019) Predicting student satisfaction and perceived learning within online learning environments. Distance Educ 40(1):133–148
20. Fenner-McAdoo E. (2021) Breakout rooms more effectively. https://www.teachhub.com/classroom-management/2020/12/how-to-use-breakout-rooms-more-effectively/
21. Reina S (2021) Proven virtual group activities to engage students in any discipline. get creative with your use of breakout rooms. https://hbsp.harvard.edu/inspiring-minds/get-creative-with-your-use-of-breakout-rooms

Rapid Transition from Offline to Online: The Teaching and Learning Experience in Engineering Education

Yit Yan Koh and Yaw Long Chua

Abstract In the pre-COVID era, many engineering education academics shared that engineering education is challenging to conduct through online learning. They felt that teaching engineering deals with laborious calculations, and the need for physical presence for experiments and online education is difficult to be achieved. Nonetheless, COVID-19 forced education institutions to adopt the online learning model, which is the only way to continue learning during the Pandemic. From rejection to reception, the institutions put on the show to convert the lectures and assessments into the online version with various challenges such as lectures, connectivity issues, submission of assessment items, and even academic integrity on formal assessments. With an available Learning Management System, lectures, tutorials, laboratory sessions and examinations were conducted through the online platform. This paper illustrates the process of the conversion from traditional face-to-face teaching to online teaching and how academics and students cope with such transformation by maintaining consistent academic performance. In addition, a comparison of students' academic performance face-to-face (pre-COVID) and online (during the Pandemic) shows the positive outcome of such conversion of teaching and learning activities. Lastly, a short discussion on the continuous improvement process was implemented to ensure that the adopted system is sustainable throughout the Pandemic.

Keywords Face-to-face teaching · Online teaching · Engineering education

Y. Y. Koh (✉)
Newcastle Australia Institute of Higher Education, Singapore, Singapore
e-mail: yityan.koh@newcastle.edu.au

Y. L. Chua
Universiti Tenaga Nasional, Kajang, Selangor, Malaysia

C. S. Choo et al. (eds.), *Proceedings of ASEAN-Australian Engineering Congress (AAEC2022)*, Lecture Notes in Electrical Engineering 1072,
https://doi.org/10.1007/978-981-99-5547-3_8

1 Introduction

Learning to the data published by UNESCO, within a short period, COVID-19 impacted global education by forcing 1.5 billion students from 184 countries to change the way they learn [16]. This reflects that 89.4% of learners in the world faced the need to change their way of learning within weeks. Under such rapid conversion, and most countries underwent the so-called "lockdown" period, online learning has turned into the primary learning mode, thanks to the advancements in technology [2, 6, 22].

To assist the schools and universities in getting ready for the flexible education systems, UNESCO introduced a complete distance learning strategy which includes eight pointers towards inclusion, equity, and maximize the quality of learning [27]. The eight pointers are illustrated as (1) fine-tune curricular objectives and focus on humanitarian and social caring; (2) confirm the readiness and select the most effective technological solutions; (3) enhance the technological and content readiness to ensure the continuity of education and learning; (4) ensure equity and inclusion; (5) protect learners' privacy and data security; (6) provide assistance to educators on distance learning, and encourage supports from parents and caregivers; (7) adopt student-centered teaching, monitoring and assessment methodologies to for effective distance learning strategies; and (8) plan for sustainability and long-term goals.

In response to the guidelines published by UNESCO, the IHLs are also expected to ensure the effectiveness and relatedness of online teaching [3]. Provided a series of guidelines to achieve such achievements. First, the IHLs need to provide a high relevance between the instructional design and student learning so that students can make sense of the content of the subjects. Furthermore, online instructional information must be delivered effectively with controlled quality. The faculty must provide sufficient support to the educators to ensure they can fully understand the online delivery philosophy and methodology. Students should also be highly encouraged the participants to improve the breadth and depth of their learning since they are not physically in the classroom. With further distractions in their learning environments, continuous motivation is needed. Finally, educators must have a contingency plan in case of any incident during online learning, which includes internet disruption, technical glitches, and students' responses.

With the motivation and guidelines mentioned above, this paper aims to review the process of the conversion of face-to-face teaching to online teaching, and steps taken by the institution to ensure students' academic performance throughout the transition.

2 To Blend or Not to Blend? A Question About Face-To-Face and Online Learning

Blended learning was introduced as early as the 2000s. The idea relates to a mixture of face-to-face teaching in the classroom and online activities (online videos, forums, discussion groups) to students' learning flexibility. During the introductory period, such methods have proved to be effective in the medicine and psychology area [4, 12]. On the contrary, students in Engineering did not seem to be able to adapt to the idea, and they felt that face-to-face learning was more effective and conducive [8].

Since then, a few different types of blended learning modes have been introduced by various researchers to explore the suitability of blended learning in the education system and its effectiveness. A few examples are provided in this paper to illustrate the variety of blended learning.

Combined asynchronous (lecture capture) and synchronous (on-campus or live broadcasted lecture) were presented by [17]. This is the method where the lecture is delivered through the Learning Management System (LMS) and classroom and recorded. In addition, the lectures are recorded where students can refer to the lecturer at the desired speed to obtain important information. Students can attend the lecture on campus online from home. Such combination is also called Hybrid Flexible (Hyflex) learning, as outlined by [15].

As presented by [14], hybrid learning is where lectures are conducted online. In addition to the lectures, additional virtual activities such as forums, videos, e-documents, and discussions are held where students can access these activities any time outside the lectures. The advantage of such hybrid learning is students do not need to be physically present in the discussions/forums but can leave comments anytime, which promotes more valuable suggestions or comments towards a topic discussed.

The Flipped Classroom was a relatively new concept in the late 2010s [26]. The learning materials are provided before the lectures, and students are requested to study and familiarize themselves with the materials. As such, the face-to-face lecture is no longer a one-way communication from the lecturer to the students but turns into a discussion forum where students present and discuss the topic under the lecturer's or tutor's guidance.

It should be noted to date there is no clear delimitation towards the use of "hybrid", "blended", and "hyflex", where researchers are using them interchangeably [18]. Hence, the main idea one should have when it comes to online or face-to-face teaching is that "hybrid", "blended", and "hyflex" could be the combination of both online and face-to-face teaching. In some cases, additional activities are introduced to enhance students' learning experience.

Even though blended learning was not welcomed in the early 2000s, the idea has gained popularity among undergraduates since then [1]. Notwithstanding the attention given to the undergraduate, online learning was still predominantly conducted at the postgraduate level [17]. Furthermore, research studies such as the interactive video presented by [20], Cloud-Computing-Learning as practiced by [23] showed the effectiveness of online learning in Engineering Education [17]. Have demonstrated that online lectures through LMS or webinar sessions are as practical as face-to-face lectures. Summarizing these reports, it is evident that while some academics were worrying about the effectiveness of online learning, research has provided the answers to confirm the successful implementation of online learning in Engineering Education.

3 Teaching and Learning During Circuit Breaker

The Circuit Breaker (lockdown) was enforced nationwide in Singapore in May 2020 to contain the spread of the COVID-19 virus in the country [11]. The COVID-19 (Temporary Measures) Act 2020 was released to enforce people's movement restrictions [24]. During this period, schools and IHE had to implement Home-based Learning (HBL) through online learning.

At the Newcastle Australia Institute of Higher Education, the delivery mode of the courses is achieved using Blackboard Collaborate, which is common in many IHEs [25]. The institution adopted Canvas as the replacement for Blackboard in early 2022, where Canvas serves a similar function. The Centre of Teaching and Learning has been providing annual training to academics on using LMS (Blackboard) for teaching. Hence, despite the Pandemic's sudden hit, the training required to teach using LMS is relatively lesser as most academics are aware of or already incorporate Blackboard Collaborate in their teachings.

Despite the familiarity with using Blackboard Collaborate to teach, academics in Engineering still face challenges where courses requiring heavy calculations were not easy to handle by just using Blackboard Collaborate. This problem was solved by immediately purchasing a graphic tablet and training provided to the academics in using the tablet for teaching courses containing heavy calculations.

Students were informed about the conversion of the classes to the online mode. Following this, a quick survey was also conducted to ensure that all students had a computer/laptop that could connect to the internet for learning. Students with technological issues were flagged and immediate assistance was provided to them to ensure they could access the online lectures without many hassles. To reduce internet bandwidth usage during the lecture, students were not required to turn on their videos during the lectures. To ensure that students were continuously paying attention, lecturers adopted frequent polls so that all students were required to cast their polls in the questions, asking open-ended questions requiring specific students to answer the question and mini quizzes requiring full-class participation.

At the beginning of the implementation, students raised their frustration that they were not able to stay focused as compared to the classroom, they found it hard to raise and explain questions during the class (especially questions on the calculations), and intermittent internet disruption deprives them of paying full attention during the lectures. These issues were slowly ironed out throughout the conduct, where students slowly got themselves adapted to the different learning environments.

The laboratory sessions were inevitable to be conducted virtually during the Circuit Breaker. Lecturers/tutors conducted experiments before the circuit breaker and video recorded. Students viewed the laboratory session as if they were experimenting, and then with the data given, they continued with the calculation and analyses. The condition was eased with the campus opening in Phase 2 on 19 June 2020, where lectures were still conducted online, and students attended physical sessions in the school [19].

In Phase 3, the hyflex teaching and learning model was adopted as some students were allowed to enter the school for the lectures. The main reason for the hyflex mode was that some students were still locked in their home country and unable to return to Singapore, so online learning mode would remain their foremost. Academics were trained to provide the hyflex teaching and learning mode to ensure they can cope with the new skills in the hyflex learning. Such training focuses on the provision of attention to both students who attend the class physically and virtually and make sure that both sides of students can read the writing or presentation of information. Furthermore, the academic will also need to ensure that students in the classroom can still interact with students in the virtual mode to ensure total inclusiveness for the lectures.

Since then, the hyflex teaching and learning mode has been the primary teaching mode. During the subsequent lockdown processes, all students and academics adopted online learning, which was already familiar.

4 Assessment and Academic Performance

Since the lectures and tutorials were shifted online, the conduct of assessments was also shifted online. Both students and academics were not facing many issues with submitting assignments, as this had been practiced most of the time, even before the Pandemic. For the written assignments, students would complete the assignments, scan the document, and upload it on the submission page. For the word-processed assessments, they would complete the assessment, save it in the Document format (.doc or.docx) or Portable Document Format (.pdf) and submit it on the assessment page. The academics downloaded the assessment in bulk and marked them as a soft copy.

The main issue faced by the IHE is the conduct of tests and examinations conventionally conducted face-to-face. The institution had only two months, from the observance of the first case in Singapore to the lockdown, to ensure that the examinations could be conducted through the LMS successfully.

As the first try by the IHE, the team decided to turn all examinations into Open Book Examinations. Furthermore, the Scan and Upload method was used for the examination. Under such a case, students are needed to answer the questions on paper, as many calculations were required, and it was not feasible to answer questions online. At the end of the examination, students scanned the solution and uploaded it to the LMS. The examination was conducted through the Blackboard collaborate, where all students were required to turn on the camera to ensure academic integrity was observed as much as possible.

For such an approach to be effective, a mock examination was conducted before the final examination to ensure that all students could follow the procedure and gain initial feedback on such implementation. The procedure was reviewed and fine-tuned with feedback from the academics and students. Students attempted their first online examination without many problems reported.

An oral examination was conducted after the examination to ensure that the students were not breaching academic integrity. A random number of students were selected for a 5-minute oral examination, where the academics asked questions about their solutions to check if they fully understood their approach to the solution. Students who could not provide a satisfactory answer to the academic would be categorized as "academic misconduct allegation" and referred to Student Academic Conduct Officer for further investigation.

With such procedures in place, it was expected that students would be given the same quality of education despite the learning environment changes. The results of such conversion should yield similar academic performance. Hence, in the analyses of the student's results, the following hypothesis was declared:

H_{01}: there is no significant difference in academic performance for students who do face-to-face teaching and online teaching.

To demonstrate the validity of this hypothesis, some academic courses were taken into consideration, and in this paper, one course, Thermodynamics, was demonstrated. As such, student performance for 2019, 2020 and 2021 was compared.

The sample courses were chosen based on the condition that they were taught face-to-face in 2019 and online in 2020 and 2021. To ensure high similarity, the lecturer and assessment components for these cohorts are consistent with the only difference in the mode of delivery. Table 1 shows the total enrolment of students in the course over the three years.

Figure 1 shows the comparison of the academic performance of Thermodynamics students across the years 2019, 2020 and 2021. By comparing results, the distribution of students' performance is consistent, and no abnormal distribution was observed.

Table 2 shows the mean and standard deviation for the marks obtained by students from 2019 to 2021. The mean obtained by the cohort that went through online learning is similar to that of face-to-face learning.

A Normality test was performed to confirm the null hypothesis for this research. The face-to-face teaching (2019) and online teaching (2020 and 2021) was performed through the two-sample *t*-Test, assuming unequal variances at the 95% confidence

Table 1 Enrolment number for thermodynamics for years 2019, 2020 and 2021

Year	Number of students
2019	40
2020	36
2021	28

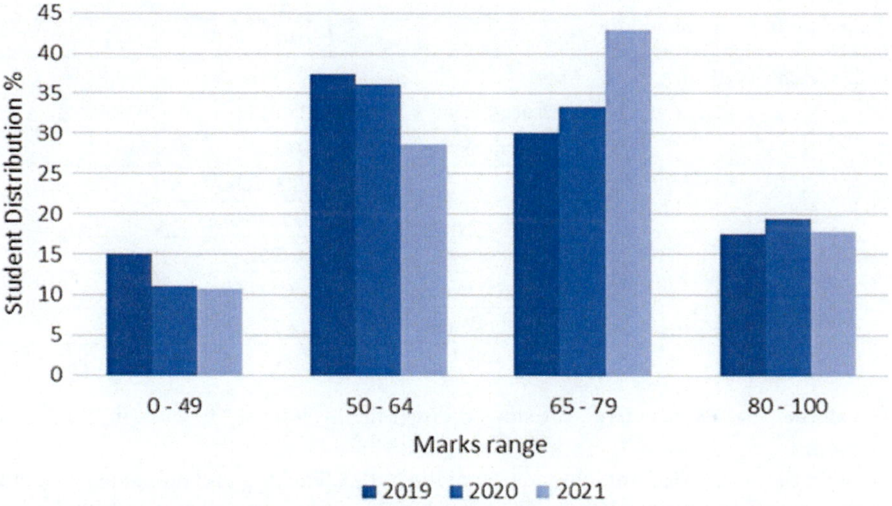

Fig. 1 Comparing students' performance for face-to-face teaching (2019) and online teaching (2020, 2021)

Table 2 The mean and standard deviation for the marks obtained by students for face-to-face teaching (2019) and online teaching (2020, 2021)

Year	Mean	Standard deviation
2019	61.72	18.07
2020	64.90	15.79
2021	63.49	19.10

level. Tables 3 and 4 show the t-Test scores for face-face teaching and online teaching for 2019 and 2020 and 2019 and 2021. The results show no significant difference between the scores of the two tests ($p > 0.05$); hence, H_{01} is accepted.

The conversion from face-to-face teaching to online teaching during the Pandemic is said to be incidental learning for global institutions, where the institutions face this complex problem and solve it quickly. A survey conducted by Harvard Business Review [13] reported that as high as 86% (with 43% agree and 43% somewhat agree) of business leaders felt that complexity affects their ability to decide and respond

Table 3 t-Test scores for face-face teaching (2019) and online teaching (2020)

	2019	2020
Mean	61.7218	64.89636
Variance	335.0696	256.3134
Observations	40	36
t Stat	−0.80643	
P(T < = t) two-tail	0.42258	
t Critical two-tail	1.992543	

Table 4 t-Test scores for face-face teaching (2019) and online teaching (2021)

	2019	2021
Mean	61.7218	63.48849
Variance	335.0696	378.2002
Observations	40	28
t Stat	−0.37766	
P(T < = t) one-tail	0.353557	
t Critical one-tail	1.672522	
P(T < = t) two-tail	0.707114	
t Critical two-tail	2.003241	

to external threats effectively. It slowed down the growth of business in the digital economy.

With the successful implementation of the hyflex teaching and online assessment, such practice has been taken in place until 2022, and both students and academics have been living in the norm of such practice. 2022 marks the ease of the restriction, where most of the country's boundaries are opened, and students are returning to Singapore from their home countries. The practice remains the default practice until all students are back in Singapore, and the IHL will review the current practice to decide if this is to be continued or if new conduct is to be implemented.

5 Conclusion

This paper presented the conversion from face-to-face teaching to online teaching for the Newcastle Australia Institute of Higher Education to address the COVID-19 Pandemic in the region. The conversion process was successful, thanks to the cooperation provided by academics and students. A continual review was in place to ensure that the practice was consistently applicable based on the country's current situation. The results show the possibility of converting from face-to-face to online teaching, which can be done, even in a short duration, thorough application of different strategies to solve this complex problem, proving the 10,000-h rule outlined by Gladwell [10]. There was no significant difference in the academic performance between students who did the face-to-face classes and students who attended the online classes.

References

1. Allen IE, Seaman J (2016) Online report card: tracking online education in the United States. s.l.:Babson Survey Research Group
2. Armin A et al (2021) Digital teaching 2020: students assess attention during an online lecture as equivalent to a face-to-face lecture. Der Ophthalmologe: Zeitschrift der Deutschen Ophthalmologischen Gesellschaft
3. Bao W (2020) COVID-19 and online teaching in higher education: A case study of Peking university. Hum Behav Emerg Technol 2(2):113–115
4. Beale EG, Tarwater PM, Lee VH (2014) Embryology instruction with online lectures in a human anatomy course. Anat Sci Educ 7:234–241
5. Bernama (2021) EMCO in several localities in five states, says Ismail Sabri. https://www.theedgemarkets.com/article/emco-several-localities-five-states-says-ismail-sabri. Accessed 4 June 2021
6. Brakora K, Day L, Choo D (2021) Succeeding in a pandemic: Online lectures and in-person laboratory teaching maintained exam scores in the anatomical sciences. FASEB J 35
7. Bunyan J (2020) The response of university towards COVID-19 Measures—Singapore. https://www.malaymail.com/news/malaysia/2020/03/16/pm-malaysia-in-lockdown-from-wed-until-march-31-all-shops-closed-except-for/1847204. Accessed 4 June 2021
8. Chan C, Koh YY (2008) Different degrees of blending benefit students differently: a pilot study. Khon Kaen, s.n.
9. Fan V, Cheong R (2021) Malaysia: MCO, CMCO, RMCO, CMCO Again: Regulations And SOPs. https://www.mondaq.com/operational-impacts-and-strategy/1022936/mco-cmco-rmco-cmco-again-regulations-and-sops. Accessed 4 June 2021
10. Gladwell M (2008) Outliers: the story of success. Little, Brown Co, San Francisco
11. gov.sg (2020) COVID-19 circuit breaker: Closure of workplace premises. https://www.gov.sg/article/circuit-breaker-extension-and-tighter-measures-what-you-need-to-know#:~:text=PM%20Lee%20on%2021%20April,1%20June%202020%20(inclusive).&text=From%20now%20till%204%20May%202020%2C%20stay,home%20as%20much%20as%20possible. Accessed 4 June 2021
12. Grabe M, Christopherson K (2008) Optional student use of online lecture resources: resource preferences, performance and lecture attendance. J Comput Assist Learn 24(1)
13. Havard Business Review (2015) The business case for managing complexity. https://hbr.org/resources/pdfs/comm/sap/19277_HBR_SAP_Report_5.pdf. Accessed 11 June 2021
14. Hoic-Bozic N, Mornar V, Boticki I, (2009) A blended learning approach to course design and implementation. IEEE Trans Educ 52(1)
15. Keiper MC, White A, Carlson CD, Lunipek JM (2020) Student perceptions on the benefits of Flipgrid in a HyFlex learning. J Educ Bus
16. Marinoni G, van't Land H & Jensen T, (2020) The impact of COVID-19 on higher education around the world: IAU global survey report. Int Assoc Univ, Paris
17. Martínez P, Aguilar FJ, Ortiz M (2020) Transitioning from Face-to-Face to blended and full online learning engineering master's program. IEEE Trans Educ 63(1)
18. McGee P, Reis A (2012) Blended course design: A synthesis of best practices. J Asynchronous Learn Netw 16(4):7–22
19. MOH (2020) Ministry of health. https://www.moh.gov.sg/news-highlights/details/moving-into-phase-two-of-re-opening Accessed 4 June 2020
20. Onime CE, Uhomoibhi JO (2013) Using interactive video for on-line blended learning in engineering education. s.l., s.n
21. Ożadowicz A (2020) Modified blended learning in engineering higher education during the COVID-19 Lockdown--Building automation courses case study. Educ Sci 10(292).
22. Pham H, Ho TT (2020) Toward a 'new normal' with e-learning in Vietnamese higher education during the post COVID-19 pandemic. Higher education research and development 39(7).
23. Porumb C, Orza B, Micu D, Porumb S (2012) Cloud computing and its application to blended learning in engineering. s.l., s.n.

24. Singapore Statues Online (2020) COVID-19 (Temporary Measures) (Control Order) Regulations 2020. https://sso.agc.gov.sg/SL-Supp/S254-2020/Published/20200407?DocDate=20200407 Accessed 4 June 2021
25. Soesmanto T, Bonner S (2019) Dual mode delivery in an introductory statistics course: design and evaluation. J Stat Educ 27(2):90–98
26. Thai NTT, De Wever B, Valcke M (2017) The impact of a flipped classroom design on learning performance in higher education: Looking for the best "blend" of lectures and guiding questions with feedback. Comput Educ 107:113–126
27. UNESCO, 2020. Distance learning strategies in response to COVID-19 school closures. https://unesdoc.unesco.org/home

Engineering Computing

A Task-Driven Visualization Framework for Diabetes Management

Lip Shen Low, Wan Tze Vong, Brian Chung Shiong Loh, and Patrick Hang Hui Then

Abstract Diabetes are chronic diseases that require patients to regularly manage their own health conditions. The goal is to have their blood glucose levels under control by making frequent changes to their diabetes therapy, diet, and lifestyle behaviour. As a result, patients are advised to use personal health devices such as glucometers and blood pressure meters to collect their own health data. These multi-source health data are presented by patients to their health care providers during clinical visits in separate visualizations and are often complicated by the variety of data sources and formats. To address this issue, TaskVisDMa conceptual task-driven visualizations framework for diabetes management—is proposed in this paper to produce advanced visualizations that can provide both informative and actionable insights for health care providers to make correct reasoning for treatment plan adjustments and to better communicate with patients. The framework is composed of a collection of data pre-processing, design choices and visualization strategies to handle multi-source and multi-dimensional patient-generated health data, and is divided into five modules: an input module that receives multiple health datasets, a data pre-processing module that processes and integrates the input datasets, an data-idiom-task linking module that uses a task base to determine the data attributes and visualization idioms that best represent the data-analysis tasks identified from the pre-processed dataset, a visualization generation module that uses a strategy base to generate a holistic visual representation of the multi-source and multi-dimensional datasets, and an output layer that presents an interactive and holistic visual representation to users. Our future work will focus on: (1) expanding the task base and strategy base to formulate if–then rules for the automated selection of design choices and visual and analytic strategies; and (2) advancing the framework as a tool for creating visualizations from patient-generated health data for diabetes management.

Keywords Information visualization · Patient-generated data · Diabetes management

L. S. Low (✉) · W. T. Vong · B. C. S. Loh · P. H. H. Then
Faculty of Engineering, Computing and Science, Swinburne University of Technology, Kuching, Sarawak, Malaysia
e-mail: llow@swinburne.edu.my

C. S. Choo et al. (eds.), *Proceedings of ASEAN-Australian Engineering Congress (AAEC2022)*, Lecture Notes in Electrical Engineering 1072,
https://doi.org/10.1007/978-981-99-5547-3_9

1 Introduction

Diabetes mellitus, a group of metabolic chronic diseases characterized by elevated blood glucose levels, can lead to a variety of complications such as cardiovascular disease, amputations, and nerve damage [1]. To prevent both short-term and long-term complications, diabetics must modify their eating habits and lifestyles to maintain healthy blood glucose levels. As a result, patients are advised to use personal health devices such as glucometer, blood pressure meters and body weight scales, and mobile applications such as MySugr, Health2Sync and Glucose Buddy to collect their own health data. These multi-source health data are then presented by patients to their health care providers (HCPs) in logbook format or in separate graphs using mobile applications during clinical visits [2]. However, these health data are frequently complicated by the volume and variety of data collected using different devices and displayed in different graphical formats and units of measurement, causing them difficult to be analysed by care providers for valuable and actionable insights [3].

A high level of continuity of care has been associated with reduced complication risks, mortality rate and health service utilization among diabetic patients [4]. The first step in achieving high continuity of care is to gather patient-generated data from different diabetic devices in one place. The second step is to visualize the data in the most effective ways possible for meaningful insights. Meaningful insights such as trends, distribution, and outliers can help HCPs in developing correct reasonings about the interplay between events such as exercise intensity, food intake and insulin administration, and blood glucose levels. For example, a blood glucose line chart can help identify both hyperglycemic and hypoglycemic trends over time [5] while multiple box plots can help show the change in carbohydrate and insulin intake before and after meals [6]. These insights can in turn help HCPs in designing effective treatment plans as well as educating and communicating with their patients in self-management of their diabetes and about behavior change for better treatment outcomes [7–9].

A considerable number of information visualization tools have been developed to support HCPs in deriving visualizations from patient-generated health data for valuable insights. Platforms such as Tidepool and Glooko aim to overcome the lack of data interoperability by combining diabetes data from multiple devices and displaying them in a uniform manner [10]. Many research prototypes have also been developed to apply different visualization strategies such as temporal event sequence alignment, temporal folding, and juxtaposed views to help HCPs make sense of complex and multidimensional diabetes data [6, 11, 12]. Despite the advances in data integration and visualizations, challenges persist. Existing diabetes information visualization tools are limited by their ability to support task-driven visualizations from multi-source patient-generated health data. In an attempt to close this research gap, VizML [6] introduced a unique task abstraction approach called hierarchical task abstraction to classify the decision-making process HCPs utilize for diabetes management and to define the design requirements for IDMVis—an interactive temporal event sequence visualization tool. However, the approach covers part of the data analysis tasks undertaken by HCPs in managing patients with Type 1 diabetes only, which in turn limits the visualizations that can be produced using IDMVis for insight extraction and for patients with other diabetes types. Moreover, instead of allowing users to specify their data analysis tasks for graph visualizations, most existing diabetes visualization tools are specifically designed to analyze certain types of patient health data with predefined visualizations and data analysis tasks. There is also a lack of study that investigates how different data analysis tasks could affect the design space, including data, view, and strategy, of an information visualization tool for diabetes management. The present study addresses these gaps by proposing a conceptual framework to support HCPs in deriving meaningful visualization insights from multi-source patient-generated health data based on their data analysis task requirements. In this paper, we contribute: (1) a task base comprising 12 data analysis tasks together with the corresponding data attributes and visualization idioms; (2) a strategy base comprising eight groups of data analysis tasks that are summarized with the visual and analytic strategies used to arrange and process different design choices; and (3) a conceptual framework for task-driven visualizations of patient-generated health data for diabetes management, TaskVisDM.

The rest of the paper is arranged as follows: Sect. 2 provides an overview of the proposed conceptual framework. Sections 3–6 describes the five modules of the framework, followed by the conclusion and future work in Sect. 7.

2 Conceptual Framework Overview

As shown in Fig. 1, the proposed conceptual framework is divided into five modules. The input module is designed to collect patient-generated data residing in different sources. As mentioned earlier in this paper, patients are advised to regularly use personal health devices to maintain healthy blood glucose levels. These multi-source patient-generated health data are stored either in paper files or electronically. Therefore, users will be able to load datasets or enter the data manually. The data preprocessing module is designed to extract dataset properties such as data type, number of attributes and number of samples from the multi-source datasets, to standardize the format of the data attributes, to transform numeric attributes to nominal, and to convert the one-dimensional datasets into a multi-dimensional dataset. Users will also have the freedom to adjust the properties of the loaded datasets. The purpose of this module is to combine multi-source datasets and display them in a uniform manner so that it is easier for HCPs to understand the interplay between different events such as exercise intensity and medication, and blood glucose levels. Following pre-processing, the data-idiom-task linking module is designed to determine the best design choices for the multi-dimensional dataset based on the tasks and columns (i.e., data attributes) selected by users. This is achieved by using a rule-based approach to

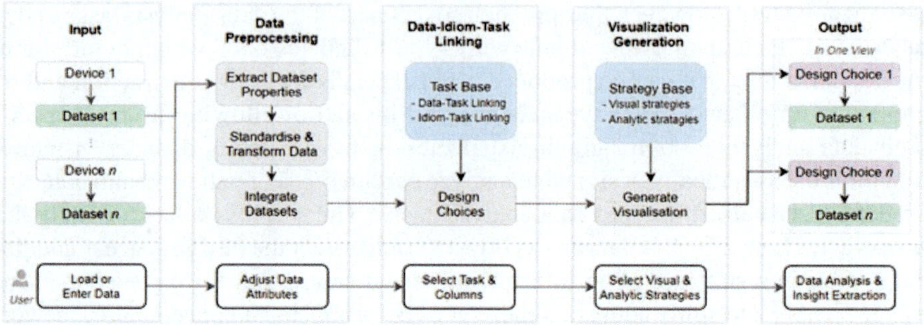

Fig. 1 A conceptual task-driven visualization framework for diabetes management, TaskVisDM

determine the visualization idioms that best represent the tasks and columns selected by users. A range of data analysis tasks involved in diabetes management such as blood glucose pattern over the past 14 days and hyperglycemic event frequency in the past one month, together with their corresponding data attributes and visualization idioms will be used in this module to determine the chart properties such as type of charts, color, and axis for the visualization generation module. The visualization generation module is designed to employ visual strategies such as juxtaposed and superimposed views and analytic strategies such as temporal event sequence alignment and temporal folding to help HPCs gain meaningful insights from visualizations for analytical reasonings. A set of hand-crafted or machine-generated rules will be used to determine the visual and analytic strategies that can be applied to the selected task and columns. Lastly, the output module displays interactive visualizations based on the task, columns and strategies selected by HCPs to perform data analysis tasks and insight extraction for analytical reasonings.

3 Input and Data Pre-processing

The input module is used to read and load multiple datasets from different personal health devices. Users will be able to import data files in common spreadsheet format such as CSV or Excel file or enter their data manually into a spreadsheet table. The datasets will then be loaded into the TaskVisDM tool and will serve as the input of the data pre-processing module. The main objective of the data pre-processing module is to process and integrate the multi-source datasets and prepare the integrated dataset for the data-idiom-task linking and visualization generation modules. The module is divided into three parts. The first part extracts the meta information about each input dataset, which includes column names, data types, and number of columns and rows. The second part processes the data from multi-source datasets into a structured and standardized format. A variety of diabetes data such as blood glucose and blood pressure readings include a date and time stamp, which can be recorded in different temporal data types such as Date, Time, and Timestamp and in different units of measurement such as mmol/L, mg/dL, and percentage. The module includes functionality to decompose time-series attributes into different temporal scales (such as from daily to weekly to monthly) and to transform numerical attributes into nominal attributes (such as from daily blood glucose readings to number of hyper and hypo events over the past 7 days). The third part aims to map the multi-source one-dimensional time series datasets to the same temporal interval in order to integrate them into a multi-dimensional time-series dataset. This helps reduce the complexity of the multi-source datasets and enable analytic strategies such as temporal sequence event alignment to be applied in the subsequent visualization generation module. Users will have the option to alter the column header and the unit of measurement of each attribute in the integrated datasets based on their preferences. The data pre-processing module produces two types of datasets: uni- or multi-dimensional time-series dataset with a date/time column and non-time series dataset.

4 Data-Idiom-Task Linking

The effectiveness of a visualization is dependent on its design and how well it supports the tasks a user wishes to complete [13]. Therefore, choosing an appropriate visual encoding method is essential to ensure that the resulting visualization is useful. As mentioned in the introduction, existing information visualization tools support part of the data analysis tasks undertaken by HCPs only and are limited by their ability to allow users to specify their intended data analysis tasks for graph visualizations. The present study sought to extend the previous research by VizML [6] to close this research gap by including more data analysis tasks that are related to diabetes management into the proposed task-driven visualization framework. The data-idiom-task linking module aims to determine the best design choices for the pre-processed datasets based on the tasks and data attributes (or columns) selected by the users. A task base comprising 12 data analysis tasks is created by identifying data analysis tasks from previous research works that focused on the effectiveness of different forms of visualization related to diabetes management activities. As shown in Table 1, each data analysis task is linked with the corresponding data attributes (such as blood glucose reading and insulin administration) and visualization idioms (line chart, violin plot and bar chart) found in the literature. The line chart idiom for example is a visual encoding of one quantitative value attribute and one ordered key attribute that displays a series of data points connected by a single line, which is commonly used in diabetes management to help identify trends in blood glucose levels over time for hyperglycemic and hypoglycemic events. Using this task base, the module determines the design choice properties such as chart type, color scheme, axis scaling and time range for the following visualization generation module. Other than that, users will be able to adjust the data analysis tasks and data attributes to generate different visualizations for various types of insights.

5 Visualization Generation

Multi-source patient-generated data stored using disparate platforms such as using mobile applications or paper-based logbooks can be challenging to analyze. Therefore, the visualization generation module aims to create a holistic visual representation of multi-source datasets with analytics elements that can support users in identifying meaningful insights from complex patient-generated health data in one view. The module provides the functionality to create, design and arrange the design choices stemmed from the data-idiom-task module. Previous work has investigated various strategies to visualize multi-dimensional diabetes data. Visual strategies that have been used to arrange design choices in a one view include tabular, juxtaposed, and superimposed views [19]. For instance, juxtaposed views allow users to easily review and digest multiple charts by placing them side-by-side in one screen without requiring them to memorize the details of multiple charts that are displayed in separate screens, while superimposed views overlay two charts on top of each other in one view to allow users to compare across different datasets without having them to jump between different parts of the screen. A variety of analytical strategies such as extraction, alignment, folding and pattern simplification have also been used to support users in identifying meaningful insights from visualizations [19]. For instances, temporal folding has been used to split long-interval time-series data into short-interval data to reduce the complexity of visualizations; aligning strategy is applied to align multiple charts by day of the week to understand before and after-effect events for cause-and-effect relationships exploration. Although prior work has investigated approaches to support the visualizations of multi-dimensional health data, there is a lack of research on how various data analysis tasks can influence the design space, which includes the data, view, and strategy, of an information visualization tool for managing diabetes. A strategy base is presented in this paper in attempt to reduce this gap. Table 2 shows eight groups of data analysis tasks. Each group is summarized with information about how visual strategies and analytic strategies are applied to arrange and design multiple design choices in one view. For example, three types of charts are used to visualize the first group of data analysis tasks in Table 2: small multiple line charts, single line chart and violin plots. Small multiple line charts are used to display two weeks of event sequence data over 24 h and users can select one of the days involved to view event details in a separate line chart. Violin plots on the other hand are used to show the distribution of insulin and carbohydrate intake at three mealtimes. The small multiple and single line charts

Table 1 Task base for data-idiom-task linking

Data analysis task	Data attribute	Idiom	References
Correlation between physical exercise and blood glucose levels	Physical exercise (NUM), blood glucose (NUM)	Scatterplot 	[14]
Compare frequency of hypoglycemic events across all mealtimes	Blood glucose (NUM), mealtimes (CAT)	Stacked bar chart 	[15]
Compare percentage of hypo- and hyper-glycemic events across all mealtimes	Percentages of hypo and hyper events (NUM), mealtimes (CAT)	Stacked bar chart 	[12]
Compare blood glucose levels by day of week	Blood glucose (NUM), day of week (CAT)	Stacked bar chart 	[15]
Compare average blood glucose levels over 7 days	Average blood glucose (NUM), 7 days (CAT)	Bar chart 	[16]
Distribution of blood glucose levels over 30 days	Blood glucose (NUM), timestamp (TEM)	Histogram 	[12]
Distribution of insulin at three mealtimes	Insulin (NUM), mealtimes (CAT)	Violin plot 	[6]
Distribution of carbohydrates at three mealtimes	Carbohydrates (NUM), mealtimes (CAT)	Violin plot 	[6]
Trend of blood glucose levels over 14 days	Blood glucose (NUM), timestamp (TEM)	Line chart 	[17]
Blood glucose reading in the hyperglycemic range (outliers) over 14 days	Blood glucose (NUM), timestamp (TEM)	Line chart 	[5]

(continued)

Table 1 (continued)

Data analysis task	Data attribute	Idiom	References
Blood glucose reading in the hypo- and hyperglycemic ranges (outliers) by time of a day and day of a week	Blood glucose (CAT), time of a day (TEM), day of a week (CAT)	Scatterplot 	[15]
Overview of HbA1c	HbA1c (NUM), timestamp (TEM)	Gauge chart 	[18]

TEM Temporal Data, *CAT* Categorical Data, *NUM* Numerical Data

are placed on top of each other and aligned by hour to support the identification of hypo- and hyperglycemic events over 24 h. Similarly, the violin plots for insulin and carbohydrate intakes are aligned by mealtimes to understand the effect of breakfast, lunch and dinner to the blood glucose levels [6]. An analysis of the strategy base shows the potential to extract features such as task type, chart type, number of charts and time scale from the visualizations for rule-based generation.

The strategy base will be further expanded in our future study by identifying visualizations that have been used to present multiple heterogenous datasets from literature. Important features will then be extracted from this strategy base to investigate the relationship between data analysis tasks and visual and analytic strategies. The findings will then be formulated into if–then rules to determine the visual and analytic strategies that should be applied to visualize multi-source and multi-dimensional datasets in one view.

6 Output

The output module is used to output a holistic visual representation that is not only interactive but also designed with visual elements that can assist users in extracting different insights for analytical reasoning. As shown in Fig. 2, the visualizations are juxtaposed and aligned by hour. Users will be able to adjust the task, the attributes from given datasets, and the duration of 1-day, 7-day, 14-day, 30-day and 60-day to generate different visualizations for insight extraction.

7 Conclusion and Future Work

In this paper, we present a conceptual framework, TaskVisDM, to support HCPs in management patients with diabetes. The framework aims to address the limitations of existing diabetes information visualization tools in supporting task-driven visualization from multi-source and multi-dimensional datasets. A task base and a

Table 2 Strategy base for visualization generation

Data analysis task (idioms)	Visual strategy	Analytic strategy	References
DC1: distribution of blood glucose levels over 24 h for 14 days (small multiple line charts) DC2: distribution of blood glucose levels over 24 h for a specific day (line chart) DC3: distribution of basal and bolus insulin (violin plots) DC4: distribution of insulin at three mealtimes (violin plots) DC5: distribution of carbohydrates at three mealtimes (violin plots)	Juxtaposed (DC1 and DC2 are placed on top of each other; DC3, DC4 and DC5 are placed on top of each other)	Alignment (DC1 and DC2 are aligned by hour; DC4 and DC5 are aligned by mealtimes) Folding (by adjusting analysis period)	[6]
DC1: distribution of blood glucose levels over 24 h in a day (line chart) DC2: comparison of bolus insulin and carbohydrate intakes over 24 h in a day (bar chart with annotation) DC3: distribution of basal rate over 24 h in a day (area chart)	Juxtaposed (DC1, DC2 and DC3 are placed on top of each other)	Alignment (DC1, DC2 and DC3 are aligned by hour) Folding (by adjusting analysis period)	[11]
DC1: distribution of blood glucose levels over the past 14 days (scatterplot) DC2: distribution of blood glucose levels over 24 h in the past 14 days (superimposed scatterplot) DC3: distribution of hyperglycemic events over 24 h (area chart) DC4: comparison of three categories of blood glucose events (donut chart)	Juxtaposed (DC1, DC2, DC3 and DC4 are placed on top of each other or side-by-side)	Alignment (DC2 and DC3 are aligned by hour) Folding (by adjusting analysis period) Pattern simplification (by grouping blood glucose levels)	[20]
DC1: distribution of blood glucose levels over 24 h (line chart) DC2: distribution of insulin delivery over 24 h (line chart) DC3: distribution of carbohydrate intake over 24 h (line chart)	Juxtaposed (DC1, DC2 and DC3 are placed on top of each other)	Alignment (DC1 and DC2 are aligned by hour)	[21]
DC1: distribution of blood glucose levels over 3 days (line chart) DC2: distribution of carbohydrate intake over 3 days (line chart) DC3: comparison of insulin delivery methods over 3 days (bar chart)	Superimposed (DC1, DC2 and DC3 are overlayed on top of each other)	Alignment (DC1, DC2 and DC3 are aligned by hour)	[17]
DC1: distribution of blood glucose levels over 24 h by day of the week (superimposed line charts)	Superimposed (line charts are overlayed on top of each other)	Alignment (small line charts are aligned by hour)	[22]

(continued)

Table 2 (continued)

Data analysis task (idioms)	Visual strategy	Analytic strategy	References
DC1: distribution of average blood glucose levels over 24 h from year 2010 to year 2013 (small multiple line charts)	Juxtaposed (small line charts are placed on top of each other)	Alignment (small line charts are aligned by hour)	[22]
DC1: distribution of temporary basal, reported sleep, skin temperature, air temperature galvanic skin response and heart rate over 24 h (superimposed line charts) DC2: distribution of basal rate, continuous glucose monitoring, meals, exercise, bolus event, and hypo event over 24 h (superimposed scatter plot and line chart)	Juxtaposed (DC1 and DC2 are placed on top of each other)	Alignment (DC1 and DC2 are aligned by hour)	[23]

DC Design Choice

strategy base are also presented in this study for future formulation of hand-crafted or machine-generated rules for the selection of design choices and the application of visual and analytic strategies for a holistic visual representation of patient-generated health data. Our future work will focus on expanding the task base and strategy based and the development of TaskVisDM as a visualization tool for diabetes management. Besides, the TaskVisDM prototype will be compared with other visualization tools such as Rath and Microsoft Power BI to evaluate its effectiveness in producing the desired design choices. Evaluation metrics such as entropy and Gini coefficient will be used to evaluate the performance of each visualization tool based on a collection of dataset-visualization pairs collected from the Plotly Community Feed.

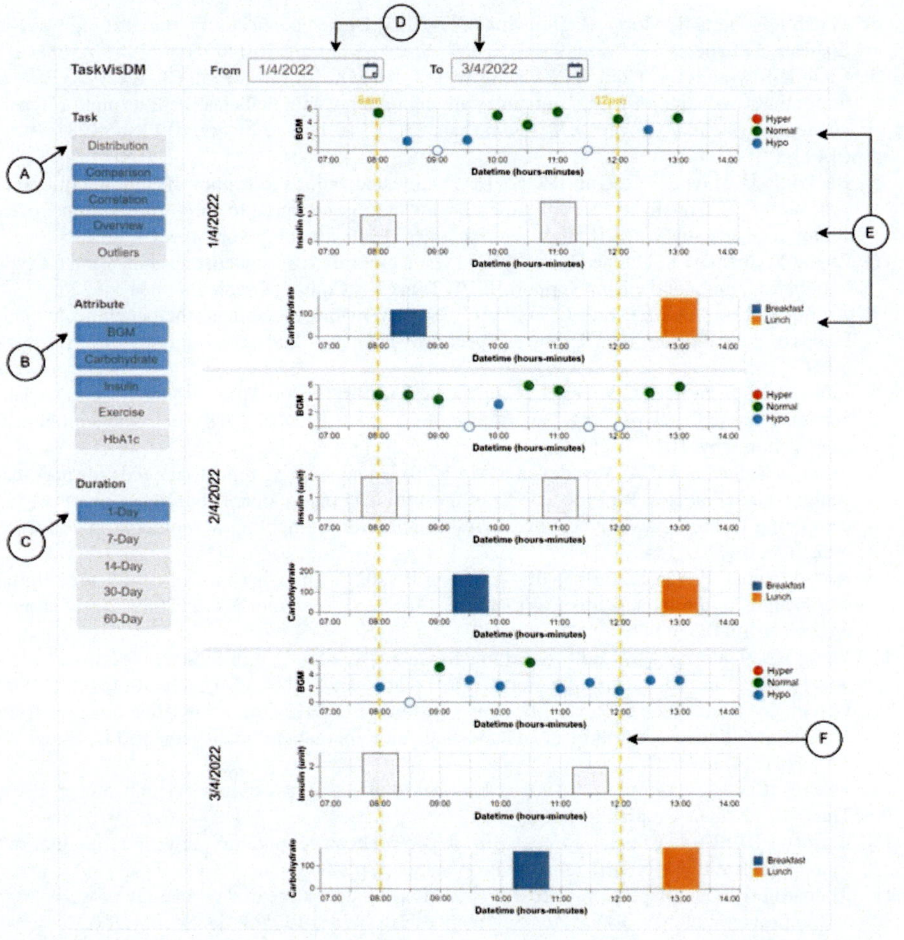

Fig. 2 A sample output of the TaskVisDM framework. **a** Task selection, **b** attributes selection, **c** duration (temporal folding), **d** dates selection, **e** juxtaposed charts, **f** alignment of timeline (alignment)

Acknowledgements This research was supported by Swinburne HDR Scholarship fund provided by Swinburne University of Technology Sarawak Campus.

References

1. Mauricio D, Alonso N, Gratacòs M (2020) Chronic diabetes complications: the need to move beyond classical concepts. Trends Endocrinol Metab 31(4):287–295. https://doi.org/10.1016/j.tem.2020.01.007
2. Wolfsdorf J (2012) Intensive diabetes management. American Diabetes Association

3. Walsh J, Roberts R, Morris R, Heinemann L (2015) Device connectivity: the next big wave in diabetes. J Diabetes Sci Technol 9(3):701–705. https://doi.org/10.1177/1932296814568806
4. Chan K-S, Wan EY-F, Chin W-Y, Cheng WH-G, Ho MK, Yu EY-T, Lam CL-K (2021) Effects of continuity of care on health outcomes among patients with diabetes mellitus and/or hypertension: a systematic review. BMC Fam Pract 22(1):145. https://doi.org/10.1186/s12875-021-01493-x
5. Backonja U, Haynes SC, Kim KK (2018) Data visualizations to support health practitioners' provision of personalized care for patients with cancer and multiple chronic conditions: user-centered design study. JMIR Hum Factors 5(4):e11826. https://doi.org/10.2196/11826
6. Zhang Y, Chanana K, Dunne C (2018) IDMVis: temporal event sequence visualization for type 1 diabetes treatment decision support. IEEE Trans Vis Comput Graph 25(1):512–522
7. Arbiter B, Look H, McComb L, Snider C (2019) Why download data: the benefits and challenges of more diabetes data. Diabetes Spectrum 32(3):221–225. https://doi.org/10.2337/ds18-0099
8. Rajabiyazdi F, Perin C, Oehlberg L, Carpendale S (2018) Personal patient-generated data visualizations for diabetes patients. Paper presented at the IEEE VIS 2018 electronic conference, Berlin, Germany, 2018-10-21
9. Prioleau T, Sabharwal A, Vasudevan MM (2020) Understanding reflection needs for personal health data in diabetes. In: Proceedings of the 14th EAI international conference on pervasive computing technologies for healthcare. Association for Computing Machinery, pp 263–273. https://doi.org/10.1145/3421937.3421972
10. Ramakrishnan P, Yan K, Balijepalli C, Druyts E (2021) Changing face of healthcare: digital therapeutics in the management of diabetes. Curr Med Res Opin 37(12):2089–2091. https://doi.org/10.1080/03007995.2021.1976737
11. Wong JC, Neinstein AB, Look H, Arbiter B, Chokr N, Ross C, Adi S (2017) Pilot study of a novel application for data visualization in type 1 diabetes. J Diabetes Sci Technol 11(4):800–807
12. Derozier V, Arnavielhe S, Renard E, Dray G, Martin S (2019) How knowledge emerges from artificial intelligence algorithm and data visualization for diabetes management. J Diabetes Sci Technol 13(4):698–707
13. Saket B, Endert A, Demiralp Ç (2018) Task-based effectiveness of basic visualizations. IEEE Trans Vis Comput Graph 25(7):2505–2512
14. Ehrmann BJ (2014) Chapter contents. In: A comprehensive guide to geriatric rehabilitation: [previously entitled Geriatric rehabilitation manual], p 329
15. Rodbard D (2009) Display of glucose distributions by date, time of day, and day of week: new and improved methods. J Diabetes Sci Technol 3(6):1388–1394. https://doi.org/10.1177/193229680900300619
16. Fareed N, Singh P, Boettcher E, Wang Y, Venkatesh K, Strouse R (2022) Developing and testing an integrated patient mHealth and provider dashboard application system for type 2 diabetes management among Medicaid-enrolled pregnant individuals based on a user-centered approach: mixed-methods study. medRxiv
17. Giordanengo A, Årsand E, Woldaregay AZ, Bradway M, Grottland A, Hartvigsen G, Granja C, Torsvik T, Hansen AH (2019) Design and prestudy assessment of a dashboard for presenting self-collected health data of patients with diabetes to clinicians: Iterative approach and qualitative case study. JMIR Diabetes 4(3):e14002. https://doi.org/10.2196/14002
18. Yandrapalli BT, Jones J, Purkayastha S (2019) Development and implementation of a dashboard for diabetes care management in OpenMRS. arXiv:191011437
19. Zhang Y, Parker AG, Dunne C (2020) Information visualization for diabetes management: a literature review. In: Proceedings of the 14th EAI international conference on pervasive computing technologies for healthcare, pp 427–430
20. Kumar RB, Goren ND, Stark DE, Wall DP, Longhurst CA (2016) Automated integration of continuous glucose monitor data in the electronic health record using consumer technology. J Am Med Inform Assoc 23(3):532–537. https://doi.org/10.1093/jamia/ocv206
21. Buckingham B (2008) Clinical overview of continuous glucose monitoring. J Diabetes Sci Technol 2(2):300–306. https://doi.org/10.1177/193229680800200223

22. Lee V, Thurston T, Thurston C (2017) A comparison of discovered regularities in blood glucose readings across two data collection approaches used with a type 1 diabetic youth. Methods Inf Med 56(S 01):e84–e91
23. Marling C, Bunescu R (PMC7881904) The OhioT1DM dataset for blood glucose level prediction: update 2020. In: CEUR workshop proceedings, Sep 2020, pp 71–74

Experimental Studies on Horizontal Shear Capacity of Prestressed Beams Casted in Stages with Multiple-Stage Tendon Stressing as Compared to Various Codes

Hin-Foo Low, Sih-Ying Kong, and Daniel Kong

Abstract In concrete composite construction, the first cast and the second cast of the concrete must be effectively connected together so that the structure behaves in a monolithic manner as a whole. If the two casts are not bonded together, slips occur at the interface under flexure and each component slides against each other at the interface joint. This will cause each component of the composite subjected to bending about their own individual centroidal axis, and eventually result in substantial reduction in stiffness and structural strength of the section. To prevent this, various codes of practice include provisions for the design of the horizontal shear across the interfacing concrete joint of the two casts in order to ensure full composite action can be developed. For practical construction of thick prestressed members at site, the structures are normally casted in few layers in two or three stages. It is not uncommon that the first cast of the concrete is designed to be prestressed to support the wet concrete weight of the subsequent casts in order to reduce the burden of the formwork. This paper studies the effect of multiple-stage tendon stressing on the horizontal shear capacity of a prestressed beam specimens casted and stressed in stages. The experimental results obtained from the specimen testing are then compared with the interfacing shear capacity computed by various codes. The major international codes of practice, like Eurocode EC2:2004, American ACI318M-14, British BS8110:1997, Australian AS3600-2009, and Chinese GB50010:2010 are considered in this research [1–5].

Keywords Interfacing shear · Horizontal shear · Interface slip capacity · Concrete bond strength between two casts · Horizontal shear strength at concrete interface · Joint strength between two casts of concrete · Concrete shear-friction capacity ·

H.-F. Low (✉)
OSD Consultants (M) Sdn Bhd, Kuala Lumpur, Malaysia
e-mail: lowhf@osdconsultants.com.my

S.-Y. Kong · D. Kong
School of Engineering, Monash University Malaysia, Subang Jaya, Malaysia

C. S. Choo et al. (eds.), *Proceedings of ASEAN-Australian Engineering Congress (AAEC2022)*, Lecture Notes in Electrical Engineering 1072,
https://doi.org/10.1007/978-981-99-5547-3_10

Interfacing shear stress · Horizontal shear stress · Effect of staged casting · Sequential tendon stressing

1 Introduction

In composite construction, the first cast and the second cast must be effectively connected together in order for the structure to behave in a monolithic manner. Horizontal shearing action develops at the interface joint between the two casts of concrete when the composite section is subjected to loading.

If the two castings are not bonded together, slips occur at the interface under flexure and both the components will be subjected to bending about their own individual centroidal axis which eventually results in substantial reduction in stiffness and structural strength [6].

Across the interface joint, horizontal shear stresses are building up under the bending action of the composition section. In an elastic uncracked section, the distribution of horizontal shear stress at the joint of two components can be calculated based on the shear flow using the following equation.

$$\tau_h = V \cdot S/I \cdot b_f \tag{1}$$

where τ_h represents the horizontal shear stress at the joint; V is the shear force; S is the first moment of area of the second cast about the centroid of the full composite section; I is the moment of inertia of the gross composite section; and b_f is the width of the contact surface.

This linear elastic shear flow will become non-linear and inelastic towards ultimate limit state after the concrete has cracked and redistribution of stress occurs in the material. Then, the average shear stress across the interfacing surface is used for the ultimate strength calculation. A simpler form of horizontal shear stress, τ^* acting on the contract surface at the ultimate load is derived in the equation below:

$$\tau^* = \beta V^*/z \cdot b_f \tag{2}$$

where V^* is the ultimate shear force; β is the ratio of the compressive force within the second cast and the total compressive force of the composite section; z is the internal moment lever arm of the section as indicated in Fig. 1; and b_f is the width of the contact surface.

In the evaluation of the interfacing slip behavior of various beam specimens in this research, Eq. 2 was used to compute the horizontal shear stress at the construction joint in order to reflect the ultimate stress distribution at failure. The computed horizontal shear stress from each category of the beam specimen with and without staged casting and stressing sequence were discussed [8, 9].

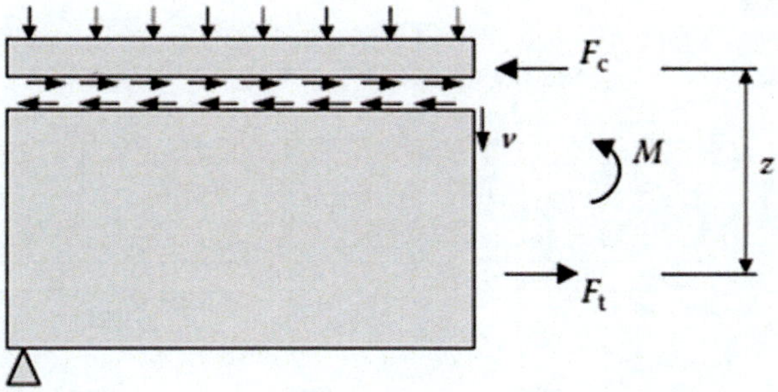

Fig. 1 Horizontal shearing action on composite beam [7]

2 Description of Test Specimens

The beam test specimens were prepared according to different staged casting and stressing sequence to study the effect of locked-in or residual stress on the cracking and ultimate behavior of the pre-stress structure. These specimens were subjected to three-point load test at quarter span to create more adverse effect to the beams in order to investigate the interfacing shear behavior when subjected to ultimate load.

Figure 2 shows the details of geometry and reinforcement of pre-stressed post-tensioned concrete beam specimens in this research. The beam specimens were prepared with length of 3.4 m. 12 mm diameter high tensile bars of yield strength 500 MPa were provided as top and bottom flexural reinforcements while 8 mm diameter shear links of mild steel with 250 MPa were used. Localized reinforcements consisted of 12 mm diameter U-shaped bars and closer links were provided at the end block to prevent bursting and spalling of concrete at the anchorage zone. Corrugated 60 mm diameter duct of straight profile was installed at a distance of 75 mm below the beam top and another one at 75 mm from the beam soffit before concrete casting. Each duct was installed with a single strand of 12.7 mm diameter and stressed to 75% of the breaking load (UTS = 184 kN). This strand was grouted before load testing.

Four types of specimens were prepared with different concrete casting and tendon stressing sequences to investigate their effects of locked in stresses on the interfacing shear strength of post-tensioned concrete beams casted in stages. The contact surface between the first and second cast was roughened to expose the coarse aggregate and it was cleaned from any loose materials before the subsequent cast. Concrete C35/45 (f_{ck}/f_{cu}) was adopted for the preparation of the beam specimens, and the average compressive strength of six concrete cylinders was 35.06 MPa with variation of 4.46%. The casting and stressing sequences for all beams are presented in Table 1.

Three-point load test setup was adopted with loading at quarter span (Fig. 3) so that higher shear force occurred at the shorter side of the beam to investigate the interface slip behavior at the beam section of lower shear resistance from concrete bonding with minimum amount of links. The beam end supports were located at 3000 mm, and steel cylinder was placed at the quarter span from one of the supports to load the beams at a constant displacement rate of 0.05 mm/s. Data logger was used to record the applied load until the failure of the beam specimens.

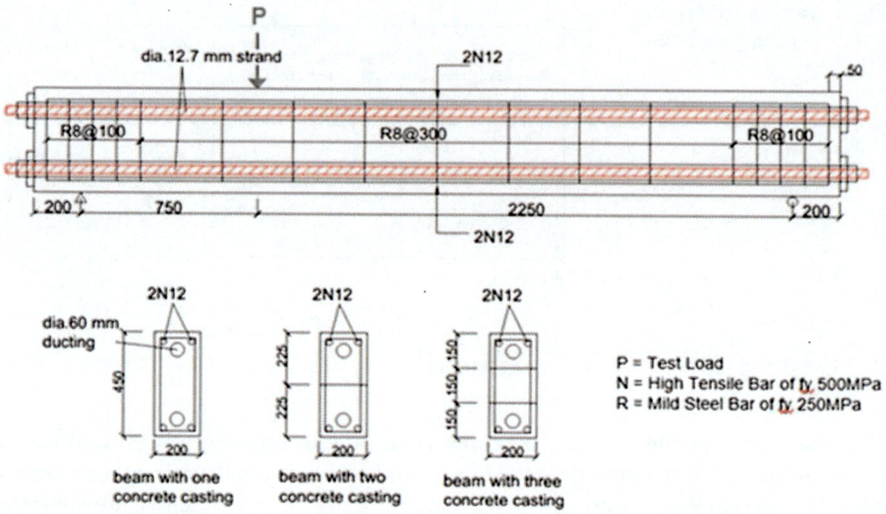

Fig. 2 Details of post-tensioned concrete beam specimens (all dimensions in mm)

Table 1 Summary of specimen type and their preparation procedures

No	Specimen	Abbreviation	Casting and stressing sequence
1	Full cast and simultaneous stressing	Control	The beams were prepared by a single concrete casting. Both top and bottom strands were stressed after 14 days of concrete casting
2	Two casts and simultaneous stressing	2C1S	The first concrete layer was cast to a height of 225 mm. After 7 days, the second layer was poured. Both top and bottom strands were stressed after 14 days of second casting
3	Two casts and sequenced stressing	2C2S	The first concrete layer was cast to a height of 225 mm. The bottom strand was stressed after 14 days of first casting. The second concrete layer was cast and the top strand was stressed after 14 days of second casting
4	Three casts and sequenced stressing	3C2S	The height of the first concrete layer was 150 mm and the bottom strand was stressed after 14 days of first casting, followed by casting of the second concrete layer and third concrete layer. The top strand was stressed after 14 days of third casting

Fig. 3 Experiment setup for three-point load test at the quarter span

3 Observation and Finding from Experiment Results

The first hairline crack appeared at the soffit of the point load at quarter span when the test load reached 115–135 kN for all the test specimens, and all the beam demonstrated flexural cracks of similar manner as the control beam. The initial flexural cracks were found propagating upwards while the crack width continued to increase. At the same time, more new cracks started to form adjacent to the initial cracks.

When the load continue to increase, it was found that the cracking pattern and their distribution were influenced by the location of the concrete casting joints. For beam specimens with two and three concrete casting (2C1S, 2C2S and 3C2S), the cracks propagated from the soffit upwards until the casting joint whereby they started to propagate along the interface surface before the cracks continued diagonally into the subsequent layer of the concrete. The observed crack patterns of the beam specimens are summarized in Fig. 4.

The control beam demonstrated a typical shear flexural crack pattern at failure as shown in Fig. 5a. For specimens with and without stressing stages, 2C1S beams demonstrated similar crack distribution (Fig. 5b) with the 2C2S beams (Fig. 5c).

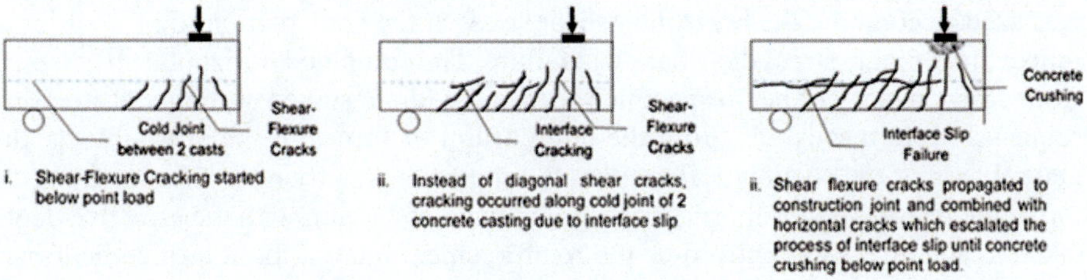

Fig. 4 Observed crack patterns on beam specimens during load test

However, more cracks were observed for beam specimens when casting increased from two casts to three casts. Beam specimen 3C2S shows more and severe concrete cracking at the interface between the bottom layer and middle layer (Fig. 5d), and at the same time, minor cracking at the interface between the middle layer and top layer of the concrete could be observed. Some cracks initiated at the soffit of the second concrete layer of beam 3C2S due to interface slip as shown in Fig. 5d. Similar results were reported by [10].

All the test specimens that involved staged casting and sequential tendon stressing demonstrated shear flexure cracks propagating to the construction joint and combined with horizontal cracks which escalated the process of interface slip until the concrete crushes below the position of the point load. The crack propagation at the contact surface was caused by the interface slip between the two concrete casting as the result of the failure of concrete adhesion (chemical bonding of concrete). Once this happened, the horizontal shear at the interface was sustained by the friction interaction between the contact surfaces, as well as the dowelling action of stirrups across the joint.

Beam specimens with two and three casting stage generally demonstrated a lower stiffness and strength as compared to the control beams as the result of the interface slip behavior at the construction joint. The test loads when the first crack of interface slip occurred, and the ultimate load at failure of the beam specimens prepared based on different casting sequences and staged tendon stressing are sumarized in Table 2.

From the test results, it was found that when the stages of construction increased, a lower failure load was generally obtained during the experiment with a failure load of 221 kN for the control beam, 216 kN for 2C1S and 204 kN for 2C2S, and 203 kN for 3C2S. Lower margin of structural reserve was observed with the increase of the number of beam casting due to the interface slip weakness across the joint. Nevertheless, it was found that the horizontal shear capacity at the interface was not affected much by the casting sequence and staged tendon stressing.

4 Evaluation of Interface Slip Resistance to Various Codes of Practice

In the evaluation of the interfacing slip behavior of various beam specimens, Eq. 2 was used to compute the horizontal shear stress at the construction joint in order to reflect the ultimate stress distribution at failure. The computed horizontal shear stress from each category of beam specimen with and without staged casting and stressing sequence were discussed. From the equilibrium of forces in the stress block, the neutral axis of the section at the ultimate limit state was found at 77 mm from the top of the beam. The lever arm, z was found to be 357.9 mm with the effective depth taken as the effective centroid of the reinforcement among the reinforcement bars and prestressing steel based on weighted average.

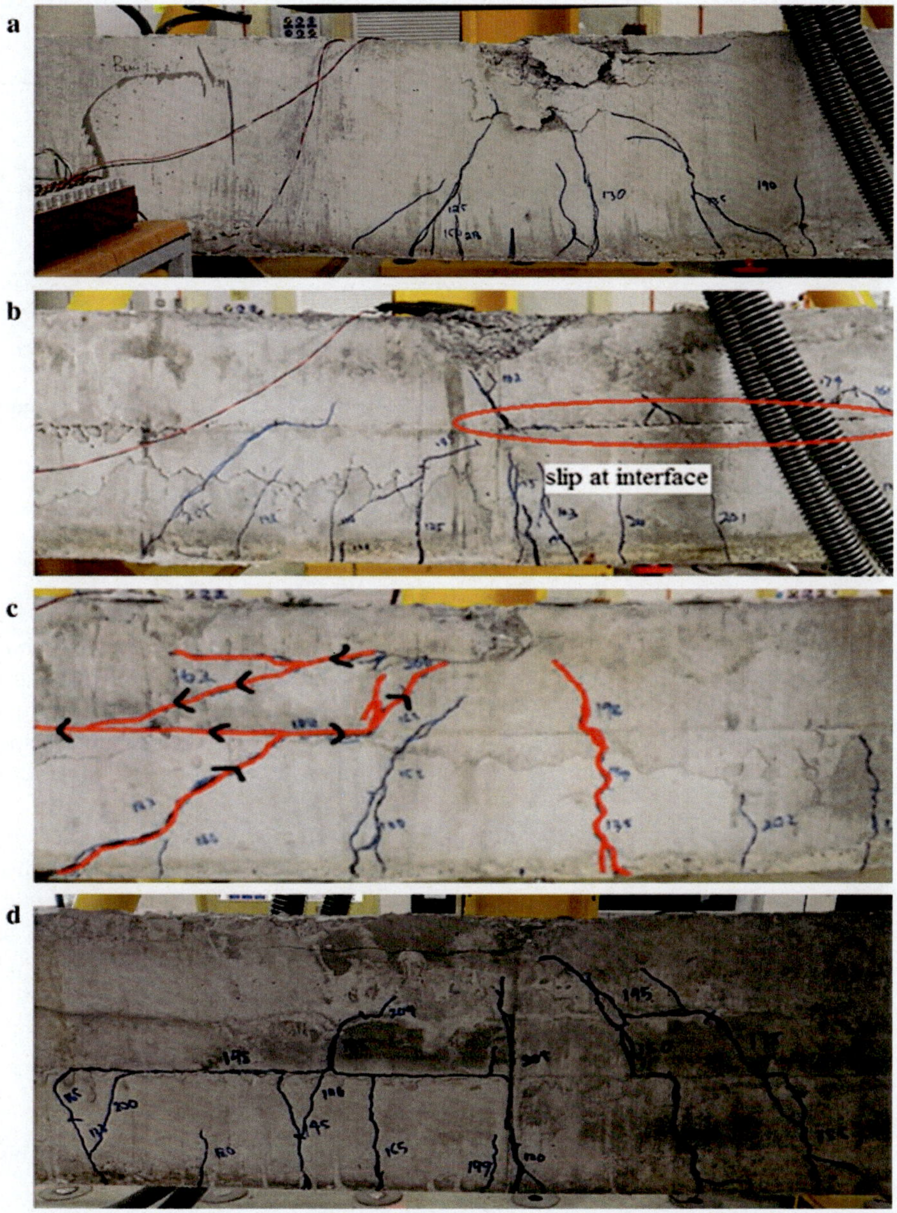

Fig. 5 Crack lines observed during experiments at the loaded quarter span **a** shear-flexural cracking and compression crushing of the control beam, **b** shear-flexural cracking, interface slip and compression crushing of the 2C1S beam, **c** shear-flexural cracking, interface slip and compression crushing of the 2C2S beam, **d** shear-flexural cracking, interface slip and compression crushing of the 3C2S beam

Table 2 Summary of load at first interface slip and load at failure of beam specimens

Beam reference	Max load at first interface slip (kN)	Max load at failure (kN)
Control	–	221
2C1S	195	216
2C2S	198	204
3C2S	190	203

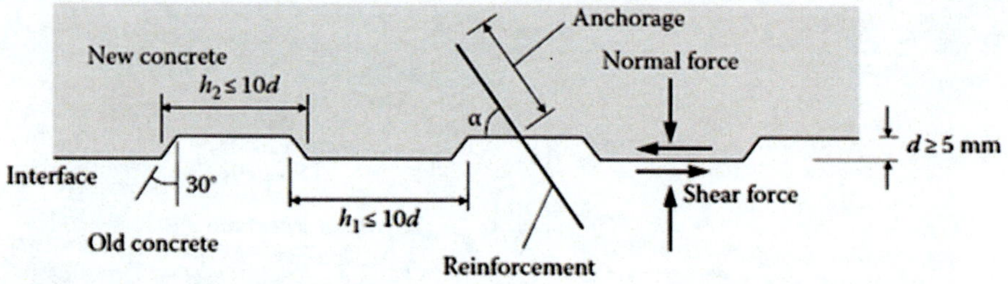

Fig. 6 Indented construction joint (EC2:2004)

Based on Eq. 2, the horizontal shear stress at the interface was calculated with β taken as 1.0 because the entire compressive stress block lies within the second cast with neutral axis is located at 77 mm from the top surface. The horizontal shear stress was calculated to be 2.07 MPa and this value shall be compared with the interfacing shear capacity of the test beam obtained using formulas in accordance with various major codes.

Most of the codes consider the interfacing shear strength of the contact surface were developed from three major components consisting of concrete chemical adhesion, interlocking action of aggregate or indented joints, and dowelling action of the reinforcement across the interface with sufficient anchorage. This is presented in Fig. 6 for illustration.

Based on Eurocode 2: 2004, the horizontal shear stress resistance, ν_{Rdi} at the interface was given as

$$\nu_{\text{Rdi}}(\text{EC}) = c\, f_{ctd} + \mu\, \sigma_n + \rho\, f_{yd}\, \mu \tag{3}$$

where c and μ are coefficients depending on the roughness of the interface. In this research, c and μ were taken as 0.4 and 0.7 respectively based on EC2 corresponding to the rough contact surface during preparation of the test specimen casting joint. f_{ctd} is the concrete tensile strength. In this research, the upper bound value of $f_{ctk,0.95}$ was adopted for evaluation purpose with the material factor, γ_m 1.0. Thus, $f_{ctd} =$ 4.16 MPa. σ_n was taken as the load at the first interface slip. ρ was calculated based on the ratio of link reinforcement (R8 at 300 mm spacing) to the contact surface area where 2 * 50 mm^2/200/300 = 0.00168. f_{yd} of mild steel link reinforcement was taken as 250 MPa with the material factor, $\gamma_s = 1.0$. Thus, ν_{Rdi} (EC) = 2.19 MPa.

AS3600-2009 adopted a similar approach in computing the horizontal shear stress capacity at the interface based on

$$\nu_{\text{Rdi}}\,(\text{AS}) = \mu[\,A_{sv}.f_{yd}/(s_v.b_f)\, + \sigma_n] + k_{co}f_{ct} \tag{4}$$

where A_{sv} and S_v are referred to the area of link reinforcement and link spacing respectively. μ and k_{co} were taken as 0.9 and 0.4 respectively from Table 8.4.3 of AS3600-2009 corresponding to a deliberately roughened surface. Instead of the lower characteristic tensile strength of the concrete f_{ct}, the upper bound value of $1.8{*}0.36\sqrt{fc} = 3.83$ MPa was used for the evaluation purpose [6].

Thus, ν_{Rdi} (AS) = 2.21 MPa.

For BS8110, ACI318 and GB50010, the contribution of friction due to normal load is not considered in the calculation of horizontal shear stress capacity in the code formula.

In BS8110-1997, the ultimate horizontal shear stress at interface was provided in Table 5.5: Part 1. Based on the concrete cube strength, f_{cu} of 45 MPa and surface treatment as brushed and roughened, horizontal shear stress capacity was found to be 2.2 MPa from the table. This had included a material factor of 1.5 according to note 4 of the table. For evaluation and comparison of this research, the upper bound value without material factor was used. Thus,

$$\nu_{\text{Rdi}}(\text{BS}) = 2.2 * 1.5 = 3.3\,\text{MPa}$$

ACI318M-14 did not address the horizontal shear stress resistance based on the type of surface treatment. A general formula of ν_{Rdi} at the interface was given as

$$\nu_{\text{Rdi}}\,(\text{ACI}) = 1.8 + 0.6A_{sv}\cdot f_{yd}/(s_v\cdot b_f) \tag{5}$$

using the same value for the parameters, ν_{Rdi} (ACI) = 2.05 MPa.

Similarly, GB50010 provided the horizontal shear stress resistance without considering the type of contact surface in the formula.

$$\nu_{\text{Rdi}}\,(\text{GB}) = 1.2f_t + 0.85A_{sv}\cdot f_{yd}/(s_v\cdot b_f) \tag{6}$$

With f_t of 1.8 MPa from Table 4.1.4-2 in GB50010, the horizontal shear stress capacity was found to be

$$\nu_{\text{Rdi}}\,(\text{GB}) = 2.52\,\text{MPa}$$

Table 3 summarizes the horizontal shear stress capacity calculated using the relevant formula given by various codes of practice in comparison with the corresponding shear stress at the interface when the first slip was observed on the test beam during testing.

The applied load when the first interface shear slip occurred was found to be constant among all the test specimens with and without the staged casting and

Table 3 Comparison of horizontal shear stress at 1st interface slip of beam specimen and calculated capacity from various codes

Beam ref.	Applied load when 1st slip occurred (kN)	Corresponding interface shear stress (MPa)	EC2-2004 (MPa)	S3600-2009 (MPa)	ACI318M-2014 (MPa)	GB50010-2010 (MPa)	BS8110-1997 (MPa)
2C1S	195	2.04	2.19	2.21	2.05	2.52	3.30
2C2S	198	2.04					
3C2S	190	2.00					

stressing sequence. Even though the interface joint between each layer of casting was located at different levels, 225 mm from soffit for test beams 2C1S and 2C2S while 150 mm from soffit for test beam 3C2S, the effect on the interface slip behavior was found not significant at the ultimate limit state where redistribution of force was expected to occur. This was supported by the fact that only minor difference of the applied load was observed when the first slip occurred between these beams during the testing.

From Eq. 2 with these loads, the corresponding horizontal shear stresses in the range of 2.00–2.08 MPa were obtained. These values were found matching the horizontal shear stress capacity calculated based on various codes of practice considering the upper bound values for concrete parameters and without any safety factor, except BS8110. Using the upper characteristic tensile strength for concrete, EC2 and AS3600 predicted interface slip failure occurred at a horizontal stress of 2.19 MPa and 2.21 MPa respectively. ACI318 calculated the horizontal shear stress capacity of 2.05 MPa which was very close to the value from the test, while GB50010 computed a higher stress of 2.52 MPa. In actual design, these values predicted by various codes were found to be safe if the lower bound values and the appropriate material factors are applied.

However, BS8110 was found to over-estimate the horizontal shear stress capacity of the contact surface at the joint by about 60%. It indicated a stress capacity of 3.3 MPa without the 1.5 concrete material factor whereas the interface joint of all the test beams slipped at the horizontal shear stress of 2.00–2.08 MPa. From the comparison among the codes, it seemed that the horizontal shear stress capacity from Table 5.5: Part 1: BS8110 might not have included the appropriate safety factor.

5 Conclusion

The interface shear behavior of the composite section depends on the location of the contact surface. The first slip at the interface joint was expected to occur at a similar test load intensity for all the beam specimens if the horizontal construction joint is located between the concrete compression stress block and the tension force of the reinforcements.

The concrete casting stages and the tendon stressing sequence were found not having significant effect on the interface slip capacity of all the beam specimens. The yielding of the interfacing shear links allowed the redistribution of the locked in stresses due to the construction stages at ultimate, and the interlocking action of the aggregate across the contact surface was found not unaffected.

From the specimen testing, it was found that various codes have sufficient design provisions to address the interfacing shear of different cast of a composite section. Even though adopting different approaches and design consideration, the horizontal shear capacity by various codes of practice were able to predict the interface slip within the upper bound approach (without material factor), except BS8110. The authors recommended to consider an appropriate reduction factor if the interface shear was designed according to BS8110 to allow for site tolerances.

Acknowledgements This work was undertaken with funding from the Monash Industry Matching Fund with OSD and Nehemiah as part of the Ph.D. research project of Ir. Dr. Low Hin Foo. The authors also would like to express their gratitude to BBR Construction Systems (M) Sdn Bhd for sponsoring the prestressing systems, and Mr. Dominic Hoy for his kind assistance in samples preparation and testing.

References

1. ACI Committee 318 (2014) ACI 318M building code requirements for structural concrete and commentary. ACI Committee 318 Structural Building Code, Sec R14, pp 212–214
2. China Academy of Building Research (2010) GB50010: Chinese codes for design of concrete structures. National Standards of the People's Republic of China, Beijing, Appendix H, pp 221–225
3. Committee BD-002 (2009) AS3600 Australian Standard: concrete structures, Chap 8.4. Standard Australia Ltd., Sydney, pp 128–130
4. Technical Committee B/525 (1997) BS8110: structural use of concrete. British Standard Institute, Sec 5.4.9, pp 121–122
5. Technical Committee CEN/TC250 (2004) Eurocode 2: design of concrete structures. European Committee for Standardisation, S6.2.5, pp 92–94
6. Gilbert RI (2016) Design of prestressed concrete to AS3600-2009. CRC Press, Florida, pp 393–398
7. Gilbert RI, Mickleborough BC (2016) Design of prestressed concrete to Eurocode 2. Taylor & Francis, Florida, pp 393–398
8. Low HF (2020) Experimental and numerical studies on a prestressed transfer plate subjected to staged casting and sequential stressing. PhD thesis, School of Engineering, University Monash Malaysia, pp 133–168
9. Low HF, Kong SY, Kong D, Paul CS (2019) Interface slip of post-tensioned concrete beams with stage construction: experimental and FE study. Comput Concr: Int J 24(2):173–183
10. Fang Z, Jiang H, Liu A (2018) Horizontal shear behaviors of normal weight and lightweight concrete composite T-beams. Int J Concr Struct Mater 12(1):55

Fabrication of Sulphonated Poly(Ether Ether Ketone) (SPEEK)–Calcium Oxide (CaO) Multilayer Membrane for Fuel Cell Application

Khairul Anwar Mohamad Said, Wee Loong Chang, and Mohamed Afizal Mohamed Amin

Abstract Demand for energy security and more has pushed researchers to look at fuel cells as an alternative energy source. The electrolyte barrier controlling the proton flow was the main factor limiting the fuel cell's effectiveness and performance. Sulphonated Poly Ether Ether Ketone (SPEEK) is an aromatic ether ketone polymer with good mechanical and thermal properties suited for use in a fuel cell. This project aims to fabricate and analyse the properties of impregnating Calcium Oxide (CaO) into the PEM material and to increase the characteristic of SPEEK for fuel cell application. Tests, including water uptake and proton conductivity, were compared to that of Nafion as the benchmark material. Performance comparison revealed that the modified membrane performed better in the proton conductivity test than Nafion. Moreover, the membrane with a low composition of calcium oxide (0.05%CaO–SPEEK) shows promising results with proton conductivity and water uptake at twice and 8.4 times more than Nafion 117.

Keywords PEEK · Calcium oxide · Multilayer · Proton exchange membrane fuel cell · Proton conductivity

1 Introduction

As an energy conversion device, a fuel cell undergoes oxidation and produces electricity from the electrochemical reaction. The main distinction between fuel cell and dry cell batteries is that they do not need to be recharged with an external source; they can be replenished by refilling the fuel component. Classification of a fuel cell is based on electrolytes, fuels, type of ion exchange, operating condition, and

K. A. Mohamad Said · W. L. Chang · M. A. Mohamed Amin (✉)
Department of Chemical Engineering, Faculty of Engineering, Universiti Malaysia Sarawak, 94300 Kota Samarahan, Sarawak, Malaysia
e-mail: mamafizal@unimas.my

C. S. Choo et al. (eds.), *Proceedings of ASEAN-Australian Engineering Congress (AAEC2022)*, Lecture Notes in Electrical Engineering 1072,
https://doi.org/10.1007/978-981-99-5547-3_11

regeneration methods. However, fuel cells are commonly known for the type of electrolyte used. Based on this definition, there are five known types of fuel cells which are, alkaline fuel cells (AFC), phosphoric acid fuel cells (PAFC), molten carbonate fuel cells (MCFC), solid oxide fuel cells (SOFC), and proton exchange membrane fuel cells (PEMFC) [6]. PEMFC is the most promising of these five fuel cell types because it has a power density of 350 mW/cm^2 with a fuel efficiency of 45–60% [16]. In PEMFC, the electrolyte is a membrane that acts as a fuel and oxidant separator that can conduct protons. Nafion is the industrial membrane produced by PEMFC [3]. Chemically, Nafion is a perfluoro-sulfonic-ethylene polymer widely known and studied for its physio-chemical stability and high proton conductivity (5×10^{-2} S/cm at 23 °C) as a solid polymer electrolyte [13]. However, several drawbacks limit the application of Nafion, such as high production cost and difficulties, high fuel crossovers, and high operating temperature, essentially creating a boundary for the commercialization of PEMFC [19]. Research on alternatives for Nafion has been ongoing for years, and a few have been identified as possible replacements. These polymers have been noted for their high thermo-mechanical stability and low cost [14]. Such polymers were Poly Ether-Ether Ketone (PEEK), Sulfonated Poly Ether-Ether Ketone (SPEEK), Chitosan (CS), Polysulfone (PSF), Polyvinyl Alcohol (PVA), Para-Toluene Sulfonic Acid (pTSA) and Polybenzimidazole (PBI) [1]. Ideally, an excellent polymer membrane suited to be used in a fuel cell is desired to have the characteristic outlined in Table 1 [8].

A sulfonated form of poly(ether ether) ketone (PEEK) functionalized by electrophilic substitution with sulphuric acid is known as SPEEK where; it will be focused on as the alternative for Nafion in this paper [2]. The polymer SPEEK has excellent mechanical and thermal properties while being low in cost was the primary reason SPEEK was chosen for this study [15]. The solubility of SPEEK depends on the amount of sulphuric groups within it whereas when the amount increases, the hydrophilic ability of the polymer increases. When SPEEK has less than 30% Degree of Sulphonation (DS), it is soluble only in concentrated sulfuric acid. Meanwhile, at 30–40% DS, it is soluble in hot dimethylformamide (DMF),

Table 1 Excellent PEM characteristics

Properties	Requirements
Proton conductivity	High
Electrical insulation	Good
Mechanical strength	High
Thermal stability	High
Oxidative and hydrolytic stability	Good
Fuel crossover	Low
Water management	Good
Cost effectiveness	Good
Swelling stress	Low
Membrane electrode assembly (MEA)	Yes

N,N-dimethylacetamide (DMAc), and dimethylsulfoxide (DMSO). At concentrations above 40–70% DS, SPEEK is soluble in the solution above at room temperature. At points above 70%, SPEEK can dissolve with methanol, and at 100% DS, dissolution in hot water is possible [17]. In this paper, the author aims to have a SPEEK/CaO multilayer composite membrane where excellent properties of SPEEK, such as high conductivity and thermo-mechanical strength, were coupled with the modified CaO membrane for increased conductivity and lowered fuel crossovers in the multilayer membrane.

2 Experimental

2.1 SPEEK Preparation

Sulphonation of PEEK began by preparing 30 g of PEEK powder, which was added gradually to a mechanically stirred 570 ml sulphuric acid (95–98%) at room temperature for 1 h and then stirred continuously for 3 h at 60 °C. After that, the reaction is stopped by pouring the mixture into an ice–water mixture. The chilled SPEEK was then washed with distilled water and repeatedly rinsed until the pH of the SPEEK was around the range of 6–7 and then oven-dried for one day at 60 °C. The dried SPEEK polymer would then have a pure white colour and be kept in an airtight bottle for later use.

2.2 SPEEK Solution Preparation

10 wt% of SPEEK solution was prepared by dissolving SPEEK flakes obtained earlier in pure N,N-dimethylacetamide (DMAc) at room temperature, similar to the method performed by Raja Rafidah et al. [12]. The solution was kept away from sunlight, heat, and moisture to minimize the deterioration of the SPEEK quality. A composite inorganic calcium oxide solution was prepared by adding five different concentrations of CaO into the SPEEK solution. The five samples were as follows; 0.05, 0.1, 0.2, 0.5, and 1.0%.

2.3 Membrane Casting

The preparation of a multilayer composite membrane starts with forming a pristine SPEEK membrane by casting unmodified SPEEK onto a glass plate. Then, the membrane surface was smoothened with a wiper, where the excess solution was scraped off. On visual confirmation that the membrane is of even surface, it was placed into the oven at 60 °C overnight and 100 °C for 2 h. The method is repeated for the second inorganic layer. Once dried into a membrane, it is removed and immersed in distilled water to remove the membrane from the glass plate. Pretreatment of the multilayer membrane is performed with immersion into 1 mol sulphuric acid solution for one day to protonate and crosslink the multilayer membrane, according to the method proposed by Raja Rafidah et al. [12]. At the end of the protonation process, the membrane is washed repeatedly with deionized water until neutral, daping with tissue to remove excess water, and kept in a plastic bag and sealed airtight.

2.4 Fourier Transform Infrared (FTIR)

FTIR tests were done from 4000 to 500 cm^{-1} to identify the membrane structure. This test was used to confirm the presence of specific chemical groups within the polymer SPEEK especially the sulphuric group. Two absorption peaks at 1024 and 1080 cm^{-1} are associated with the symmetric and asymmetric stretch vibration. The intensity of the absorption peaks is associated with the sulphonation of PEEK.

2.5 Hydrogen Nuclear Magnetic Resonance (^{1}H-NMR)

^{1}H-NMR tests were conducted to identify sulphuric groups within the SPEEK samples. The test was conducted under the frequency of 500 MHz, where the results were used to calculate the DS using the relation shown below [20].

$$\frac{DS}{12 - 2DS} = \frac{A_{H_E}}{\sum A_{H_{A,A',B,B',C,D}}} \quad (1)$$

The DS can be obtained by determining the ratio of the peak area of H_E to that of the other hydrogen atoms located next to the aromatic ring, as shown in Fig. 1.

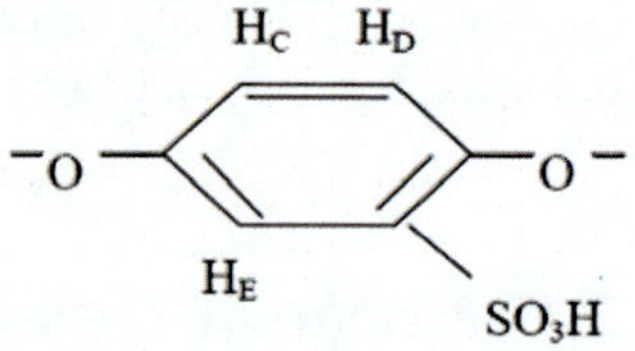

Fig. 1 Sulphuric acid group ring

2.6 Morphology Study

The cross-sectional area morphology of the membrane was investigated under Scanning Electron Microscopy (SEM) with a magnification of 500× and 2000× to identify the pores, calcium oxide distribution, and separation between the unmodified SPEEK and the modified SPEEK layer. The cryogenic breaking method was used to split the membrane for cross-sectional area morphology to observe the membrane's pure, undisturbed state.

2.7 Water Uptake

Water uptake tests were conducted at room temperature to measure the membrane's ability to absorb water. This ability of the membrane is highly affected by the DS within the membrane and the surface morphology, where it could affect the results of proton conductivity and tensile strength. The following relation is used to calculate the amount of water absorbed.

$$W_{H_2O} = \frac{W_{WET} - W_{DRY}}{W_{DRY}} \tag{2}$$

2.8 Proton Conductivity

Proton conductivity was conducted using the impedance test, which operates from 1 Hz to 10 MHz. Before the test, the membrane was immersed in deionized water for 1.5 h and removed, then dap dried to remove water from the surface. After that, the membrane was placed into the equipment cell to measure the room temperature. The formula shown is used to obtain the membrane's conductivity, is measured in Siemens per square centimetre. τ represents the thickness, and Ψ is the area of the membrane. R is the resistivity value obtained from the AC impedance spectroscopy machine.

$$\sigma = \frac{\tau}{R \cdot \Psi} \tag{3}$$

Table 2 FTIR absorption wavelength table

Frequency (cm^{-1})	Group	Compounds
613	Ca-O	Calcium oxide
675	$CaCO_3$	
851	S-OH	Sulphuric
1018	O–S–O	
1078		

3 Results and Discussion

3.1 FTIR Analysis

In Table 2, one can identify that the peak shown at 1018 and 1078 cm^{-1} is similar to the peak reported in the literature (Al [7]. These peaks were due to the symmetric vibration and asymmetric stretching of the O=S=O of the sulphuric groups within the SPEEK. The availability of these two peaks within the test samples indicates that PEEK was successfully transformed into SPEEK. For the presence of calcium oxide inside the membrane structure, the peaks at 613 and 675 cm^{-1} represent the Ca-O symmetric vibration and lattice vibration of $CaCO_3$, respectively. [9] investigate the application of eggshells for biodiesel production. The eggshell, which mainly consists of calcium oxide (CaO), was characterized using FTIR to identify the CaO. Their findings indicate that the CaO presence inside the eggshell can be divided into Ca-O symmetric vibration and lattice vibration of $CaCO_3$ that peaked at 550 and 613 cm^{-1} [9]. Although both CaO peak assignments in our study shifted upwards compared to [9], it is still relevant to justify the presence of CaO inside the SPEEK membrane.

Figure 2 illustrates the FTIR spectra for Nafion and SPEEK membranes. The peak associated with sulphuric compound appears to be shifted in Nafion at 842 cm^{-1}. A similar trend was observed for both O–S–O functional groups in which the peak was shifted to 1022 and 1053 cm^{-1}. Hence, it can be concluded that SPEEK and Nafion possess sulphuric bonds in their structure.

3.2 H-NMR Analysis

The amount of confirmed sulphur within the SPEEK membrane was identified in this test, known as the degree of sulphonation (DS). The findings from the analysis show three peaks occurring in 7.53, 7.22, and 7.12 ppm, representing the resonance signal of H_E, H_C, and H_D, respectively, as described in Sect. 2.5. The signal of 7.53 ppm corresponds with other researchers to that of the sulphuric group [11]. Calculating the amount of DS based on the relation available in Sect. 2.5, it was determined that the amount of DS is 61.655%.

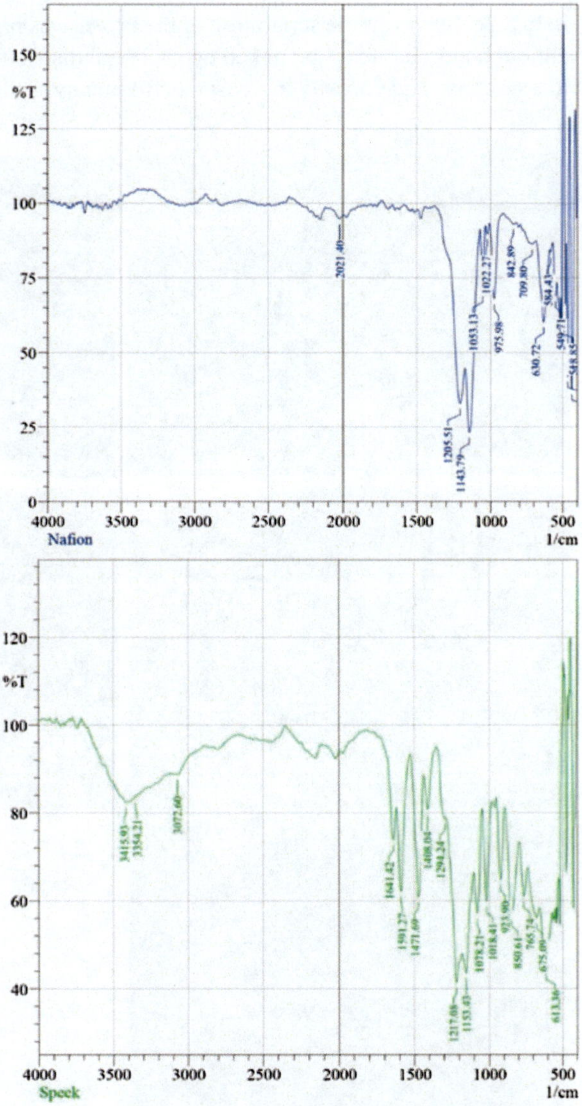

Fig. 2 FTIR spectra of Nafion (above) and SPEEK membrane (below)

3.3 *Morphology Study*

Cross-sectional areas of the membrane were studied to identify the boundary of the two layers, determine whether these layers adhere to each other firmly, and investigate the crystalized formation and porosity within the membrane. The results of this study are shown in Fig. 3, in which Fig. 3a–d show the dense structure and tightly adhered layers. This observation shows that the composite membrane adheres well to the unmodified SPEEK membrane with visible boundary layers at these concentrations.

In Fig. 3e, however, the separation of the layers was profoundly observed in the SEM, although not visible to the naked eye. Crystal distribution was uneven throughout all five samples, and porosity increased in frequency as the CaO concentration increased.

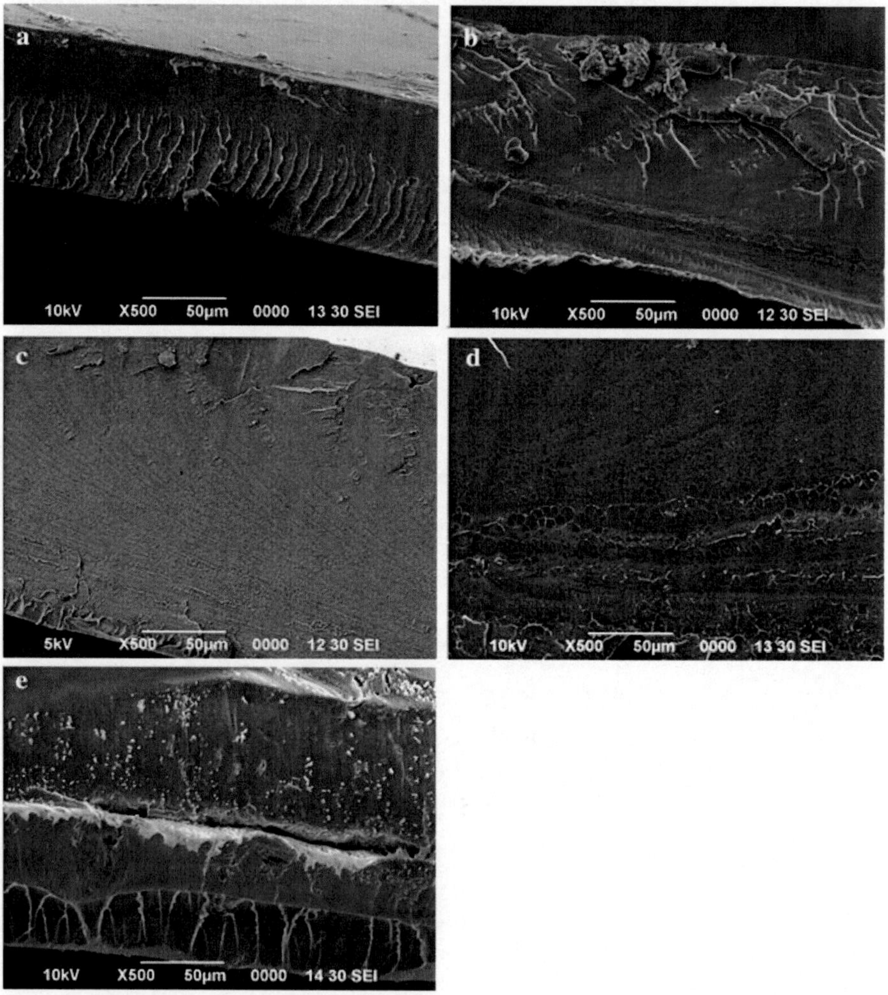

Fig. 3 Cross-sectional morphology pictures of **a** 0.05% CaO; **b** 0.10% CaO; **c** 0.20% CaO; **d** 0.50% CaO; **e** 1.00% CaO

Table 3 Water uptake test results

Membrane designation		Water uptake (%)
SPEEK	0.05% CaO	48
	0.10% CaO	40
	0.20% CaO	37
	0.50% CaO	36
	1.0% CaO	31

3.4 *Water Uptake Analysis*

The ability of the membrane to absorb water predominantly affects the proton conductivity of the membrane and its mechanical strength. Table 3 shows the results obtained through calculation as described in Sect. 2.7 previously. Of the five samples conducted, it can be deduced that the water uptake of the membrane reduces as the concentration of CaO increases. This reduction in water uptake is probably due to the effect of CaO within the membrane and, on the surface, blocks out effective water sorption [10].

3.5 *Proton Conductivity Analysis*

The membrane conductivity can be determined once resistivity values are available from the AC impedance spectroscopy. Figure 4 illustrates the value of the membrane conductivity.

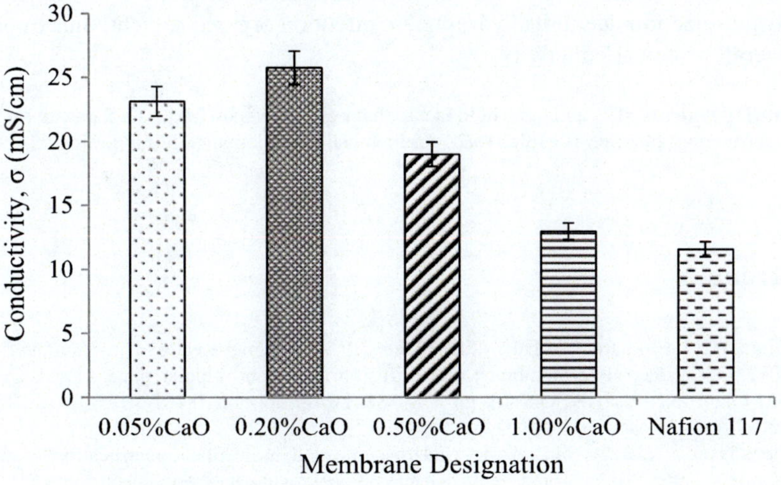

Fig. 4 Proton conductivity test results

Figure 4 shows that increasing the concentration of the CaO within the membrane initially produced a good result with the proton conductivity test. However, once the concentration exceeds 0.20% of CaO, the performance of the membrane reduces. These findings relate to the water uptake value, where a lower value contributes to lower proton conductivity [5, 7]. Testing was also performed for Nafion 117, where it can be observed that proton conductivity values of the modified multilayer membrane have better performance than its commercially available counterpart. The proton conductivity of 0.20%CaO at 27 mS/cm was twice that of a Nafion 117 at 12 mS/cm. Comparison with other researchers showed that the multilayer composite membrane had outperformed Nafion 112 at 18 mS/cm [18] and unmodified SPEEK 55%DS (5 mS/cm) and 70%DS (8 mS/cm) [4].

4 Conclusions and Recommendations

Fabrication of an inorganic impregnated membrane was successfully performed, where calcium oxide (CaO) was used to modify the existing SPEEK membrane as PEM. Nafion 117 was used as a reference point to compare since Nafion was typically used in a fuel cell. Besides, Nafion possesses excellent proton conductivity, high thermo-chemical resistivity, and high mechanical strength. Comparison of CaO effect on a membrane with Nafion is crucial if it were to be applied in fuel cell application. The water uptake test shows that only samples with 0.05%CaO were on par with Nafion, where the proton conductivity was much higher. The remaining samples failed to achieve better or even at par with the standard set by Nafion. Nevertheless, the modified multilayer membrane excels only in 1 test, which is the proton conductivity test. It is recommended to utilize calcium carbonate, a clay derived from calcium oxide. The inclusion of calcium carbonate will improve the methanol rejection due to its hydrophobic effect on organic solvent, thus improving the overall proton conductivity.

Acknowledgements The authors would like to thank the Universiti Malaysia Sarawak for Special Short Term Grant F02/SpSTG/1382/16/24 for financially and logistically supporting this project.

References

1. Branco CM, El-kharouf A, Du S (2022) Materials for polymer electrolyte membrane fuel cells (PEMFCs): electrolyte membrane, gas diffusion layers and bipolar plates. In: A-GBT-E of SM Olabi (ed) Elsevier, Oxford, pp 378–388. https://doi.org/10.1016/B978-0-12-815732-9.00011-5
2. Devi N et al (2019) Non-invasive macroscopic and molecular quantification of water in Nafion® and SPEEK proton exchange membranes using terahertz spectroscopy. J Membr Sci 588:117183. https://doi.org/10.1016/j.memsci.2019.117183

3. Duan Y et al (2022) Crosslinked PAEK-based nanofiber reinforced Nafion membrane with ion-paired interfaces towards high-concentration DMFC. J Membr Sci 655(120589). https://doi.org/10.1016/j.memsci.2022.120589
4. Erce Şengül et al (2009) Effects of sulfonated polyether-etherketone (SPEEK) and composite membranes on the proton exchange membrane fuel cell (PEMFC) performance. Int J Hydrog Energy 34(10):4645–4652. https://doi.org/10.1016/j.ijhydene.2008.08.066
5. Gebel G (2000) Structural evolution of water swollen perfluorosulfonated ionomers from dry membrane to solution. Polymer 41(15):5829–5838. https://doi.org/10.1016/S0032-3861(99)00770-3
6. Godula-Jopek D-IA,Westenberger AF (2022) Fuel cell types: PEMFC/DMFC/AFC/PAFC//MCFC/SOFC/. In: LFBT-E of ES Cabeza (ed) Oxford, Elsevier, pp 250–265. https://doi.org/10.1016/B978-0-12-819723-3.00056-1
7. Al Lafi AG (2015) The sulfonation of poly(ether ether ketone) as investigated by two-dimensional FTIR correlation spectroscopy. J Appl Polym Sci 132(2). https://doi.org/10.1002/app.41242
8. Liu Y, Tu Z, Chan SH (2022) Performance enhancement in a H_2/O_2 PEMFC with dual-ejector recirculation. Int J Hydrog Energy 47(25):12698–12710. https://doi.org/10.1016/j.ijhydene.2022.02.023
9. Mamat FM, Yacob AR (2014) Kinetic study of biodiesel using egg shell for base transesterification reaction. In: Chemistry undergraduate final year project symposium 2014/2015. Department of Chemistry, Faculty of Science, Universiti Teknologi Malaysia, pp 133–139
10. Mazinani V et al (2014) Synthesis and characterization of nafion-ZrOH-CaO hybrid membrane for proton exchange membrane fuel cell. J Fuel Cell Sci Technol 11(3). https://doi.org/10.1115/1.4026142
11. Paik Y et al (2009) Influence of water and degree of sulfonation on the structure and dynamics of SPEEK studied by solid-state 13C and 1H NMR. Polymer 50(12):2664–2673. https://doi.org/10.1016/j.polymer.2009.03.039
12. Raja Rafidah RS et al (2020) Recent progress in the development of aromatic polymer-based proton exchange membranes for fuel cell applications. Polymers. https://doi.org/10.3390/polym12051061
13. Smitha B, Sridhar S, Khan AA (2005) Solid polymer electrolyte membranes for fuel cell applications—a review. J Membr Sci 259(1):10–26. https://doi.org/10.1016/j.memsci.2005.01.035
14. Vidhyeswari D, Surendhar A, Bhuvaneshwari S (2022) Enhanced performance of novel carbon nanotubes—sulfonated poly ether ether ketone (speek) composite proton exchange membrane in mfc application. Chemosphere 293:133560. https://doi.org/10.1016/j.chemosphere.2022.133560
15. Wei P et al (2022) Sandwich-structure PI/SPEEK/PI proton exchange membrane developed for achieving the high durability on excellent proton conductivity and stability. J Membr Sci 644:120116. https://doi.org/10.1016/j.memsci.2021.120116
16. Xiao F et al (2021) Solid-state synthesis of highly dispersed nitrogen-coordinated single iron atom electrocatalysts for proton exchange membrane fuel cells. Nano Lett 21(8):3633–4363. https://doi.org/10.1021/acs.nanolett.1c00702
17. Yagizatli Y, Sahin A, Ar I (2022) Effect of thermal crosslinking process on membrane structure and PEM fuel cell applications performed with SPEEK-PVA blend membranes. Int J Hydrog Energy [Preprint]. https://doi.org/10.1016/j.ijhydene.2022.04.183
18. Yang J et al (2009) Nafion/polyaniline composite membranes specifically designed to allow proton exchange membrane fuel cells operation at low humidity. J Power Sources 189(2):1016–1019. https://doi.org/10.1016/j.jpowsour.2008.12.076
19. Yang L, Wang H, Zhang X (2021) Recent progress in preparation process of ionic polymer-metal composites. Results Phys 29:104800. https://doi.org/10.1016/j.rinp.2021.104800
20. Yee RSL, Zhang K, Ladewig BP (2013) The effects of sulfonated poly(ether ether ketone) ion exchange preparation conditions on membrane properties. Membranes. https://doi.org/10.3390/membranes3030182

Modification of Sulfonated Poly Ether Ether Ketone (SPEEK) Proton Exchange Membrane with Sodium Chloride for Fuel Cell Application

Mohamed Afizal Mohamed Amin, Noor Najlaa Asyura Sulaiman, Khairul Anwar Mohamad Said, Dayang Norafizan Awang Chee, and Ibrahim Yakub

Abstract Sulfonated polyether ether ketone (SPEEK) is a promising material in making good proton exchange membrane due to their high proton conductivity and chemical stability as well as lower-cost production compared to commercially available perfluoro sulfonated polymers. Modification on this with an additional of organic and inorganic salts material become one of the best approaches in improving the performance of the membrane. In this study, the influence of ionic salt which is sodium chloride in SPEEK matrix membrane towards membrane characteristic and performance was investigated. Based on the result, the modified membrane shows significant improvement in water uptake and proton conductivity, upon addition of sodium chloride as compared with the neat and commercial membranes reported in other studies. Further increase in the concentration of salts in the membrane matrix increases the membrane proton conductivity up to fourfold.

Keywords Proton exchange membrane · Sulfonated poly ether ether ketone · Sodium chloride

1 Introduction

Nowadays, there are several types of energy conversion technology. One of them is a fuel cell. A fuel cell is a widely used next-generation power generator and applied in everyday applications such as electric vehicles and mobile telephones [24]. Electric

M. A. Mohamed Amin (✉) · N. N. A. Sulaiman · K. A. Mohamad Said · I. Yakub
Faculty of Engineering, Universiti Malaysia Sarawak, 94300 Kota Samarahan, Sarawak, Malaysia
e-mail: mamafizal@unimas.my

D. N. Awang Chee
Faculty Resource Science and Technology, Universiti Malaysia Sarawak, 94300 Kota Samarahan, Sarawak, Malaysia

C. S. Choo et al. (eds.), *Proceedings of ASEAN-Australian Engineering Congress (AAEC2022)*, Lecture Notes in Electrical Engineering 1072,
https://doi.org/10.1007/978-981-99-5547-3_12

power was generated by an electrochemical reaction between fuels such as hydrogen gas and methanol with oxygen gas in which water and heat are produced as a by-product of the reaction. Several types of fuels can be used in the fuel cell, such as hydrogen gas, methanol, and ethanol. The electrolyte used in the fuel cell is essential to indicate the type of fuel cell. Generally, there are five types of fuel cells, which are alkaline fuel cells (AFC), phosphoric acid fuel cells (PAFC), solid proton exchange membrane (PEMFC), molten carbohydrate fuel cell (MCFC), and solid oxide fuel cells (SOFC). Fuel cell advantages include highly efficient proton conductivity, no mechanical or moving part involved, and environmental friendly [22].

As stated by Khomein et al. [16], the fuel cell produces electricity without affecting the environment. A fuel cell system can generate power with efficiency reaching 60% more than the conventional combustion, which only exhibits 30–35% efficiency [19]. This efficiency shows that the fuel cell has more advantages in terms of performance and environmental aspects. The mechanism in all fuel cell types is similar where ions released from the anode are transferred to the potential difference [29].

SPEEK has been the promising contender of PEM for various reasons, most notably its low cost, susceptibility to thermal degradation, and excellent mechanical strength [5]. Besides, due to SPEEK chemical structure that directly combined the polyaromatic backbone, which has hydrophobic characteristics with the hydrophilic group of sulfonic acid, it promoted SPEEK with a unique capability to adjust its degree of sulfonation (DS) [14]. Hence, SPEEK with an optimized degree of sulfonation could possess similar performance to the Nafion membrane. Despite the SPEEK potential of surpassing the Nafion membrane, SPEEK as a sulfonated-based membrane was prone to chemical degradation by hydroxyl radicals formed during the oxygen reduction reaction [20]. As a result of the hydroxyl radicals' attack, the main polymer branch could experience chain scission that leads to the elimination of hydrophobic sulfonic acid which over time would inflict damages to the PEM [15]. Therefore, in mitigating hydroxyl radicals, the SPEEK membrane can retain its mechanical strength and prevent chemical degradation by introducing a dual-layer composite membrane.

The high resistance of SPEEK membrane to methanol crossover relied mostly on the degree of sulfonation therefore high DS would increase the methanol permeability. Several approaches have been adopted to reduce the methanol crossover such as surface grafting, crosslink and blending. For instance, blending polyethersulfone (PES) into SPEEK would create a composite membrane that would improve its hydrolysis stability but has poor proton conductivity. In contrast, blending SPEEK with poly(amide imide) (PAI) would improve the proton conductivity as a function of methanol permeability yet at high PAI composition, the methanol permeability was reported to increase. Besides blending SPEEK with long chain polymer, inorganic material has also been reported to address the methanol crossover and proton conductivity issue in SPEEK membrane. In general, the incorporation of inorganic particles was to fill the pore and crack within the membrane thus preventing methanol permeation. Moreover, the inclusion of inorganic materials such as montmorillonite into SPEEK membrane was found to be beneficial in warding the methanol from crossover while improving the proton conductivity.

Although blended method produces a SPEEK membrane with acceptable performance, the multilayer membrane will have separated layer with unique properties. Hence, the multilayer membrane has a more focused characteristic and is easier to design. For instance, SPEEK membrane layered with Nafion on both side has higher proton conductivity than the pristine SPEEK membrane and it was prepared by dip coating the SPEEK membrane in thick Nafion coating solution. One of the inorganic material that was known for its good conductivity was salt or sodium chloride (NaCl). Besides, the reaction between sodium and methanol will result in the formation of sodium methoxide, a type of salt that may increase the layer which prevents methanol crossover and induces proton conductivity. Therefore, in this study, dual layer SPEEK membrane will be used in Proton Exchange Membrane (PEM). The second layer of SPEEK will be equipped with sodium chloride while the primary layer was the pristine SPEEK membrane. The NaCl content was varied to investigate the influence of NaCl to membrane strength, methanol crossover, proton conductivity, thermal stability and its structure.

2 Experimental

2.1 *Material and Chemicals*

Poly ether ether ketone (PEEK) in powder form (Victrex US Inc. Ltd.) and sulfuric acid (H_2SO_4) (95–98%) were used to synthesized solid SPEEK. Dimethylacetamide (DMAc) is the solvent used in preparing the membrane dope solution. Sodium chloride (NaCl) (99%) in a powdered form is used as a dope solution modifier.

2.2 *Sulfonation of PEEK*

SPEEK was synthesized through the reaction between PEEK powders in a concentrated H_2SO_4 solution. Prior to use, PEEK was dried overnight in the oven to remove the moisture. The synthesis procedure begins by mixing 30 g of PEEK into 570 ml concentrated H_2SO_4 in the round bottom flask at room temperature for 1 h. After the mixture was homogenous, the bottom flask was heat up at 60 °C for 2 h. Then, the SPEEK solution was poured into ice-cold water with vigorous stirring to terminates the sulfonation reaction. Afterwards, the solid SPEEK was washed using tap water until it reached a pH of 6–7. Finally, it was dried in an oven at 100 °C overnight [10]. The dried SPEEK polymer was stored at room temperature.

Table 1 Membrane designation

Membrane	Weight of the composition at the top layer (g)		
	SPEEK	DMAc	NaCl
$SP_{NC(0.1)}$	10	90	0.1
$SP_{NC(0.2)}$	10	90	0.2
$SP_{NC(0.3)}$	10	90	0.3
$SP_{NC(0.4)}$	10	90	0.4
$SP_{NC(0.5)}$	10	90	0.5

2.3 Membrane Preparation

Solid SPEEK was dissolved in the DMAc solvent to produce 10 wt% of the SPEEK dope solution. The mixture was stirred until the SPEEK was fully dissolved. Then, the solution was cast on a clean glass plate. The casted dope solution was undergoing a drying process at 60 °C for 8 h in removing the excess solvent. Next, the membrane was further dried at 80 °C overnight in removing the remaining solvent. Now, the neat resultant membrane on top of the class will become a new base in making a membrane second layer. For this layer, different amount of NaCl (0.1, 0.2, 0.3, 0.4, and 0.5 g) was mixed in the prepared SPEEK dope solution. Then, it was casted on top of the prepared SPEEK layer followed by the drying process under the same previous procedure. The prepared multilayer SPEEK-NaCl membrane was fully immersed in the distilled water until it was peeled off from the glass plate. Afterwards, the membrane was treated with 1M of sulfuric acid overnight. Finally, the membrane was rinsed with DI water and dried at room temperature. Membranes designated are tabulated in Table 1.

2.4 Characterization of SPEEK Polymer

2.4.1 Fourier Transform Infrared (FTIR)

A Fourier Transform Infrared (FTIR) spectrum was obtained in a scanning range of 400–2000 cm^{-1}. It was used to detect the component present in the substance, which exhibits the fuel cell component [23]. It also utilizes in determining the chemical structure of the molecule, although qualitatively [1]. The peaks at specific wavelengths were identified.

2.4.2 Nuclear Magnetic Resonance (H^1 NMR)

Hydrogen Nuclear Magnetic Resonance (H^1 NMR) was used to determine the structure of the SPEEK compounds. The NMR was conducted at 500 MHz, and dimethyl sulfide (DMSO) was used as the solvent. The results obtained were used to calculate the degree of sulfonation (DS) of the SPEEK polymer.

2.5 Membrane Analysis

2.5.1 Scanning Electron Microscopy (SEM) Analysis

The SEM analysis was conducted to study the membrane morphology due to the presence of NaCl. The samples were prepared by freezing the membrane in liquid nitrogen to obtain clean fractures, thus minimizing the damage to its cross-section. Then, the samples were sputtered using ion sputtering to improve conduction before analysis. The magnification of this analysis was within 500 to 5000×.

2.5.2 Thermal Stability by Using Thermogravimetric Analysis (TGA)

The TGA was conducted to measure the thermal stability of the SPEEK membranes. Before TGA analysis, the membrane needs to be dried to remove the water present in the membrane. The sample was prepared by cutting the membrane into small pieces approximately of weight 7 mg.

2.5.3 Tensile Strength

The tensile strength was conducted by exerting pressure on the membrane until the breaking point at a predetermined force. Besides the breaking point, the time for membrane elongation was also observed. The samples were cut into rectangular shapes of 5 mm width × 50 mm length. Each membrane ends were stick with adhesive tape to attach it to the tensile machine. The gauge lengths for all the samples were measured before testing, including the membrane thickness using a micrometer screw gauge. Thickness and gauge length was measured before running the experiment and was used as input data. The crosshead speed was set to 5 mm/min.

2.5.4 Water Uptake Test

The water uptake test was performed based on the weight of the membrane before and after the hydration in which the SPEEK membrane was cut in square size (2 × 2 cm). All the samples were dried in an oven at 60 °C for 48 h. The weight of the dried membrane was taken. The analysis continues by immersing the dried membrane into the deionized water overnight at room temperature. Before weighing the samples, the water on the surface of the sample was wiped out with tissue paper. The percentage of water uptake was calculated using the formula, as shown in Eq. (1) [2].

$$\text{Water uptake} = \frac{W_{\text{wet}} - W_{\text{dry}}}{W_{\text{dry}}} \times 100\% \tag{1}$$

where, W_{wet} is the weight of the wet membranes and W_{dry} the weight of the dry membranes.

2.5.5 Proton Conductivity Measurement

The proton conductivity of the membrane was analyzed by AC impedance technique using a 1260 Impedance instrument over a frequency range of 1–10^7 Hz with 100–500 mV oscillating voltage. The samples of the membrane were immersed in the deionized water overnight before experimenting. Proton conductivity of membrane, σ (Siemens per centimeter (S/cm)), was calculated according to Eq. (2), where σ is defined as the reciprocal of *R*.

$$\text{Protonconductivity}, \sigma = \frac{d}{RS} \tag{2}$$

where *d* and *S* are the thickness and area of the membrane, respectively. *R* was derived from the low intersection of the high frequency semi-circle on a complex impedance plane with the Re(Z) axis.

3 Results and Discussion

3.1 Functional Group Analysis

Figure 1 shows the wavelengths present in the SPEEK polymer. The peak at various wavelengths suggests that the chemical structure possess a dried SPEEK polymer. The wavelength of 1032 and 1082 cm^{-1} was referred to as the symmetric and asymmetric stretching vibration of sulfonic groups (SO_3H). The presence of SO_3H was significant to determine the sulfonation process occurring. The result shows the presence of SO_3H in the dried SPEEK polymer, where SO_3H has successfully substituted into the aromatic ring of the Poly-Ether Ether Ketone (PEEK). The peak at wavelength 1220 and 1469 cm^{-1} indicated the aromatic C–O–C and C–C aromatic ring, respectively. Thus, the SPEEK polymer was confirmed to have an aromatic ring. The

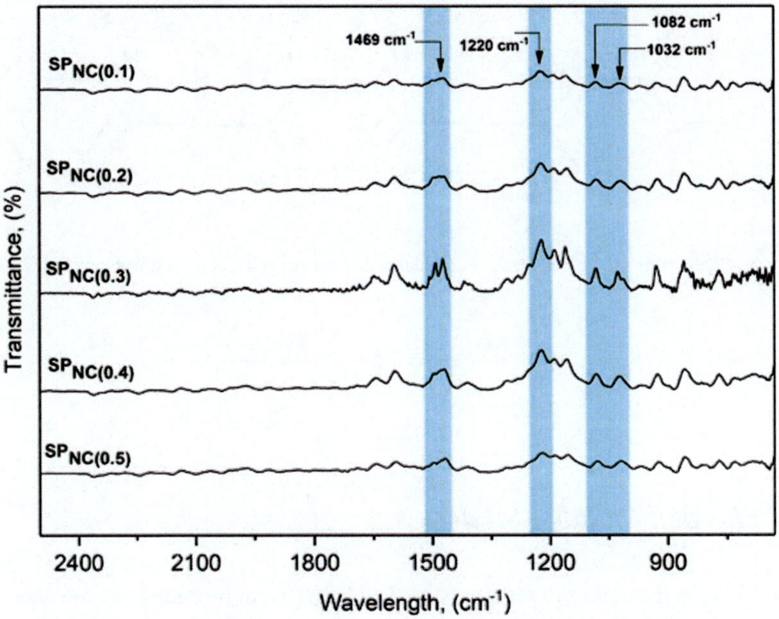

Fig. 1 FTIR spectra of the modified SPEEK membrane at different loadings of NaCl

process of sulfonation occurred due to the presence of a sulfonic group in the SPEEK polymer.

3.2 Degree of Sulfonation Study

The H^1 NMR was used to determine the degree of sulfonation (DS) of dried SPEEK polymer. Figure 2 shows the nomenclature of the SPEEK polymer. H_E intensity represents the sulfonic (SO_3H) group content in the SPEEK polymer chain while H_A, $H_{A'}$, H_B, $H_{B'}$, H_C, and H_D correspond to all the aromatic hydrogens. The ratio between the peak areas of the H_E and other aromatic hydrogen signals are calculated using Eq. (3) where the n represents the number of H_E per repeat unit. The value of n must be between 0 and 1 ($0 \leq n \leq 1$) [25] in which n is the DS, H_E is the sulfonic acid (SO_3H) group content, and $H_{A,A',B,B',C,D}$ is aromatic hydrogen. According to Trindade et al. [8], the DS indicates the concentration of sulfonic acid group imparted within the SPEEK during the sulfonation reaction, where high DS means that the SPEEK has a high concentration of sulfonic acid group. The concentration of sulfonic acid groups affects the physical and chemical properties of dried SPEEK polymer. The DS process changes the properties of PEEK and affects the solubility of the polymer. Based on the calculation, the sulfonic group concentration in the SPEEK polymer was approximately 7.5 ppm. The DS of dried SPEEK polymer obtained was 62%.

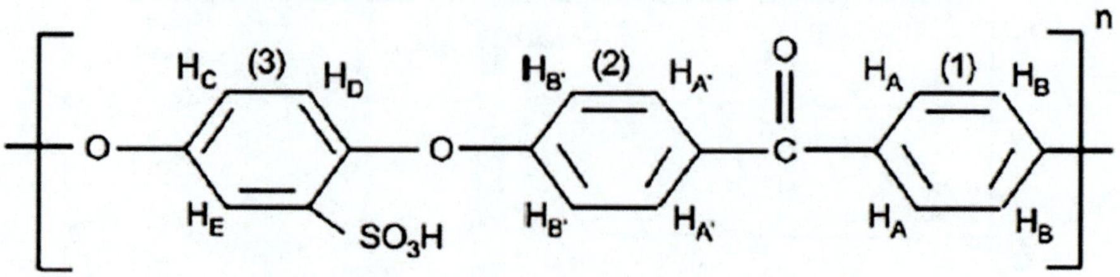

Fig. 2 Nomenclature of SPEEK polymer obtained from [6] with permission from Elsevier

$$\frac{(n)}{(12-2n)} = \frac{\sum H_E}{\sum H_{A,A',B,B'C,D}} \tag{3}$$

3.3 *Thermal Stability Study*

Figure 3 shows the stages of weight loss of all SPEEK membranes and the decompose temperature. $SP_{NC(0.2)}$ has undergone two stages where the first weight loss occurred at the temperature between 30 and 200 °C then starts to decompose at 600 °C. Sample $SP_{NC(0.3)}$, and $SP_{NC(0.4)}$ possess almost the same pattern where weight loss starts at 600 °C. $S/S_{0.5NaCl}$ shows different thermographic patterns where the first weight loss was within 40–250 °C, which was due to the vaporization from within the membrane. The second weight loss was from 300 to 500 °C that occurred when the sulfonic groups start to decompose. The third weight loss refers to membrane decomposition, which usually take place at a temperature of more than 600 °C [18]. Hence, the higher the amount of NaCl in the membrane, the higher the decomposition temperature. The modified SPEEK membrane has high thermal stability where all the samples except for $S/S_{0.1NaCl}$ remain with 40% of its weight, which could achieve high decomposition temperature if the samples were fully decomposed.

3.4 *Membrane Morphology*

Figure 4 illustrates the surface of the SPEEK membrane with magnifications of 2 and 5 k×. The porous structure of $SP_{NC(0.2)}$ was due to NaCl presence, as shown in Fig. 8. The crystalline structure of NaCl was seen on the membrane surface due to partial NaCl dissolution. The increasing amount of NaCl, leads to a more porous membrane structure. $SP_{NC(0.4)}$ has more and bigger pores on the surface membrane, and several crystalized NaCl particles were noticed scattered across the membrane surface. Based on the micrograph, the surface of $S/S_{0.5NaCl}$ was different from the other with an apparent roughness throughout the entire membrane surface. Hence, the presence of NaCl has affected the morphology of the SPEEK membrane surface.

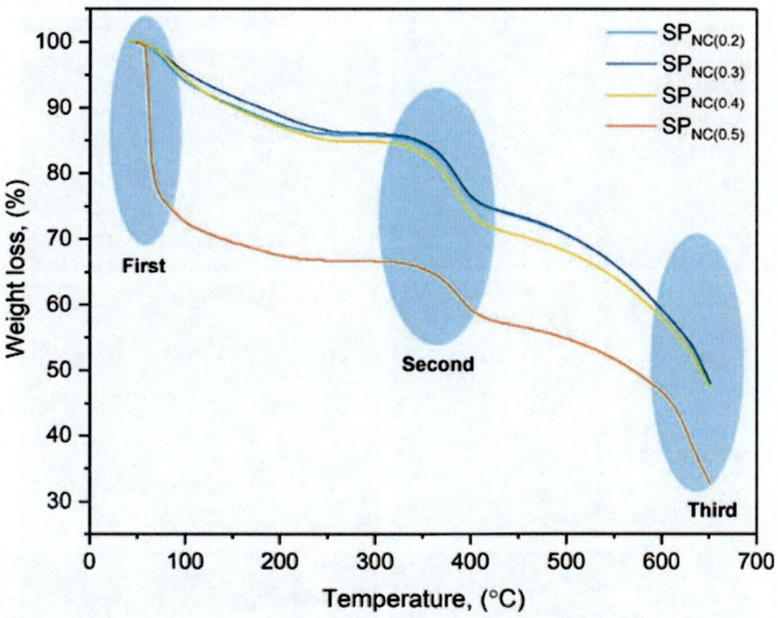

Fig. 3 TGA analysis for all prepared membrane

It can be concluded that the higher amount of NaCl leads to a more porous membrane structure. In contrast, the SPEEK membrane with a minimal additive was usually free from pore defects [28]. This morphology also affects the result of the water uptake and tensile strength of the membrane where the result of water uptake increased, and the tensile strength of the membrane decreased as the NaCl increased.

Figure 5 depicted the cross-section of the SPEEK membrane with different NaCl loadings. From this figure, it shows no apparent phase separation for $S/S_{0.1NaCl}$ and $S/S_{0.2NaCl}$, which indicated that both layers were attached seamlessly. $S/S_{0.3NaCl}$ has a slightly different structure between the layers where the rough layer was identified as a pure SPEEK membrane, and the smooth layer is SPEEK with NaCl. Moreover, a similar observation was exhibited in the cross-section of the $S/S_{0.4NaCl}$. Both membranes have the pore formed at the smooth layer (bottom layer). At 2 k× magnification, the pores were noticeable for $S/S_{0.5NaCl}$ due to the high amount of salt content in the membrane. Apart from that, the separation layer was observed between the layers. To conclude, the structure of SPEEK membranes was influenced by NaCl, and a higher NaCl loading structure would produce a highly porous membrane.

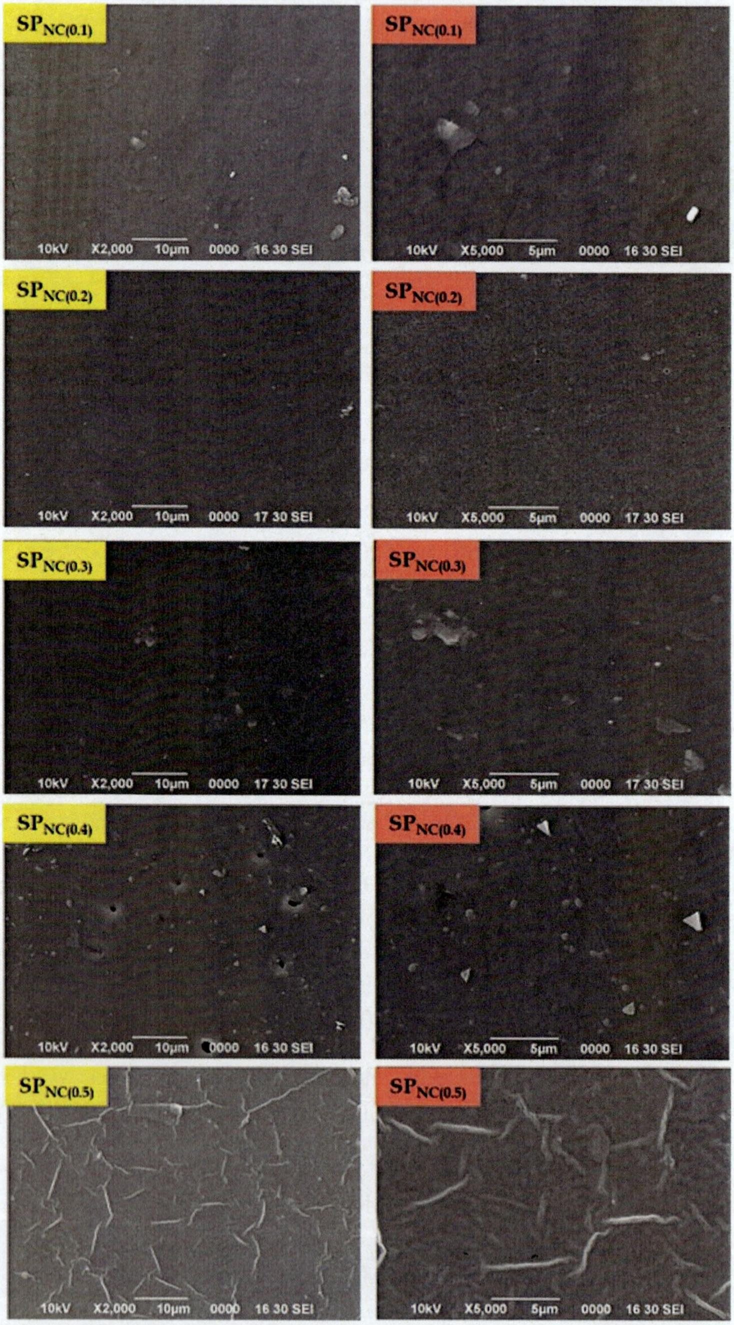

Fig. 4 Top surface morphology of the SPEEK membrane embedded with different concentrations of NaCl (0.1, 0.2, 0.3, 0.4 and 0.5 wt/wt%) at low (2 k×) and high (5 k×) magnifications

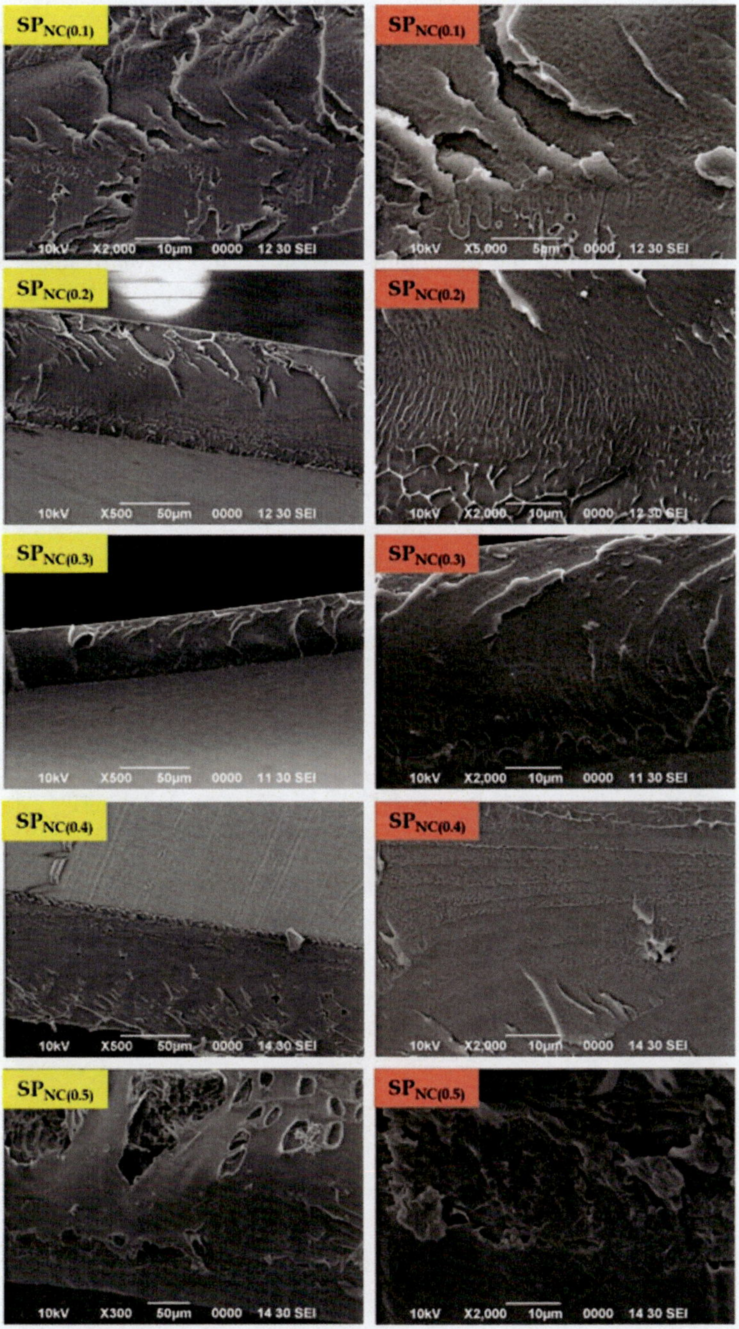

Fig. 5 Cross-sectional morphology of the SPEEK membrane embedded with different concentrations of NaCl (0.1, 0.2, 0.3, 0.4 and 0.5 wt/wt%) at low and high magnifications

3.5 Tensile Strength

Figure 6a shows the tensile strength of the SPEEK membrane at different amounts of NaCl, and the trends show the membrane loses strength at higher loading of NaCl. This weaker structure might due to the pore defects on the membrane. The morphology obtained from the SEM analysis confirms the inference that the membrane has more pores as the NaCl loading increases.

The breaking point of the membrane was shown in Fig. 6b. It takes a shorter time for the membrane to break for the sample that has more NaCl. According to Gashoul et al. [11], the membrane strength is directly related to the water uptake. A high-water uptake would reduce the tensile strength of the membrane due to membrane deterioration. For instance, the tensile strength of the SPEEK membrane was 39.9, and compared to the $S/S_{0.5NaCl}$, its tensile strength was lower than the

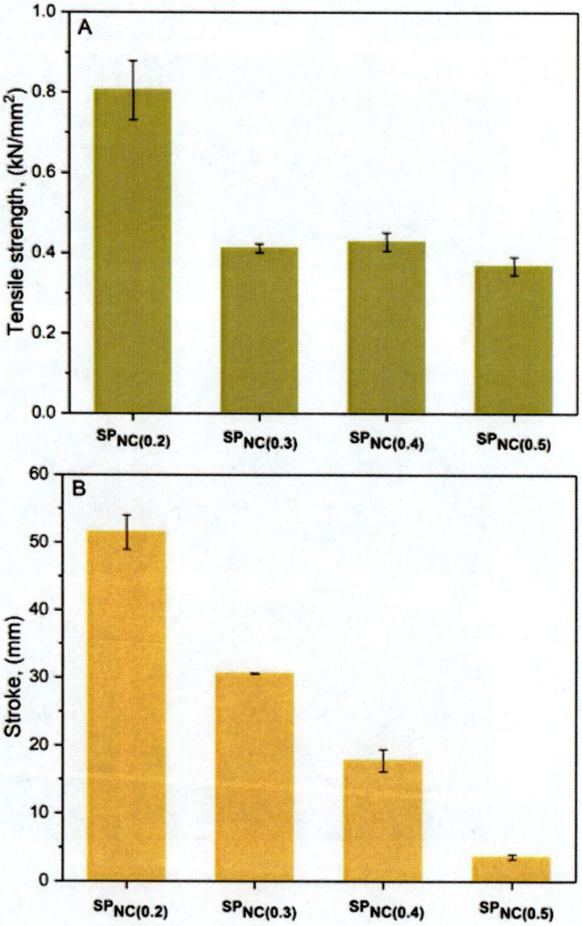

Fig. 6 Mechanical property of the modified membrane incorporated with different loadings of NaCl **a** tensile strength and **b** breaking point

SPEEK membrane. Thus, NaCl was identified as a substance that can reduce the strength of the membrane.

3.6 Water Uptake

Figure 7 shows an increment in water uptake for the membrane when the NaCl loading was increased. The increase of water uptake could be due to the stronger hydrogen bond between a water molecule and the hydrophilic groups [4]. The water uptake is one of the critical parameters to determine the performance of the membrane. Mohtar et al. reported that the water uptake influences the proton conductivity and mechanical strength of the membrane. For instance, a pristine SPEEK membrane was reported to exhibit up to 31% water uptake lower than the Nafion membrane [21]. This low water uptake was supported by Huang et al. [13], which states that the SPEEK membrane has a low percentage of water uptake as compared to the modified composite membrane. As depicted in Fig. 3, water uptake increases were recorded as the NaCl amount increased from 0.1 to 0.5 g. The highest recorded water uptake was 43%, while SPEEK exhibited the lowest with 0.1 g NaCl at 38%. Each has surpassed the water uptake of the pristine SPEEK membrane at 30.8%. Upon adsorbing water into the membrane matrix, a proton bridged channel would be formed, which proves vital in proton conductivity, although excessive water uptake would cause downsides such as reducing mechanical strength and methanol permeability. Wang et al. [26] simulate the SPEEK and Nafion membrane water state through radial distribution functions (RDF) by investigating the relation between $-SO_3^-$ with H_2O and H_3O^+. The relation was represented by O_s–O_h and O_s–H_h in which the subscript s and h refer to $-SO_3^-$ and hydronium. Hence, by increasing the hydration, O_s–O_h and O_s–H_h intensities in Nafion were shown to decrease while SPEEK was exhibiting a favourable relationship. Furthermore, the average hydronium amount was reported to increase surrounding the O_s of SPEEK, which reveal better interaction in SPEEK at high water content [26]. Hence, blending SPEEK with NaCl has shown to be effective in curbing the water uptake as shown in Fig. 7 and when combined with SPEEK resistance to water would result with membrane that exhibit hydrophobic characteristics. Although at higher NaCl composition, the water uptake was found to be drastically increased up to ~48%. However, a further increase in NaCl loading appears to be devastating to the SPEEK membrane due to NaCl solubility in water. Thus, at a higher amount of NaCl, NaCl probability of interacting with water was high, thus contributing to high water intake. Thus, the reaction that produces sodium methoxide was not as rapid as the interaction between NaCl and water except at low NaCl content.

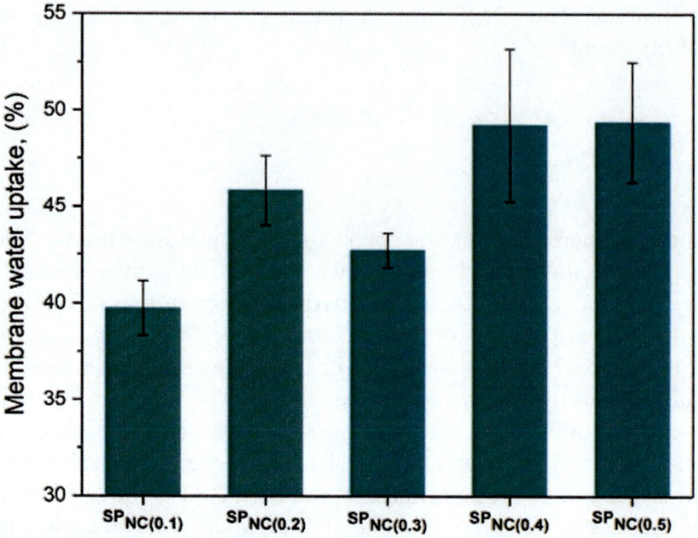

Fig. 7 Trend of water uptake with the addition of NaCl

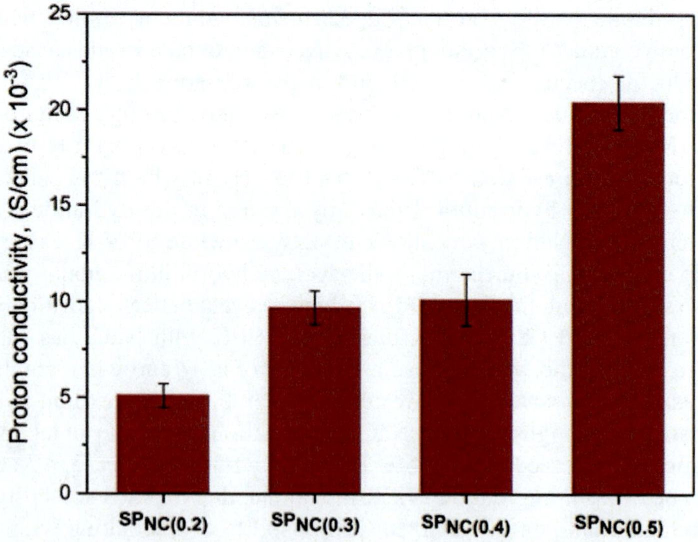

Fig. 8 Membrane proton conductivity at different loadings of NaCl

3.7 Proton Conductivity

Proton conductivity for SPEEK modified membrane was shown in Fig. 8. From this figure, the addition of NaCl significantly affects the proton conductivity of the SPEEK membrane. The result increased significantly when the loading was kept increasing. It shows that the proton conductivity exponentially increased as more NaCl was added. SPEEK membrane with 0.5 g of NaCl has 0.0196 S/cm, while the least NaCl added SPEEK membrane only produced 0.0067 S/cm. Generally, NaCl is a poor electric conductor in powder form due to the strong electrostatic forces between the ions that hold the ions to each other, causing the ions to be unable to move freely. However, when dissolved in water, the NaCl will separate into ions, allowing them to move freely and thus conduct the electricity. Hence in our case, the increase in sodium and chloride ions leads to a higher electricity generation. The increase of proton conductivity was due to the proton transfer from anode to cathode. As water content is present in the modified SPEEK membrane, NaCl was dissociated into ions of Na^+ and Cl^-. The presence of water also breaks the bonding of the sulfonic acid group, where more H^+ ions were transferred since the sulfonic acid group is a charge carrier. Hence, the water content in the modified SPEEK membrane helps the ions to transfer quickly.

In theory, the proton transfer was driven by two widely accepted mechanisms (i) Grotthus mechanism and (ii) vehicle mechanism. Grotthus mechanism revolved with the idea of water decomposing into two charged particles which possess different polarities [9]. The two charged particles were then hopping along with the water molecules that produced hydrogen and oxygen gases. In 1806, this could be the first time molecules referred to as particles with distinct opposite polarities [7]. Grotthus mechanism was specified to occur at higher elevated temperature with low humidity.

Meanwhile, the vehicular mechanism occurs at a lower temperature with high humidity, opposite to the Grotthus mechanism. The vehicular mechanism stipulated that the water molecule was assimilated with the electrolyte membrane and formed a proton-conducting channel [17]. Here, unlike the Grotthus mechanism that infers proton hopping, the vehicle mechanism uses water to act as the medium of proton transportation. For instance, in the SPEEK membrane, the polymer membrane was occupied with sulfonic group intermittently bonded across its backbone [3]. The sulfonic group on the PEEK backbone was recognized to have a high polarity, thus improving the PEEK backbone polarizability significantly, creating water clusters [27]. In our case, we concur that the mechanism which drives the proton conductivity of NaCl equipped SPEEK membrane was according to the vehicle mechanism. The proton conductivity analysis was conducted at relatively low room temperature (28–30 °C) at high humidity of a tropical climate. The dissociation of H^+ bonded with the oxygen group formed Zundel cation $(H_5O_2)^+$, a protonated water dimer. In the mid-1990s, the vehicle mechanism was first to introduce H^+ mobility in water that was caused by water molecule rotation. As a consequence of water rotation, the low operating temperature of the vehicle mechanism required the hydrogen bond on H_3O^+ to be broken. The H-bonds cutoff simultaneously have reduced the activation energy required for water molecule rotation, leading to the rate-limiting step. Therefore, there must be a rate-limiting in place that regulates the mobility of H^+ in water. Thus, even one H-bond in H_3O^+ causes a disruption, thus creating an ideal condition for H^+ to be transferred between the oxygen in Zundel cation. Therefore, with the presence of NaCl, one of the SPEEK layers will contribute to the presence of Na^+ and Cl^-

ions. According to Hellström and Behler [12], proton transfer involving Na^+ and OH^- is possible which is also known as the proton transfer pathway. A water ligand whose presence surrounding the Na^+ ion can be considered as a weak acid. However, the existence of OH^- ion will induce thermal variations, allowing the water ligand to start the proton transfer onto the OH^- ion, which contribute to the formation Na^+–OH^-. The direct contact ion pairs will dissociate and exhibit proton transfer in which the hydroxide ligand will transform to water ligand during the proton transfer in a reversible reaction as follows:

$$Na^+ - O^{\alpha}HH \cdots O^{\beta}H^- \rightleftharpoons Na^+ - O^{\alpha}H^- \cdots HO^{\beta}H \quad (4)$$

The vehicular mechanism assisted by direct contact ion pairs has bestowed the SPEEK membrane with 0.5 g NaCl with the highest proton conductivity.

4 Conclusion

A multilayer SPEEK/SPEEK-NaCl membrane was fabricated with different NaCl loadings. In general, the addition of NaCl does affect the performance of the membrane. SPEEK added with 0.5 g NaCl has increased water uptake, promoting higher proton conductivity due to the water content in the membrane. The water in the membrane helps to transport the ions. However, the drawback of high NaCl loading in the SPEEK membrane was the pore defects analyzed by its morphology. The improved porosity has reduced the membrane tensile strength.

Acknowledgements The authors would like to acknowledge Ministry of Higher Education Malaysia (MOHE) and Research, Innovation & Enterprise Centre (RIEC) Universiti Malaysia Sarawak (UNIMAS) for the financial support through a grants Research Acculturation Grant Scheme RAGS/c(13)/947/2012(48) and Special Short Term Grant F02/SpSTG/1382/16/24 respectively.

References

1. Abu-Saied MA, El-Desouky EA, Soliman EA, El-Naim GA (2020) Novel sulphonated poly (vinyl chloride)/poly (2-acrylamido-2-methylpropane sulphonic acid) blends-based polyelectrolyte membranes for direct methanol fuel cells. Polym Test 89:106604. https://doi.org/10.1016/j.polymertesting.2020.106604
2. Aiswarya SK, Joseph S (2020) Synthesis of methanol blocking PVA-TiO_2 cation exchange membrane for direct methanol alkaline fuel cell. Synth Metals 266:116442. https://doi.org/10.1016/j.synthmet.2020.116442
3. Changkhamchom S, Sirivat A (2010) Synthesis and properties of sulfonated poly(ether ketone ether sulfone) (S-PEKES) via bisphenol S: effect of sulfonation. Polym Bull 65(3):265–281. https://doi.org/10.1007/s00289-010-0243-8

4. Charradi K, Ahmed Z, Aranda P, Chtourou, R. (2019) Silica/montmorillonite nanoarchitectures and layered double hydroxide-SPEEK based composite membranes for fuel cells applications. Appl Clay Sci 174:77–85. https://doi.org/10.1016/j.clay.2019.03.027
5. Chia MY, Thiam HS, Leong LK, Koo CH, Saw LH (2020) Study on improvement of the selectivity of proton exchange membrane via incorporation of silicotungstic acid-doped silica into SPEEK. Int J Hydrog Energy 45(42):22315–22323. https://doi.org/10.1016/j.ijhydene.2019.11.010
6. Choi J, Chen Y, Abbel R, Visagie I, Parker K (2020) Flexible humidity sensors for wireless monitoring based on electrospun sulfonated polyether ether ketone (SPEEK) nanofibres. Sens Actuators B Chem 324:128704. https://doi.org/10.1016/j.snb.2020.128704
7. Cukierman S (2006) Et tu, Grotthuss! and other unfinished stories. Biochim Biophys Acta Bioenergetics 1757(8):876–885. https://doi.org/10.1016/j.bbabio.2005.12.001
8. da Trindade LG, Borba KMN, Zanchet L, Lima DW, Trench AB, Rey F, Martini EMA (2019) SPEEK-based proton exchange membranes modified with MOF-encapsulated ionic liquid. Mater Chem Phys 236:121792. https://doi.org/10.1016/j.matchemphys.2019.121792
9. de Grotthius CIT (1806) LI. Memoir upon the decomposition of water, and of the bodies which it holds in solution, by means of galvanic electricity. Philos Mag 25(100):330–339. https://doi.org/10.1080/14786440608563457
10. Gao H, Dong C, Wang Q, Zhu H, Meng X, Cong C, Zhou Q (2018) Improving the proton conductivity of proton exchange membranes via incorporation of HPW-functionalized mesoporous silica nanospheres into SPEEK. Int J Hydrog Energy 43(48):21940–21948. https://doi.org/10.1016/j.ijhydene.2018.08.214
11. Gashoul F, Parnian MJ, Rowshanzamir S (2017) A new study on improving the physicochemical and electrochemical properties of SPEEK nanocomposite membranes for medium temperature proton exchange membrane fuel cells using different loading of zirconium oxide nanoparticles. Int J Hydrog Energy 42(1):590–602. https://doi.org/10.1016/j.ijhydene.2016.11.132
12. Hellström M, Behler J (2017) Proton-transfer-driven water exchange mechanism in the Na+ solvation shell. J Phys Chem B 121(16):4184–4190. https://doi.org/10.1021/acs.jpcb.7b01490
13. Huang T, Song J, He H, Zhang, YB, Li XM, He T (2020) Impact of SPEEK on PEEK membranes: demixing, morphology and performance enhancement in lithium membrane extraction. J Membr Sci 615:118448. https://doi.org/10.1016/j.memsci.2020.118448
14. Iulianelli A, Basile A (2012) Sulfonated PEEK-based polymers in PEMFC and DMFC applications: a review. Int J Hydrog Energy 37(20):15241–15255. https://doi.org/10.1016/j.ijhydene.2012.07.063
15. Jiang Y, Hou M, Hao J, Yi B, Wang Z, Song W, Shao Z (2017) Enhanced durability of sulfonated poly (ether ether ketones)-based polymer electrolyte membranes by a multi-layer composite technology. Solid State Ion 309:33–40. https://doi.org/10.1016/j.ssi.2017.07.003
16. Khomein P, Ketelaars W, Lap T, Liu G (2020) Sulfonated aromatic polymer as a future proton exchange membrane: a review of sulfonation and crosslinking methods. Renew Sustain Energy Rev 110471. https://doi.org/10.1016/j.rser.2020.110471
17. Kreuer KD (1996) Proton conductivity: materials and applications. Chem Mater 8(3):610–641. https://doi.org/10.1021/cm950192a
18. Li Z, Guan Z, Wang C, Quan B, Zhao L (2020) Addition of modified hollow mesoporous organosilica in anhydrous SPEEK/IL composite membrane enhances its proton conductivity. J Membr Sci 118897. https://doi.org/10.1016/j.memsci.2020.118897
19. Mahapatra MK, Singh P (2013) Fuel cells. Energy conversion technology. In: Future energy: improved, sustainable and clean options for our planet. https://doi.org/10.1016/B978-0-08-099424-6.00024-7
20. Mikhailenko SD, Celso F, Kaliaguine S (2009) Properties of SPEEK based membranes modified with a free radical scavenger. J Membr Sci 345(1):315–322. https://doi.org/10.1016/j.memsci.2009.09.017
21. Mohtar SS, Ismail AF, Matsuura T (2011) Preparation and characterization of SPEEK/MMT-STA composite membrane for DMFC application. J Membr Sci 371(1):10–19. https://doi.org/10.1016/j.memsci.2011.01.009

22. Nguyen HQ, Shabani B (2020) Proton exchange membrane fuel cells heat recovery opportunities for combined heating/cooling and power applications. Energy Convers Manage 204(October):112328. https://doi.org/10.1016/j.enconman.2019.112328
23. Simari C, Lo Vecchio C, Baglio V, Nicotera I (2020) Sulfonated polyethersulfone/polyetheretherketone blend as high performing and cost-effective electrolyte membrane for direct methanol fuel cells. Renew Energy 159:336–345. https://doi.org/10.1016/j.renene.2020.06.053
24. Verma S, Mishra S, Gaur A, Chowdhury S, Mohapatra S, Dwivedi G, Verma P (2021) A comprehensive review on energy storage in hybrid electric vehicle. J Traffic Transp Eng (English Edn) 8(5):621–637. https://doi.org/10.1016/j.jtte.2021.09.001
25. Wan Mohd Noral Azman WNE, Jaafar J, Salleh WNW, Ismail AF, Othman MHD, Rahman MA, Rasdi FRM (2020) Highly selective SPEEK/ENR blended polymer electrolyte membranes for direct methanol fuel cell. Mater Today Energy 17:100427. https://doi.org/10.1016/j.mtener.2020.100427
26. Wang R, Liu S, Wang L, Li M, Gao C (2019) Understanding of nanophase separation and hydrophilic morphology in Nafion and SPEEK membranes: a combined experimental and theoretical studies. Nanomaterials 9(6). https://doi.org/10.3390/nano9060869
27. Wei J (2019) Chapter 9—Proton-conducting materials used as polymer electrolyte membranes in fuel cells. In: Song K, Liu C, JZBT-P-BMN, Guo TA (eds). https://doi.org/10.1016/B978-0-12-815067-2.00009-3
28. Zeng H, Li Y, Wang C (2020) Using zeolite imidazole frameworks as ionic cross-linkers to improve the performance of composite SPEEK membrane. Sep Purif Technol 236:116258. https://doi.org/10.1016/j.seppur.2019.116258
29. Zhu LY, Li YC, Liu J, He J, Wang LY, Lei JD (2021) Recent developments in high-performance Nafion membranes for hydrogen fuel cells applications. Pet Sci. https://doi.org/10.1016/j.petsci.2021.11.004

Performance of Palm Oil Fractions as Rejuvenators in Bitumen

Sharon Gooi and Daniel Kong

Abstract Vegetable oils, such as sunflower oil, have been found to effectively rejuvenate aged bitumen and heal cracks. In this study, palm oil fractions (POF), specifically palm cooking oil (PCO), palm mid fraction (PMF), and palm stearin (PS), are evaluated as potential rejuvenators for short-term aged bitumen (RTFOT), with sunflower oil (SO) a control. Thin-film bitumen samples were produced using 3D-printed moulds. The healing rates of the rejuvenators were obtained by measuring the reduction in crack width over time. PMF showed an equivalent healing speed as SO, with the crack closing at 20 min, followed by PCO, PS, and RTFOT at 30, 120, and 270 min, respectively. The healed samples were analyzed using Fourier Transform Infrared (FTIR) spectroscopy to identify the rejuvenators' peaks at 1740 cm^{-1}. The areas of the rejuvenator, carbonyl, and sulfoxide peaks were found to determine their respective indices, which were then used to evaluate the anti-aging performance of the POF in aged bitumen. The results showed that PS had the highest reduction in the carbonyl and sulfoxide indices, at 74%, followed by PMF, SO, and PCO at 62%, 56%, and 53%, respectively. Based on these results, it can be concluded that the POF, particularly PMF, have great potential as rejuvenators in aged bitumen.

Keywords Palm oil fractions · Rejuvenator · Aged bitumen · Healing efficiency

---

S. Gooi
Department of Civil Engineering, School of Engineering, Monash University Malaysia, Jalan Lagoon Selatan, 47500 Bandar Sunway, Selangor, Malaysia

D. Kong (✉)
Monash Construction Circularity Node (Monash-ConCerN), School of Engineering, Monash University Malaysia, Jalan Lagoon Selatan, 47500 Bandar Sunway, Selangor, Malaysia
e-mail: daniel.kong@monash.edu

C. S. Choo et al. (eds.), *Proceedings of ASEAN-Australian Engineering Congress (AAEC2022)*, Lecture Notes in Electrical Engineering 1072,
https://doi.org/10.1007/978-981-99-5547-3_13

1 Introduction

Bitumen aging through oxidation is the main cause of cracking and rutting damage in asphalt pavements. Currently, some pavement repair methods include fog sealing and spray treatments that require road closures to ensure the repairs are not affected [17]. An alternate method that has recently gained interest is the enhancing of bitumen's self-healing abilities through rejuvenation. Rejuvenation uses materials such as vegetable oils, waste oils, aromatic extracts, or designed bio-oils to restore the brittle and stiff bitumen to its original colloidal balance by increasing the maltenes to asphaltenes ratio [1]. In the past decade, there has been a drastic increase in studies investigating the utility of vegetable oils as rejuvenators for bitumen. Vegetable oils are ideal as rejuvenators due to their low cost, non-toxicity, and sustainability (depending on production processes). In addition, many vegetable oils possess antioxidant properties that could reduce the formation of oxidation products such as carbonyls and sulfoxides in aged bitumen. Sunflower oil and waste cooking oil, in particular, have been widely tested for their effect on aged bitumen.

Palm oil has been tested as an additive for bio-asphalt development [2, 8], but only recently has it received attention as a potential rejuvenator. Rafiq et al. [13] and Devulapalli et al. [7] used crude palm oil and palm oil sludge, respectively, to restore reclaimed asphalt pavement (RAP). The crude palm oil increased the RAP percentage to 76% while maintaining the target specifications required by Jabatan Kerja Raya Malaysia [13]. Bilema et al. [5] utilized virgin and waste palm cooking oil to rejuvenate laboratory-aged bitumen. The samples containing virgin palm oil showed good storage stability, similar to the virgin bitumen sample, indicating that the palm oil had sufficient homogeneity after mixing with bitumen, with no phase separation [5]. Recently, Sánchez et al. [15] evaluated the performance of crude palm oil as well as soft and hard stearin in comparison with commercially available rejuvenators. These rejuvenators increased the resistance of the RAP blend asphalt to fractures and showed excellent adhesion with the aggregates used in the asphalt concrete, comparable to the commercial rejuvenators [15]. Kargari et al. [9] encapsulated refined palm oil and included steel wool fibers in the asphalt mixture to allow for both rejuvenation and induction heating. The palm oil capsules achieved a healing index of 32% in the aged asphalt samples [9]. In conclusion, palm oil has excellent prospects as a rejuvenator in its various forms.

However, some POF, such as palm mid fraction (PMF), have not yet been evaluated as rejuvenators, especially in comparison to sunflower oil (SO). SO is an oil frequently used as a rejuvenator due to its non-toxicity and low cost [4, 11, 14]. Additionally, there has not yet been a defined method to assess the changes caused by the rejuvenators diffusing out from the crack zone of bitumen samples on the level of carbonyls and sulfoxides found in aged bitumen.

Therefore, the primary objective of this study was to evaluate the healing efficiency and anti-aging performance of three palm oil fractions on aged bitumen: virgin palm cooking oil (PCO), PMF, and palm stearin (PS), with sunflower oil (SO) as a control

Table 1 Properties of PMF and PS

Properties	PMF	PS
Free fatty acid (% as palmitic)	0.115	0.090
Color (5.25″ Lovibond cell)	2.1R	2.9R
Moisture and impurities (%)	0.055	0.043
Iodine value (Wijs)	43.67	35.36
Slip melting point (°C)	28	51.0

rejuvenator. A novel method utilizing a 3D-printed grid mold for the bituminous thin film samples was used to analyze the anti-aging ability of the rejuvenators.

2 Materials and Methods

The SO and PCO used in this study were generic brands commonly found in supermarkets, while the PMF and PS were obtained from Mewah Oils Sdn. Bhd. The properties of the PMF and PS are listed in Table 1. Bitumen with a penetration grade 60/70 was obtained from Shell Bitumen. The virgin bitumen had a penetration of 64 $\times 10^{-1}$ mm at 25 °C and softening point at 48.2 °C. The bitumen was aged according to the Rolling Thin-Film Oven Test specified in ASTM D2872 [3].

3 Healing Rate of POF

The rejuvenating ability of the POF was evaluated through the healing of cracked, aged bitumen samples using a procedure similar to that proposed by Norambuena-Contreras et al. [12]. Thin film bituminous samples were prepared in square 3D-printed molds measuring 40 $\times$ 40 mm with a height of 1.0 mm placed atop silicone paper (Fig. 1). The molds were designed with grids to separate the bitumen into 16 sections, with the top left grid being circular in shape to mark the A1 coordinate grid. The thin film samples were conditioned at room temperature for two hours, after which a standardized heated blade was used to form a crack across the center of the sample. A micropipette was then used to release the rejuvenator onto the crack, which represented 2% of the bitumen weight.

Five types of bituminous samples were produced, including an unmodified aged bitumen sample (RTFOT) and aged bitumen samples with liquid SO, PCO, PMF, and PS, respectively. Three samples were prepared for each type of bituminous sample. The samples were then allowed to heal at room temperature until crack closure was observed. The image processing software ImageJ was used to measure the crack width, with five measurements taken for each sample. The change in crack width over time was used to quantify the healing efficiency (HE) of the bitumen samples, and the slope was used as an indicator of the healing rate.

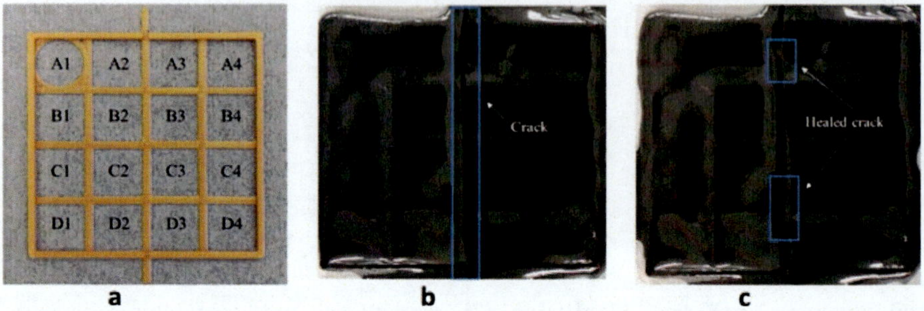

Fig. 1 Bitumen samples at different stages: **a** an empty mold, **b** bitumen thin film with crack, **c** bitumen thin film with closed crack

4 Fourier Transform Infrared Spectroscopy (FTIR) Analysis of Bitumen

FTIR was used to conduct a chemical analysis on the aged bitumen, the thin film bitumen samples, the rejuvenators, and the encapsulated rejuvenators. After the crack-healing was completed in the thin film bitumen samples, the absorbance spectra were obtained for each of the 16 sections formed by the square grid mold. The FTIR absorbance spectra of the bitumen and rejuvenators were normalized and overlaid to identify distinguishing peaks (Fig. 2). It was found that the bitumen samples lacked a peak at 1740 cm^{-1}, which was present in the PCO, PMF, and SO rejuvenators. As such, the presence of these peaks could indicate the diffusion of the rejuvenator into the bitumen. The FTIR software OMNIC was used to obtain the peak areas at the relevant bands. The quantification of the rejuvenator's diffusion away from the crack can be seen in Eq. 1.

To quantify the anti-aging properties of the POF, the carbonyl index (I_C) and sulfoxide index (I_S) of the samples were calculated according to the method proposed by Cong et al. [6]. The FTIR peak areas at 1700 cm^{-1} and 1030 cm^{-1} represent I_C and I_S, respectively (Eqs. 2 and 3). The FTIR peaks at 1375 cm^{-1} and 1457 cm^{-1} represent the –CH3 and –CH2– vibration bands that are inert even after the aging process, allowing them to be used as reference points to remove any interference by the bitumen film thickness [6]. The changes in these indices represent the anti-aging performance of the different rejuvenators. I_R represents the rejuvenator index, I_C is the carbonyl index, I_S is the sulfoxide index, and A_{1740}, A_{1700}, A_{1030}, A_{1375}, and A_{1457} are the peak areas at 1740 cm^{-1}, 1030 cm^{-1}, 1375 cm^{-1}, and 1457 cm^{-1}, respectively.

$$I_R = \frac{A_{1740}}{A_{1375}+A_{1457}} \quad (1)$$

$$I_C = \frac{A_{1700}}{A_{1375}+A_{1457}} \quad (2)$$

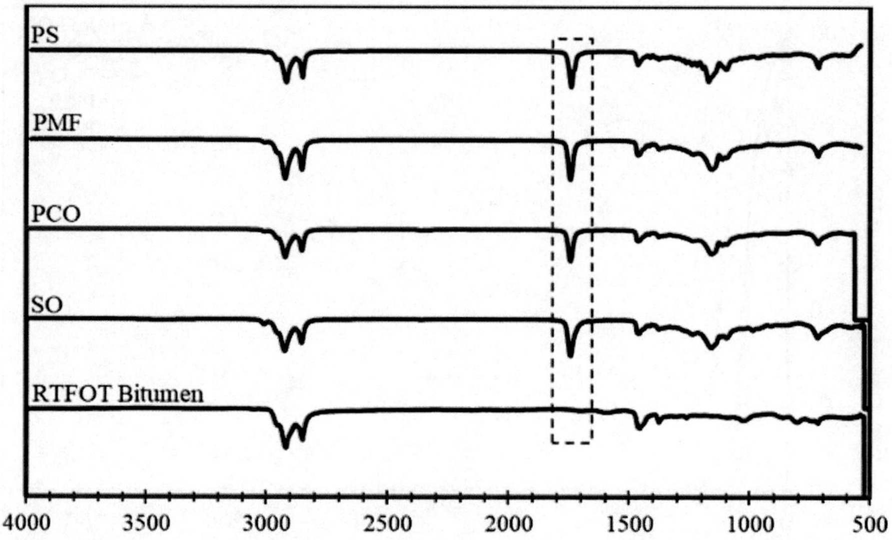

Fig. 2 FTIR spectra of bitumen and rejuvenators, with the peak at 1740 cm^{-1} unique to rejuvenators highlighted

$$I_S = \frac{A_{1030}}{A_{1375} + A_{1457}} \tag{3}$$

5 Results and Discussion

5.1 Healing Efficiency of POF

The healing efficiencies of the unmodified RTFOT, SO, and POF bitumen thin films were obtained by plotting the change in crack width over time (Fig. 3). The thin films with PMF and SO had the fastest healing rate, with the crack closing within 20 min, followed by the PCO, PS, and RTFOT films at 30, 120, and 270 min, respectively. It can be concluded that the presence of rejuvenator can dramatically improve the healing efficiency of aged bitumen. Although the PMF has a slip melting point of 28 °C, above the room temperature of 26 °C, after being heated at 40 °C for an hour beforehand, it maintained the liquid state throughout the healing process. This allowed for the quick diffusion of PMF into the aged bitumen, resulting in a high healing rate.

On the other hand, despite heating it for an hour at 60 °C, the PS solidified within minutes after being dripped into the crack due to its high melting point. However, despite solidifying quickly, it could diffuse into the bitumen during the initial liquid

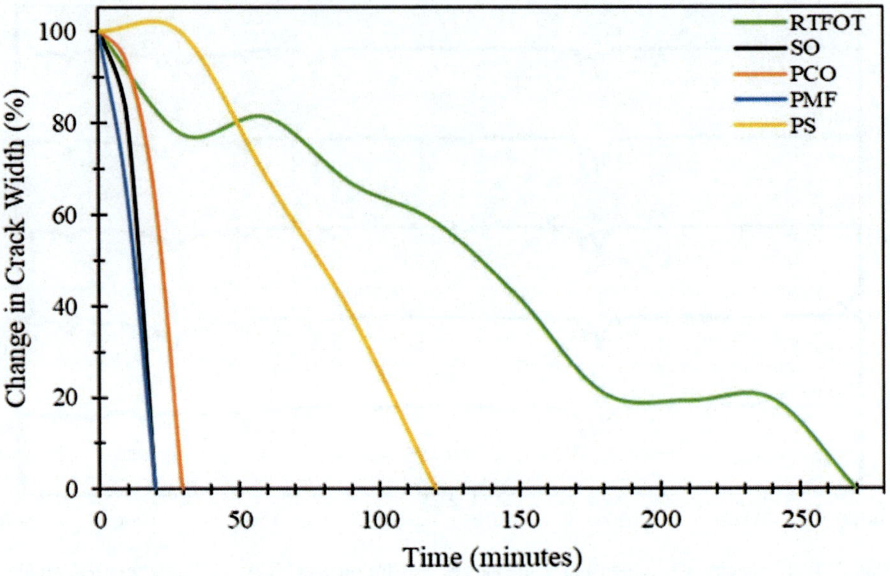

Fig. 3 POF's healing rate compared to SO and unmodified RTFOT

state, allowing it to heal faster than the RTFOT bitumen. All the rejuvenators successfully repaired the thin crack faster than the unmodified RTFOT bitumen, with the SO and PMF having an average healing time of 20 min, followed by PCO at 30 min, PS at 120 min, and the unmodified RTFOT samples at 270 min. At higher temperatures, it can be assumed that the healing rate would only increase, particularly for PS, once it has achieved its slip melting point at 51 °C.

5.2 *Anti-Aging Performance of POF*

In the cracked bitumen thin film samples, the rejuvenator was dripped across the center of the crack. To identify the anti-aging performance of the rejuvenator throughout the thin film, FTIR spectra were obtained for each of the sixteen grids (labeled A1 to D4) available in the customized 3D-printed mold. In Fig. 4, the clustered column charts show the calculated rejuvenator, carbonyl, and sulfoxide indices, which were calculated from Eqs. 1–3, using the aliphatic groups at 1375 cm^{-1} and 1457 cm^{-1} as reference peak areas. A higher level of I_R rejuvenator index is preferred to indicate the presence and diffusion of the rejuvenator from the crack. Conversely, lower values of the I_S sulfoxide index and I_C carbonyl index indicate an increased amount of anti-aging action.

The control RTFOT samples displayed high readings across the thin film for the carbonyl and sulfoxide indices, which are characteristic of aged bitumen. The I_{SO}

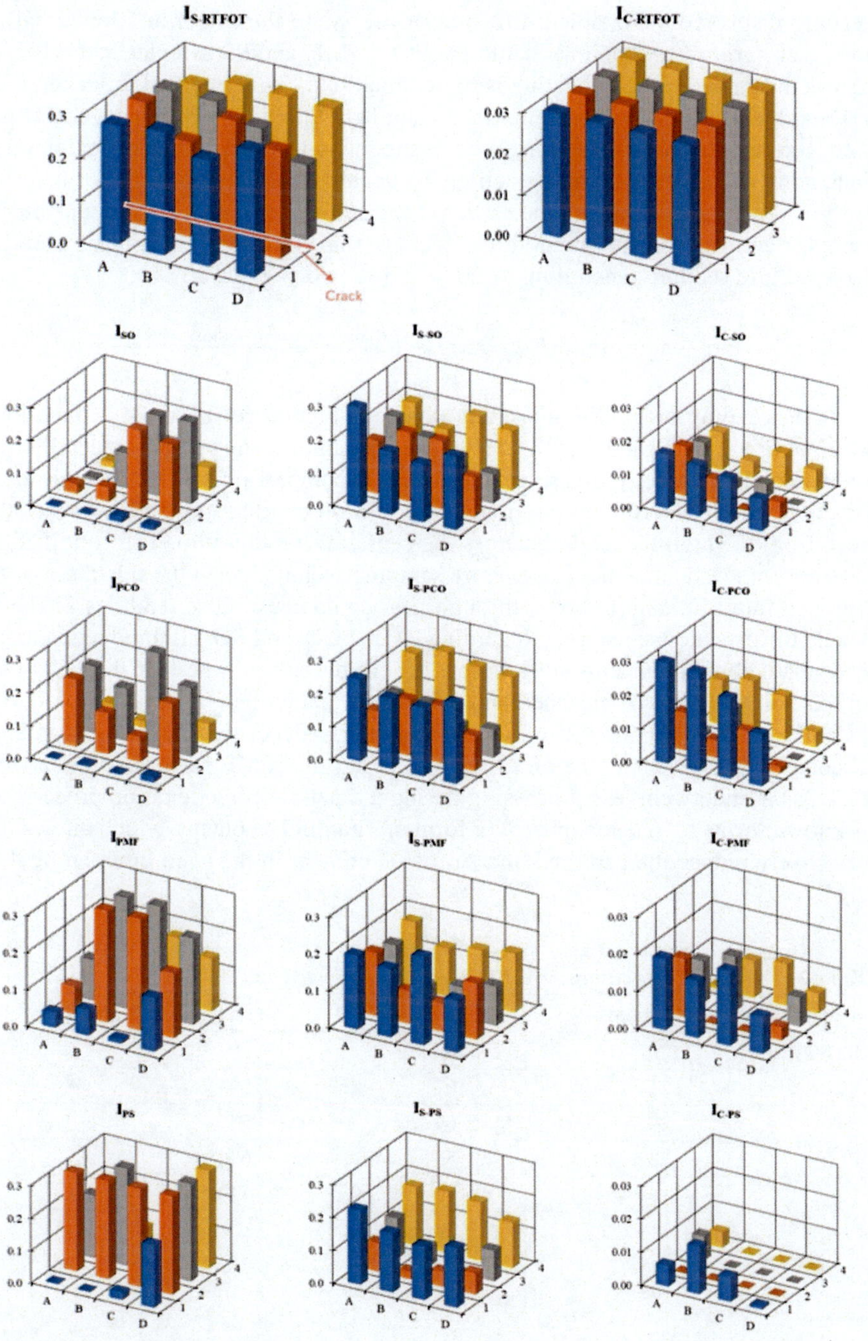

Fig. 4 The rejuvenator, sulfoxide, and carbonyl index obtained from the FTIR spectra of the RTFOT, SO, PCO, PMF, and PS thin film grid samples

reading displayed a contained diffusion of the SO to the center and lower half of the crack (grid rows 2 and 3), while the PCO, PMF, and PS samples had diffusion across the entire crack, indicating better compatibility with bitumen. In general, the carbonyl and sulfoxide indices in the SO and POF thin films exhibited a decreasing trend throughout the samples corresponding to sections with a high rejuvenator index, indicating successful anti-aging action. To quantify the anti-aging performance of the rejuvenators, the ratios of the carbonyl and sulfoxide contents of the rejuvenated samples compared to the unmodified RTFOT thin film were calculated, averaged from all grid sections, according to Eq. 4 and shown in Table 2.

$$\text{Anti}-\text{aging}\,\text{ratio}(\%) = 100 - \left(\frac{\text{Carbonyl content}+\text{Sulfoxide content}}{2}\right) \quad (4)$$

In descending order, the anti-aging ratio was highest for PS at 74%, then PMF at 62%, SO at 56%, and PCO at 53%. The PS samples showed an especially high reduction in the carbonyl and sulfoxide indices in comparison with SO, but this might be due partly to interference from solidified PS on the thin film surface, as can be seen from the high I_{PS} reading across the center of the thin film. The solidified PS was not removed since the removal by scraping would change the thickness of the bitumen thin film, and removal through heat could cause false readings as the PS would diffuse into the sample. Regarding PMF, PCO, and SO, all three rejuvenators had relatively similar anti-aging ratios. It is important to note that the PCO used in this study is refined, bleached and deodorized palm olein, not red palm oil. The PCO has thus undergone the refining process that reduces the carotene and vitamin E content [10, 18]. It is therefore not surprising that PMF performed better than PCO, as it underwent less processing during the palm oil fractionation process. SO is known for its high α-tocopherol (a form of vitamin E) content, which allows it to effectively reduce the carbonyl and sulfoxide content in the aged bitumen by 44% [16].

Table 2 The anti-aging performance of POF and SO compared to RTFOT

Thin Film	Carbonyl (%)	Sulfoxide (%)	Anti-Aging Ratio (%)
RTFOT	100	100	0
SO	21	67	56
PCO	30	64	53
PMF	23	53	62
PS	6	46	74

6 Conclusions

This study explored the healing rates and anti-aging ability of PCO, PMF, and PS against SO as a control rejuvenator and unmodified RTFOT bitumen providing a baseline comparison. It can be concluded from the results that at room temperature, PMF and PCO are viable rejuvenators, with PMF healing the crack formed as rapidly as SO. The FTIR analysis of the healed samples displayed significant drops in the sulfoxide and carbonyl levels compared to the unmodified RTFOT samples, confirming that the rejuvenators can successfully decrease the amount of oxidation products in aged bitumen through the crack healing process. Overall, palm oil fractions show great potential for further exploration as rejuvenators, with PMF performing the best with the fastest healing rate at 20 min and an anti-aging ratio of 62%.

Acknowledgements The authors express their gratitude to the Civil Engineering and Chemical Engineering Laboratories of Monash University Malaysia for providing access to their facilities throughout the course of this study.

References

1. Abadeen AZU, Hussain A, Veerappan SK et al (2022) Comprehensive self-healing evaluation of asphalt concrete containing encapsulated rejuvenator. Materials 15:3672. https://doi.org/10.3390/ma15103672
2. Al-Sabaeei AM, Napiah M, Sutanto M et al (2020) Prediction of rheological properties of bio-asphalt binders through response surface methodology. IOP Conf Ser: Earth Environ Sci 498:012012. https://doi.org/10.1088/1755-1315/498/1/012012
3. ASTM D2872–19 (2019) Standard Test Method for Effect of Heat and Air on a Moving Film of Asphalt (Rolling Thin-Film Oven Test). ASTM, West Conshohocken, PA
4. Bao S, Liu Q, Li H et al (2021) Investigation of the release and self-healing properties of calcium alginate capsules in asphalt concrete under cyclic compression loading. J Mater Civ Eng 33:04020401. https://doi.org/10.1061/(ASCE)MT.1943-5533.0003517
5. Bilema M, Aman YB, Hassan NA, et al. (2021) Performance of aged asphalt binder treated with various types of rejuvenators. Civil Eng J 7:502–517. https://doi.org/10.28991/cej-2021-03091669
6. Cong P, Chen B, Zhao H (2020) Coupling effects of wasted cooking oil and antioxidant on aging of asphalt binders. Int J Pav Res Technol 13:64–74. https://doi.org/10.1007/s42947-019-0086-0
7. Devulapalli L, Kothandaraman S, Sarang G (2021) Fracture characterization of stone matrix asphalt mixtures incorporating reclaimed asphalt pavement and rejuvenator. J Mater Civ Eng 33:04021320. https://doi.org/10.1061/(ASCE)MT.1943-5533.0003945
8. Flávia Justino Uchoa A, da Silva RW, Peter Macedo Feitosa J et al (2021) Bio-based palm oil as an additive for asphalt binder: Chemical characterization and rheological properties. Constr Build Mater 285:122883. https://doi.org/10.1016/j.conbuildmat.2021.122883
9. Kargari A, Arabani M, Mirabdolazimi SM (2022) Effect of palm oil capsules on the self-healing properties of aged and unaged asphalt mixtures gained by resting period and microwave heating. Constr Build Mater 316:125901. https://doi.org/10.1016/j.conbuildmat.2021.125901
10. Keat OC, May CY, Hock AOS (1991) Recovery of carotenoids

11. Micaelo R, Al-Mansoori T, Garcia A (2016) Study of the mechanical properties and self-healing ability of asphalt mixture containing calcium-alginate capsules. Constr Build Mater 123:734–744. https://doi.org/10.1016/j.conbuildmat.2016.07.095
12. Norambuena-Contreras J, Arteaga-Pérez LE, Concha JL, Gonzalez-Torre I (2021) Pyrolytic oil from waste tyres as a promising encapsulated rejuvenator for the extrinsic self-healing of bituminous materials. Road Mater Pav Des 1–17. https://doi.org/10.1080/14680629.2021.1907216
13. Rafiq W, Napiah M, Habib NZ et al (2021) Modeling and design optimization of reclaimed asphalt pavement containing crude palm oil using response surface methodology. Constr Build Mater 291:123288. https://doi.org/10.1016/j.conbuildmat.2021.123288
14. Ruiz-Riancho N, Garcia A, Grossegger D et al (2021) Properties of Ca-alginate capsules to maximise asphalt self-healing properties. Constr Build Mater 284:122728. https://doi.org/10.1016/j.conbuildmat.2021.122728
15. Sánchez DB, Saldarriaga JF, Caro S (2022) Chemical and thermodynamic properties of palm oil-based materials and their impact on recycled binder blends. Constr Build Mater 348:128490. https://doi.org/10.1016/j.conbuildmat.2022.128490
16. Schmidt Š, Pokorný J (2011) Potential application of oilseeds as sources of antioxidants for food lipids - a review. Czech J Food Sci 23:93–102. https://doi.org/10.17221/3377-CJFS
17. Staver M, Podolsky J, Williams RC et al (2022) Soybean oil-derived additives evaluated for use in bio-cutback and bio-fog seal treatment. Int J Pavement Res Technol. https://doi.org/10.1007/s42947-022-00199-2
18. Tan CH, Lee CJ, Tan SN et al (2021) Red palm oil: a review on processing, health benefits and its application in food. J Oleo Sci 70:1201–1210. https://doi.org/10.5650/jos.ess21108

Simulation on Application of EMG Signals for DC Motor Control

Evon Wan Ting Lim and Almon Chai

Abstract Electromyography (EMG) signals allow the prediction of neuromuscular activities and are commonly used to control robotic arm or exoskeleton to mimic human movements. A neural network is trained to classify EMG signals to distinguish hand grasp with different levels of pressing magnitude. The classification result achieved is approximately 95%. The aim of this study is to achieve better accuracy with less occurrence of error during continuous actuation of actuators when the classification rate of EMG data is not 100%. A SIMULINK model is created to simulate the continuous actuation of DC motor based on classified EMG signals for a total duration of 26.5 s. Different sample times are tested during the simulation to determine the best sample time to reduce the occurrence of error. The speed plot of the simulation with sample time of 250 ms shows the best performance. The total duration of the lead and lag time due to delay in response to control the actuation of the DC motor is <1 s. The maximum lead or lag time is approximately 0.138 s. These short lead and lag are acceptable when considering the efficiency in reducing the occurrence of errors to produce a smoother actuation with less sudden peak due to the misclassification of EMG patterns. With the combination of software and hardware, the actuation of DC motor can be controlled according to the EMG signals classified based on the sample time with the best performance.

Keywords Control · Classification · Electromyography · Neural network · Simulation

E. W. T. Lim (✉) · A. Chai
Swinburne University of Technology Sarawak Campus, Jalan Simpang Tiga, 93350 Kuching, Sarawak, Malaysia
e-mail: ewlim@swinburne.edu.my

A. Chai
e-mail: achai@swinburne.edu.my

C. S. Choo et al. (eds.), *Proceedings of ASEAN-Australian Engineering Congress (AAEC2022)*, Lecture Notes in Electrical Engineering 1072,
https://doi.org/10.1007/978-981-99-5547-3_14

1 Introduction

Electrography (EMG) signals, or commonly known as muscle signals, are generated when muscles are contracted. EMG signals have been broadly used in a variety of applications, such as power-assist exoskeleton [17], the control of prosthetic hand [1] and robotic manipulator [7], rehabilitation therapy [9], automated diagnosis of neuromuscular diseases [13, 14], and personal authentication [12].

The analysis of neuromuscular activities can be done through the evaluation of EMG signals. The prediction of human movements via classification of EMG signals has been introduced widely to replace the need of calculating complex human kinematics. Hence, EMG signals are used in the control of robot arm or exoskeleton to mimic the movement of the wearer. Actuators such as DC motor are attached to the joints of the exoskeleton robot to actuate the movement of the joints. By monitoring the EMG data of the wearer, the controller of the robot offers help to the wearer when it detects that the wearer is unable to perform the required movement during rehabilitation therapy. Besides rehabilitation, EMG signals are also used in the control of slave manipulator or powered exoskeleton to mimic the movement of the human operator. The intention of the human operator can be evaluated through the EMG signals collected via surface electrodes attached to the muscles of the human arm.

In order to increase the feasibility of EMG signals in robotic or exoskeleton control, the accuracy and consistency of the EMG pattern recognition is very important. Selecting appropriate signal conditioning, filtering and processing approaches, feature extraction approaches and classifier have always been a challenge to researchers. Current research studies focus more on proposing and identifying the appropriate feature extraction methods and the correct classifier to improve classification accuracy. Several research studies have been conducted to compare the different types of time domain features such as the studies done by Shi et al. [11], Wang and Wichakool [15]. There are also research studies that utilized multiple feature extraction methods for classification and applied principal component analysis (PCA) for feature reduction to achieve higher classification accuracy [5]. Research studies on the combination of various features to identify the best combination of feature groups to enhance the classification performance of the classifier have also been reviewed by Purushothaman and Ray [8], Zhang et al. [16]. Devaraj and Nair [2] managed to achieve an accuracy of 93% for the classification of seven hand gestures of 37 subjects via k-nearest neighbor (kNN) classifier. The methodology proposed by Khairuddin et al. [4] to discriminate elbow flexion and extension managed to achieve a classification accuracy of 96.4%. In another study regarding the classification of three kinds of hand movements via the combination of discrete wavelet transform and artificial neural network is reported to be 93.25% [6].

From the studies reviewed, different datasets, feature extraction methods, classifiers and algorithm to control robotic arm or slave exoskeleton have been utilized by different authors. Hence, it is very hard to compare the results achieved based on different methods or algorithm utilized by the different authors. However, it can be noticed that it is not easy to achieve classification with 100% accuracy especially when there are comparisons of EMG signals between different individuals. The EMG datasets used in this study is also different from those used by others. Therefore, this study will mainly focus on evaluating the feasibility regarding the application of EMG signals to control robotic arm or slave exoskeleton continuously based on the prediction of different levels of grasp forces when the classification accuracy is not 100%. Five sets of EMG signals from five sample subjects with duration of 26.5 s each are used to simulate the control of a DC motor continuously to evaluate the feasibility of this application.

2 EMG Data Preparation

EMG signals acquired while performing hand grasp with different levels of pressing magnitude from 43 participants are downloaded from the online database provided by Kaczmarek et al. [3]. During signal acquisition, the participants are asked to follow the force trajectory provided that are of different combinations of duration, pressure magnitude, and shape. A total of 24 channels of EMG electrodes are used to collect EMG signals from the right forearm of the participants. The EMG data are divided into three classes, according to the level of pressing magnitude, as presented in Table 1.

Before the EMG signals can be classified, proper signal processing and feature extraction technique need to be applied to extract a useful feature for pattern recognition. Raw EMG signals are of double polarities. Hence, during the first step of signal processing, the EMG signals are rectified via full-wave rectification to preserve all components of the signals. In this study, comparison of EMG signals between different individuals is desired. Hence, normalization is required in order to reduce the variability between individuals and at the same time raises the output reliability [10]. Normalization is done based on the Maximum Voluntary Contraction (MVC) method. After normalization, feature patterns are extracted from the EMG signals via the Root Mean Square (RMS) method as shown in Eq. (1). The window size selected is 250 ms, with an overlapping window size of 125 ms. 'N' denotes the window size.

$$\mathrm{RMS} = \sqrt{\frac{1}{N}\sum_{i=1}^{N}\mathrm{EMG}(i)^2} \tag{1}$$

Table 1 Division of EMG data into three classes according to the level of pressing magnitude

Class	Level
1	Idle state—hand grasp but relaxing (Level 0)
2	Hand grasp with pressing magnitude 1 (Level 1)
3	Hand grasp with pressing magnitude 2 (Level 2)

The feed-forward back-propagation neural network from MATLAB neural network toolbox is selected as the classifier to classify EMG feature patterns to distinguish hand grasp with three level of pressing magnitude. The neural network consists of three layers: one input layer, one hidden layer and one output layer. Weights and biases are assigned between two adjacent layers. A total of two sets of EMG data are collected from each of the 43 participants. During the training process, 70% of the EMG data are used as the training dataset and 15% of the EMG data are used as the testing dataset. The remaining 15% of the EMG data are used for validation. The number of hidden neurons is selected based on trial and errors. A total of 10 trials of training are performed by varying the number of hidden neurons, starting from 15 hidden neurons to 50 hidden neurons, with an increment of five hidden neurons each time. The highest classification accuracy among all trials is 95.7% for neural network with 45 hidden neurons. Hence, the neural network trained with 45 hidden neurons with 95.7% classification rate is saved for further usage in the next section.

3 SIMULINK Simulation

Classification of EMG signals allows the prediction of human movements. The classified EMG patterns can be applied to robotic arm or slave exoskeleton for manipulation to perform movements to closely mimic those of a human user. However, it is not easy to achieve a perfect classification rate, especially when comparison between different individuals is required. To evaluate the feasibility of using EMG signals that did not achieve 100% classification accuracy to control actuators of an exoskeleton or robotic arm, a SIMULINK model is created to simulate the actuation of a DC motor based on actual EMG signals.

EMG signals from five sample subjects are selected randomly. The EMG signals with a total duration of 26.5 s are taken from each subject. The whole duration includes the three different levels of hand grasp. The plot regarding the different levels of hand grasp for the whole duration of 26.5 s is shown in Fig. 1. The 26.5 s of EMG signals are segmented and processed, and then feature patterns are extracted for classification via the trained neural network with 45 hidden neurons. The classification rates computed from the EMG signals of the five sample subjects are more than 94%. The detailed results are discussed in Sect. 4.

The SIMULINK model is divided into two main parts, the classification of EMG signals and the actuation of DC motor based on the outcome of the classification. The actuation of the DC motor is based on actual EMG signals. The DC motor is actuated continuously based on the classification outcome of the EMG patterns extracted from the 26.5 s of EMG signals. Different applied voltages are assigned to the DC motor to represent the three different classes of hand grasp (i.e. 0 V for Class 1; 12 V for Class 2; 18 V for Class 3). The flowchart is shown in Fig. 2.

The neural network is trained to classify EMG patterns extracted with a moving window of 250 ms and an over-laping of 125 ms. Hence, every 125 ms, an EMG pattern is produced. The SIMULINK model is used to simulate the actuation of

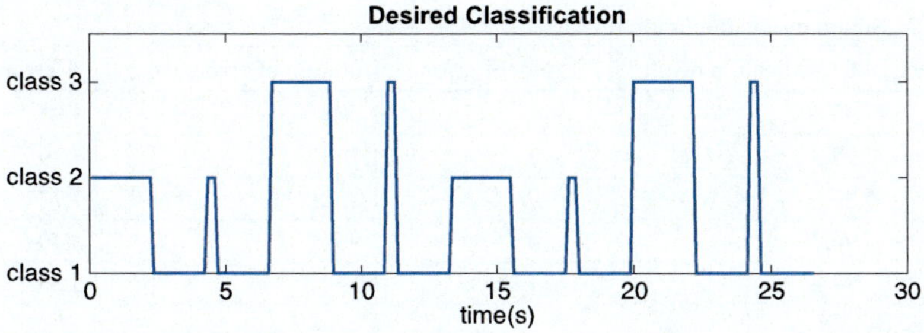

Fig. 1 The desired classification output with 100% classification accuracy

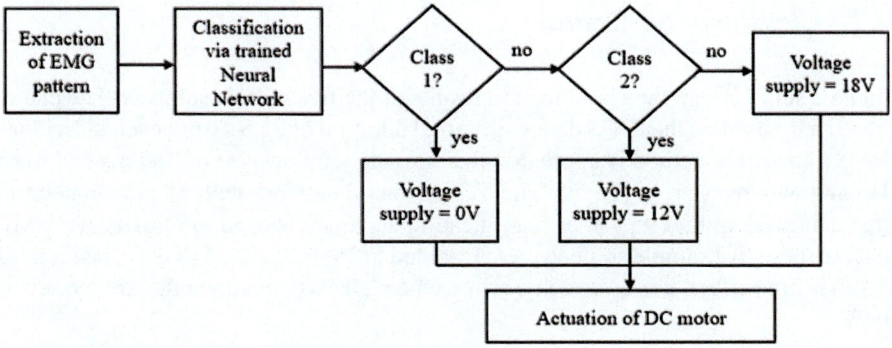

Fig. 2 Flow chart of the Simulink model

the DC motor based on the classified EMG patterns for a whole duration of 26.5 s. When the SIMULINK model is simulated at 125 ms, the actuation of the DC motor responded according to EMG patterns classified by the neural network every 125 ms. However, since the trained neural network did not achieve 100% classification accuracy, some EMG patterns are classified wrongly and hence the DC motor is actuated wrongly too according to the misclassified EMG patterns. Hence, when the speed plot obtained from the simulation of the DC motor is compared with the desired speed plot, errors are present at the times when misclassification of EMG patterns occurs.

A total of 10 experiments are conducted to analyse the effect of varying the sample time of the SIMULINK model on the speed plot of the simulation, starting from 150 to 375 ms, with an increment of 25 ms each time. The aim of these experiments is to find out the best sample time of simulation to reduce the occurrence of errors caused by misclassification of EMG patterns. When the sample time is increased, some lead and lag times are observed when compared to the speed plot simulated at 125 ms. However, if the total lead and lag time is very small, changing the sample time to reduce the occurrence of error can be taken into consideration to achieve a better DC motor actuation outcome that behaves closely to the desired outcome.

Table 2 Summary of simulation result

Subject	Classification result (%)	No. of error	Duration of errors (sec)	% error (%)
1	95.8	8	1.309	4.94
2	95.3	7	1.269	4.79
3	95.8	7	1.252	4.72
4	96.7	4	0.900	3.40
5	94.8	11	1.544	5.83

4 Results and Discussions

4.1 Classification Results

Table 2 summarized the classification results of the five sample subjects. The classification is based on the EMG data with a total duration of 26.5 s from each individual. RMS extraction method is applied with a moving window size of 250 ms and overlapping window size of 125 ms. The trained neural network with 45 hidden neurons that achieved around 95.7% of classification accuracy is used to classify the EMG data of these five sample subjects. As presented in Table 2, although the classification result is not perfect, it is quite convincing, where all classification rates are more than 94%.

4.2 Simulation Results

Figure 3 illustrates the speed plot simulated based on the EMG data of the sample subject 2. The total duration of the simulation is 26.5 s and the sample time is 125 ms. According to the classification outcome, 95.3% of the EMG data from sample subject 2 are classified correctly. Hence, the misclassified EMG patterns affect the actuation of the DC motor. Differences can be spotted from the speed plot simulated based on the predicted data when compared with the desired outcome. Based on the lower plot of Fig. 3, a total of seven errors are spotted by comparing the predicted and the desired speed plot. The largest error happens between 10 and 15 s, with duration of 512 ms. The rest of the errors have duration between the ranges of 120 ms to 140 ms. The occurrences of these errors are due to misclassified EMG patterns.

The summary on the classification accuracy, number of errors spotted and total duration of error computed from the simulation outcomes of the five sample subjects are shown in Table 2. The EMG data of sample subject 5 has the lowest classification rate among the others, which is approximately 94.8%. The number of errors spotted is 11, which is also the highest among the others. The duration of errors computed from the simulated result of sample subject 5 is also the highest among the others.

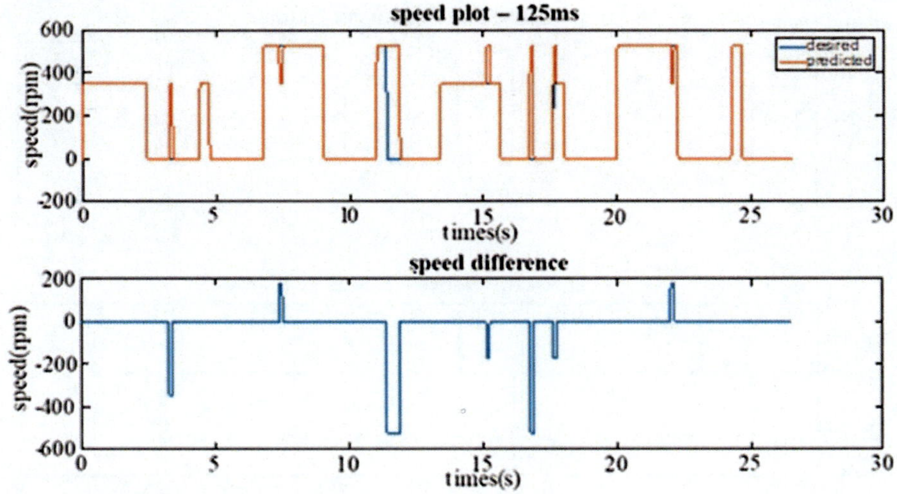

Fig. 3 Speed plot simulated based on EMG data of sample subject 2 (125 ms sample time)

From the summarized results, it can be noticed that when the classification accuracy is higher, the number of errors spotted and the duration of errors are lesser.

The occurrences of errors due to misclassified EMG patterns affect the smoothness of the speed plot, where sudden peaks are spotted at some point where misclassifications of EMG patterns happen. Although the duration of errors is not large, reduction in occurrences of error (sudden peak) is desired if possible. In order to reduce the occurrences of error (sudden peak), the sample time of the simulation is adjusted, aimed to evaluate the possibility of reducing the occurrences of error. However, when the sample time is adjusted, it is important to take into consideration the possible delay in the response of the DC motor.

A total of 10 experiments are conducted by adjusting the sample time of the simulation, starting from 150 to 375 ms with an increment of 25 ms each time. Initially, the experiments are conducted by simulating the DC motor according to the desired classification outcome (100% accuracy) to study the effect of adjusting the sample time on the actuation. During the experiment, some lead or lag time is observed to happen when the motor is actuated or stopped due to the delay in response that is present when compared to 125 ms. Figure 4 illustrates the comparison of the speed plot of DC motor simulated at 250 and 125 ms. As can be seen from the plot, some leads and lags are observed. This is because when the sample time is set as 250 ms, the respond of the DC motor will only update after each 250 ms interval.

The summary of the total lead and lag time, the maximum lead or lag time and the number of occurrences of the lead and lag time are summarized in Table 3. From the results computed from the simulation with sample time of 150–300 ms, the total lead and lag time error are less than or around 1 s, except for 275 ms. For a total duration of 26.5 s, a total lead and lag time error of 1 s is around 3.77% of the overall duration, which is acceptable. The number of occurrences of lead and lag for 250 ms

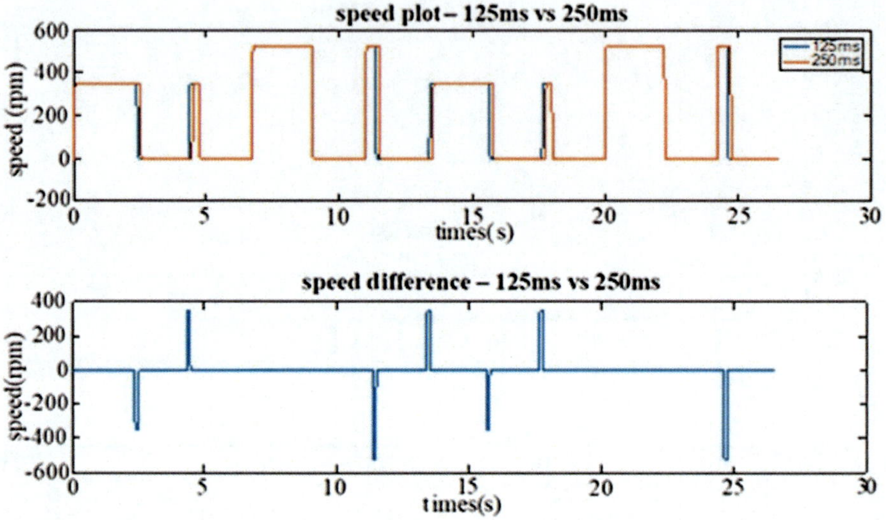

Fig. 4 Comparison of speed plot of DC motor simulated at 125 and 250 ms

sample time is 7, which is the least among the others. The maximum lead or lag time observed for simulation with sample time of 150, 175 and 250 ms are around 140 ms, which is approximately 0.53% of the overall duration only.

Table 3 Summary on the total lead and lag time, maximum lead/lag time and the number of occurrences

Sample time (ms)	Total lead and lag time		Max lead/lag		No. of occurrences
	Seconds (s)	% out of 26.5 s (%)	Seconds (s)	% out of 26.5 s (%)	
150	0.944	3.56	0.133	0.502	12
175	0.928	3.50	0.139	0.525	13
200	0.733	2.77	0.183	0.691	11
225	1.065	4.02	0.157	0.592	12
250	0.938	3.54	0.138	0.521	7
275s	1.632	6.16	0.257	0.970	15
300	1.052	3.97	0.161	0.608	13
325	2.040	7.70	0.283	1.068	14
350	3.036	11.46	0.315	1.189	15
375	2.107	7.95	0.263	0.992	11

4.3 Simulation Based on EMG Signals from Five Sample Subjects

After analyzing the occurrence of possible lead and lag time of the simulation with the various sample times, the simulation is conducted based on the actual classification results of the EMG signals of the five sample subjects for a duration of 26.5 s each. The occurrence of errors and the total duration of errors are computed from the simulation results, as summarized in Table 4. During the simulation, the lead and lag time due to changes in sample time are not taken into consideration here. The results show that the occurrences of error have been reduced during the simulation with sample time between 150 and 375 ms. The reduction in occurrences of error is obvious for sample time of 250, 300 and 325 ms. The total duration of errors for sample time of 150, 250 and 375 ms are among the lowest. By considering both the occurrences of errors and total duration of error, the sample time of 250 ms has the best performance. Besides, the total lead and lag time for the simulation with sample time of 250 ms is 0.938 s (3.54% of the total duration), and the maximum duration of lead or lag time is 0.138 ms (0.521% of the total duration). The number of times lead and lag occurs during the simulation with sample time of 250 ms is also the lowest as compared to the others.

Table 4 Summary on the occurrence and the total duration of error with different sample time

Sample time (ms)	Subject 1		Subject 2		Subject 3		Subject 4		Subject 5	
	Err	Time (s)	Err	Time (s)	Err	Time (s)	Err	Time (s)	Err	Time (s)
125	8	1.309	7	1.269	7	1.252	4	0.900	11	1.544
150	5	1.239	5	1.255	8	1.701	4	0.935	8	1.545
175	5	1.263	5	1.452	7	1.615	4	1.070	8	2.178
200	5	1.239	4	1.029	8	1.854	2	0.417	8	1.842
225	4	1.156	4	1.168	7	1.840	2	0.688	4	1.359
250	5	1.288	4	1.277	2	0.755	3	0.773	6	1.532
275	5	1.408	4	1.406	7	2.238	3	0.853	5	1.973
300	4	1.229	3	1.229	4	1.524	2	0.614	4	1.224
325	4	1.332	4	1.667	4	1.961	1	0.334	4	1.334
350	4	1.431	4	1.446	5	2.128	3	1.085	6	2.160
375	4	1.528	2	0.756	2	1.118	3	1.136	3	1.149

Err Error

5 Conclusion

The EMG data of 43 subjects recorded via 24 channels of electrodes for hand grasp with different levels of pressing magnitude are classified with a neural network. The EMG data are rectified and normalized before RMG features are extracted. During feature extraction, a moving window of 250 ms is selected, with an overlapping window of 125 ms. The classification rate achieved via the proposed method is around 95%. Although the classification rate did not achieve 100%, the feasibility on the application of these EMG data to control the actuation of actuators of an exoskeleton or robotic arm is evaluated. A SIMULINK model is created to simulate the continuous actuation of a DC motor based on the classified EMG data for a total duration of 26.5 s.

The sample time of the simulation is adjusted, starting from 125 to 400 ms, with an increment of 25 ms each time. By considering the efficiency in reducing the occurrence of errors in the speed plot, the total lead and lag time due to delay in response to control the actuation of the DC motor and the maximum lead and lag time, it is concluded that the sample time with the best performance is 250 ms. The speed plot on the actuation of the DC motor simulated with a sample time of 250 ms shows lesser error when compared to the initial sample time of 125 ms. Some lead and lag time are observed at seven locations of the speed plot, due to the delay in response to control the actuation of the DC motor when the sample time is changed to 250 ms instead of 125 ms. However, the total duration of these lead and lag times is <1 s, and the maximum lead or lag time is approximately 0.138 s.

The sample time to control the actuation of the DC motor can be easily adjusted via software. Since classification of EMG signals needs to be done via software, a Graphic User Interface (GUI) that can be built to configure some required settings such as the sample time. With the combination of software and hardware, the actuation of the DC motor can be controlled according to the EMG signals classified based on the sample time with the best performance.

Acknowledgements The authors would like to extend appreciation of support to Swinburne University of Technology Sarawak Campus.

References

1. Ahmed R, Halder R, Uddin M, Mondal PC, Karmaker AK (2018) Prosthetic arm control using electromyography (EMG) signal. IEEE. https://doi.org/10.1109/ICAEEE.2018.8642968
2. Devaraj A, Nair AK (2020) Hand gesture signal classification using machine learning. In: 2020 international conference on communication and signal processing (ICCSP), 28–30 July 2020, pp 0390–0394. https://doi.org/10.1109/ICCSP48568.2020.9182045
3. Kaczmarek P, Mánkowski T, Tomczýnski J (2019) PutEMG—a surface electromyography hand gesture recognition dataset. Sensors (Switzerland) 19(16). https://doi.org/10.3390/s19163548

4. Khairuddin IM, Sidek SN, Majeed APPA, Puzi AA (2019) Classifying motion intention from EMG signal: a k-NN approach. In: 2019 7th international conference on mechatronics engineering (ICOM), 30–31 October 2019, pp 1–4. https://doi.org/10.1109/ICOM47790.2019.8952042
5. Kunapipat M, Phukpattaranont P, Neranon P, Thongpull K (2018) Sensor-assisted EMG data recording system. In: 2018 15th international conference on electrical engineering/electronics, computer, telecommunications and information technology (ECTI-CON), 18–21 July 2018, pp 772–775. https://doi.org/10.1109/ECTICon.2018.8619880
6. Mane SM, Kambli RA, Kazi FS, Singh NM (2015) Hand motion recognition from single channel surface EMG using wavelet & artificial neural network. Procedia Comput Sci 49:58–65. https://doi.org/10.1016/j.procs.2015.04.227
7. Praveen LS, Nagananda SN, Shankapal P (2018) Design and development of real time bionic hand control using EMG signal. In: 2018 IEEE international conference on electronics, computing and communication technologies (CONECCT), 16–17 March 2018, pp 1–4. https://doi.org/10.1109/CONECCT.2018.8482393
8. Purushothaman G, Ray KK (2014) EMG based man–machine interaction—a pattern recognition research platform. Robot Auton Syst 62(6):864–870. https://doi.org/10.1016/j.robot.2014.01.008
9. Rahman MH, Rahman MJ, Saad M, Ochoa-Luna C, Ferrer SB (2013) Archambault PS control of an upper extremity exoskeleton robot to provide active assistive therapy. In: 2013 5th international conference on modelling, identification and control (ICMIC), 31 August–2 September 2013, pp 105–110
10. Rota S, Rogowski I, Hautier C, Champely S (2013) Reliability of EMG normalisation methods for upper-limb muscles. J Sports Sci 31(15):1696–1704–1704. https://doi.org/10.1080/02640414.2013.796063
11. Shi W-T, Lyu Z-J, Tang S-T, Chia T-L, Yang C-Y (2018) A bionic hand controlled by hand gesture recognition based on surface EMG signals: a preliminary study. Biocybern Biomed Eng 38(1):126–135. https://doi.org/10.1016/j.bbe.2017.11.001
12. Shin S, Jung J, Kang M, Kim YT (2018) A study on EMG signal acquisition modules and artificial neural networks for personal authentication. IEEE. https://doi.org/10.1109/CSCI46756.2018.00289
13. Tengku Zawawi TNS, Abdullah AR, Sudirman R, Shair EF, Mohd Saad N, Too J (2019) Classification of EMG signal for health screening task for musculoskeletal disorder. Int J Eng Technol 8:219–226. https://doi.org/10.14419/ijet.v8i1.7.25980
14. Ullah S, Iqbal K (2020) A preliminary review on EMG signals for assessment of diabetic peripheral neuropathy disorder. In: 2020 7th international conference on electrical and electronics engineering (ICEEE), 14–16 April 2020, pp 42–46. https://doi.org/10.1109/ICEEE49618.2020.9102488
15. Wang J, Wichakool W (2017) Artificial elbow joint classification using upper arm based on surface-EMG signal. IEEE. https://doi.org/10.1109/ICETSS.2017.8324198
16. Zhang J, Ling C, Li S (2019) Human movements classification using multi-channel surface EMG signals and deep learning technique. IEEE. https://doi.org/10.1109/CW.2019.00051
17. Zhichuan T, Kejun Z, Shouqian S, Zenggui G, Lekai Z, Zhongliang Y (2014) An upper-limb power-assist exoskeleton using proportional myoelectric control. Sensors 14(4):6677–6694. https://doi.org/10.3390/s140406677

Smart Energy and Cities

Building Information Modeling (BIM) for Construction and Demolition Waste (CDW) Management: Scientometric and State-of-the-Art Review

Samwel A. Mrema, Hieng H. Lau, San C. Liew, Palaneeswaran Ekambaram, Morshed Alam, and V. C.-C. Lee

Abstract Construction and Demolition Waste Management has received significant research attention due to the rising worldwide awareness of green environment. This review-based study analyzed 75 papers published in BIM for CDW Management through a scientometric analysis and an in-depth qualitative discussion. The scientometric analysis revealed the most influential journals, keywords, articles, and countries active in BIM for CDW Management research since 2011. Keyword analysis identified the mainstream research topics like Sustainability, Waste minimization, and Waste reuse and recycling. The follow-up qualitative discussion identified and analyzed the main research themes (e.g., quantification and estimation of CDW generation) and research gaps (e.g., lack of reliable regional dataset of CDW generation factors) and proposed the future research direction (e.g., development of reliable CDW database for CDW estimation and quantification) in BIM for CDW Management research.

Keywords Building Information Modeling (BIM) · Construction and Demolition Waste (CDW) · Science mapping · Literature review · Qualitative review

S. A. Mrema (✉) · H. H. Lau · S. C. Liew
Department of Civil Engineering, Swinburne University, Kuching, Sarawak, Malaysia
e-mail: smrema@swinburne.edu.my

P. Ekambaram · M. Alam
Department of Civil Engineering, Swinburne University, Melbourne, VIC, Australia

V. C.-C. Lee
Faculty of Engineering and Science, Curtin University Malaysia, Miri, Sarawak, Malaysia

C. S. Choo et al. (eds.), *Proceedings of ASEAN-Australian Engineering Congress (AAEC2022)*, Lecture Notes in Electrical Engineering 1072,
https://doi.org/10.1007/978-981-99-5547-3_15

1 Introduction

Construction and Demolition Waste (CDW) refers to any material produced as waste from construction, renovation and demolition activities of buildings, roads, bridges, and other structures [50]. CDW typically includes concrete, asphalt wood, steel, glass, and hazardous materials like asbestos and mercury. According to a report by the world bank group, CDW is the largest waste stream after industrial and agricultural waste globally [33]. CDW accounts for 30–40% of the total waste generated in a country [37]. CDW is associated with various negative impacts, including diminishing landfill space [41], material wastage [48], increased GHG emissions [63], and increased project costs [57]. CDW generation is still rapidly growing due to urbanization and will cause additional negative impacts if not properly managed. With such background, the research on CDW Management has attracted major attention worldwide. Current research themes in CDW Management include estimation and quantification of CDW generation, CDW diversion research in developing economies and design and planning of waste diversion [32]. Newly emerging technologies like Building Information Modelling (BIM), Geographic Information System (GIS) and Big Data (BD) have also been gradually adopted to improve the efficiency of CDW Management. Among these technologies, BIM has received considerable attention from researchers [37]. BIM significantly reduces CDW through BIM-based design validation (clash detection and design review) thus reducing design errors, change orders and reworks in design and construction stage. According to Won et al. [67] BIM can prevent up to 15.2% of construction waste through BIM-based design validation. Numerous studies have been conducted in the BIM for CDW Management field. Cheng and Ma [15] developed a BIM-based system that automatically estimates and plans for demolition and renovation waste. Wang et al. [65] combined BIM and LCA to assess the carbon emissions over the life cycle of building demolition waste. Liu et al. [39] developed a design decision making framework for improving construction waste minimization performance through BIM. Guerra et al. [23] proposed temporal-based algorithms integrated with 4D-BIM for the planning of CDW reuse and recycling throughout construction projects. Akinade and Oyedele [2] developed a BIM-based computational tool for building waste analytics and reporting in the construction supply chains to aid circular economy. Zoghi and Kim [72] developed a system dynamic (SD) model to analyze the economic aspects of CDW from using BIM. Akinade et al. [4] developed a BIM-based Deconstructability Assessment Score (BIM-DAS) for determining the extent to which a building could be deconstructed right from the design stage. The aforementioned work have demonstrated significant advancement of BIM in CDW Management. Nevertheless, there remains critical gaps in this research area that need to be filled. This paper aims to provide insights into the research gaps through (1) analysis of current state-of-art of the research area (2) identification of critical areas for near future research, and (3) formulation of framework for future research. To fill this gap, this research aims to conduct a scientometric analysis and an in-depth qualitative discussion. The scientometric analysis is used to identify the main research topics and trends within BIM for CDW Management field. It involves a bibliometric literature search and a scientometric analysis. This technique has been shown to reduce subjectivity and biasedness when selecting articles are to be included in review studies [28, 59]. The in-depth qualitative discussion evaluates the identified research topics and trends, identifies gaps and proposes framework for future research direction [32].

The remainder of this paper is structured as follows: Sect. 2 describesthe need for this research; Sect. 3 describes the methodology used; Sect. 4 presents the results of scientometric analysis; Sect. 5 discusses the research trends, gaps, and future needs; Sect. 6 highlights the research contribution of this paper.

2 Previous Reviews and Gap

This section discusses the existing literature review on BIM for CDW Management to clarify the need for a scientometric analysis and state of the art review on BIM for CDW Management are presented in this paper. Ahankoob et al. [1] and Rajendran and Gomez [56] assessed BIM's potential to minimize CDW through an in-depth literature review, and reported that BIM enhances the quality and accuracy of design and construction, detects clashes and improves collaboration between stakeholders, thus reducing design errors and rework, consequently reducing CDW. Similarly, Won and Cheng [66] developed approaches to address the current limitations of CDW Management using BIM, and reported that BIM uses like design review, 3D coordination, quantity take-off, and phase planning can address the current limitation of CDW and minimize CDW. Lu et al. [41] reported that the application of BIM for CDW Management is still rhetoric and some major hurdles need to be adequately overcome before it's a reality. A computational BIM for CDW Management is then developed by learning from the historical development of data infrastructure in the field of BIM-based cost management. Akinade et al. [5], reviewed the performance of 32 CDW Management tools using 6 Evaluation criteria obtained from conducting 4 Focus Group Interviews, and concluded that the existing tools aren't robust enough to tackle CDW Management in the design stage. Akinade et al. [3], identified limitations of existing CDW Management tools and suggested factors to consider for implementation of BIM for CDW Management. Gupta et al. [25] reviewed CDW Management tools and strategies and identified their limitations to aid in the future development of CDW Management tools and strategies. Nikmehr et al. [46], performed a scoping review on the current BIM-based CDW Management tools to identify their intellectual weakness. Han et al. [27] identified the barriers of CDW Management in Australia, and challenges for BIM based Demolition Waste Management (DWM) and presented future research directions for BIM based DWM. Although BIM has been advocated for its potential to facilitate CDW Management and various tools have been developed and reviewed, limited research efforts have attempted to synthesize research trends and to identify research directions in BIM for CDW Management. In particular, no review was carried out to delineate the state-of-the-art development of BIM and CDW Management and their interconnections.

3 Research Methodology

The overall workflow of this research is depicted in Fig. 1, which is divided into three stages: bibliometric search, science mapping and qualitative discussion.

3.1 Bibliometric Search

The bibliometric search for this research was performed in Scopus. The search phrase "TITLE-ABS-KEY ((bim OR "BUILDING INFORMATION MODEL*") AND waste)" was then input. The document type was set to articles and conference paper only. The search was performed on 26th April 2022, and there was no date limit set. A total of 415 papers were obtained from the search. First round screening was then performed to remove papers which are out of the wider scope of BIM for CDW Management. For example, paper which on other kinds of waste like data waste, transportation waste, information waste, time and cost waste. A total of 215 papers were retained. Second round screening was performed to remove papers whose objective didn't focus on BIM for CDW Management. Studies whose objectives were focused on sustainability, lean construction, Off Site Construction, Prefabrication were also removed. Finally, a total of 75 articles and conference papers were included in this review.

3.2 Scientometric Analysis

VOSViewer is a powerful text-mining tool used to analyze the structure of scientific fields. It can be used for example to analyze and visualize bibliometric network. VOSviewer has also been widely used in construction management review studies [20, 28, 47]. In this study, VOSViewer was used to visualize journal sources, keywords, articles, and countries.

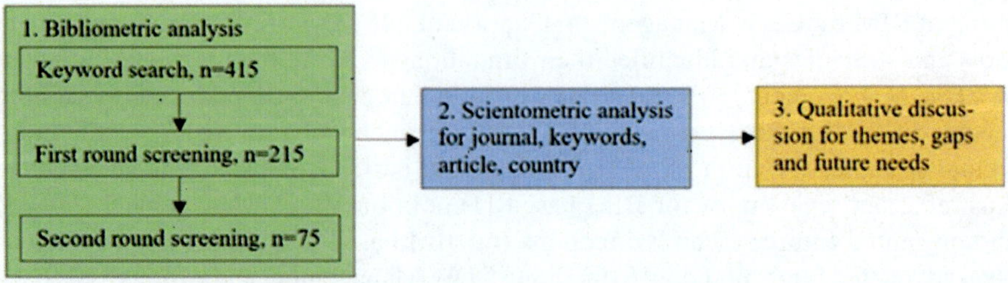

Fig. 1 The three-step workflow of the methodology adopted in this research

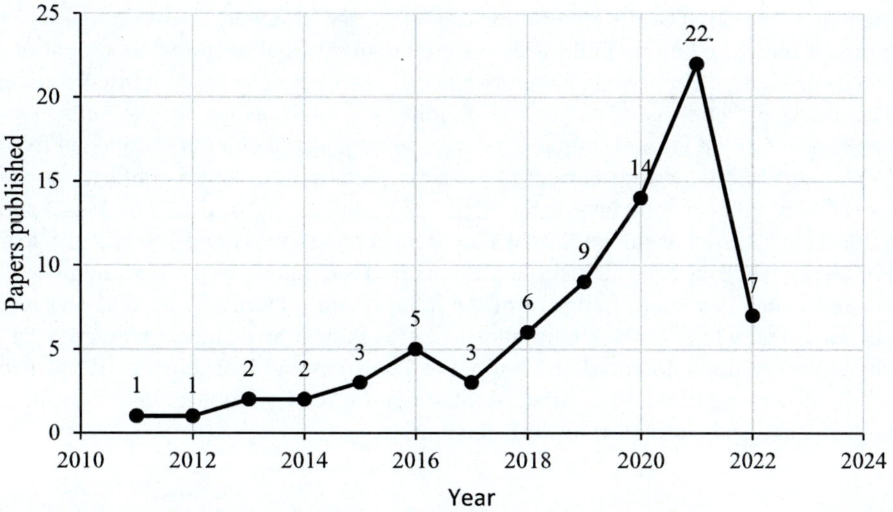

Fig. 2 Yearly publications of BIM for CDW management research (Scopus)

3.3 Qualitative Discussion

Finally, an in-depth qualitative discussion was carried out after bibliometric search and scientometric analysis. As seen in Fig. 1, the objectives of a qualitative discussion were to analyze the main research theme, identify gaps, and identify future needs in BIM for CDW Management field.

4 Result of Scientometric Analysis

4.1 Overview of Literature Sample

Figure 2 shows all the articles published in BIM for CDW Management field until April 2022. The data shows that the annual number of publications has generally been increasing from 1 in 2011 to 22 in 2021, except for 2022, which has incomplete data. This shows increasing interest in BIM for CDW Management field.

4.2 Science Mapping of Journal Sources

Figure 3 shows the science map of all journal source with a minimum of three publications. The node size represents the number of total citations received by a particular journal. The different colors represent the closeness between journal in terms of mutual citation. Table 1 shows the number of publications, total citations received, average citation, and average normalized citation by each journal source. The average citation is calculated by dividing the total citation by the number of publications of the journal source. The average normalized citation is obtained from VOSviewer and is used to correct the misinterpretation that older publications are more likely to receive higher citation than recent publications. Based on Fig. 3 and Table 1, it can be concluded that Waste Management (WM) and Journal of Clear Production (JCP) are the influential journals in the field due to the high number of citation which is a major indicator of the influence of a journal [18]. WM and JCP also have a very high average normalized citation indicating that their average yearly citation is very high. International Journal of Construction Management (IJCM) also has high average normalized citation although the total citation is low, indicating high recent research interest towards the journal.

Table 1 Journal sources with their citations

Source	Number of publications	Total citation	Average citation	Average Norm. Citation
Waste Management (WM)	6	365	61	1.7
Journal of Cleaner Production (JCP)	7	272	39	1.3
Automation in Construction (AC)	6	199	33	1.1
Resources, Conservation and Recycling (RCR)	5	140	28	0.95
Journal of Building Engineering (JBE)	3	93	31	1.14
Sustainability (Switzerland) (SS)	6	51	9	1.08
International Journal of Construction Management (IJCM)	4	41	10	2.02

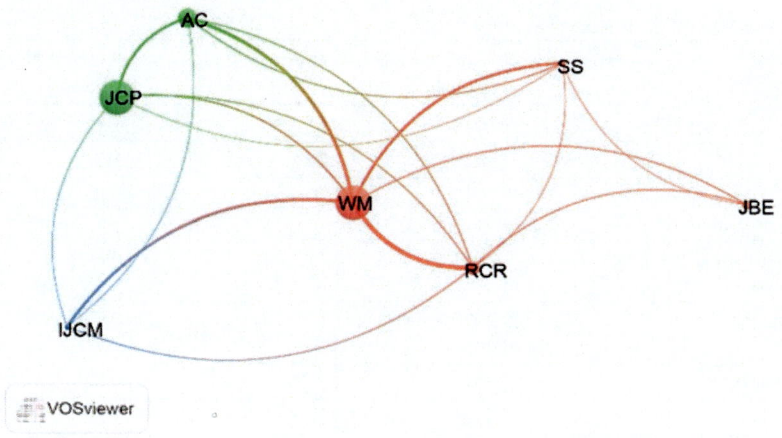

Fig. 3 Mapping of journal sources

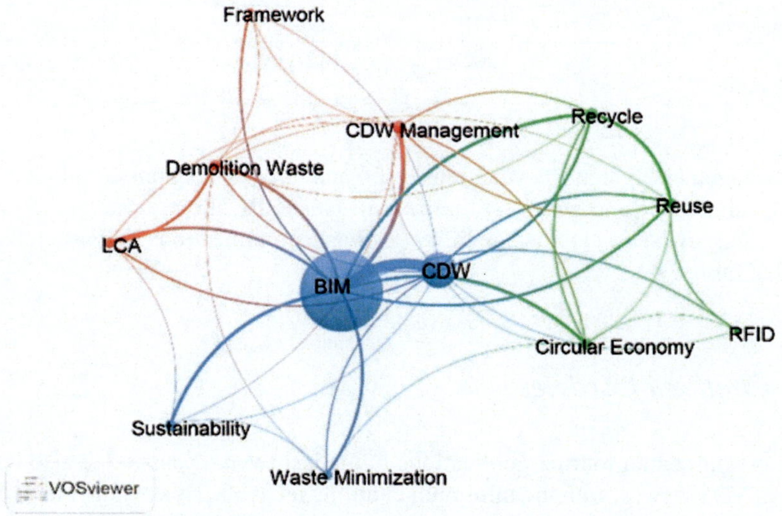

Fig. 4 Co-occurrence of author keywords in BIM for CDW management field

4.3 Co-occurrence of Keywords

A network of keywords displays the knowledge among their relationships and intellectual organization of research themes [18]. A total of 12 author keywords with a minimum occurrence of 3 were mapped by VOSViewer as seen in Fig. 4. The size of the node represents the number of occurrence of a keyword and the color of the nodes represents how strongly related the keywords are. The colors of nodes divide the keywords into 3 different clusters: red (LCA, demolition waste, framework and

Table 2 Keyword occurrences

Keyword	Occurrences	Average year published	Average citations	Average Norm. Citation
BIM	56	2019	25	1.1
CDW	24	2019	24	1.2
CDW management	9	2019	20	0.9
Circular economy	8	2021	11	0.8
Demolition waste	7	2020	17	0.9
Framework	3	2018	24	0.8
LCA	7	2020	17	1.5
Recycle	8	2020	12	1.2
Reuse	7	2020	21	1.4
RFID	3	2019	20	1.0
Sustainability	8	2019	27	1.0
Waste minimization	7	2020	12	0.8

cdw management) green (recycle, reuse, rfid and circular economy) and blue (bim, cdw, sustainability and waste minimization). Generally, the mainstream keywords can be categorized as: (1) Sustainability (2) Waste minimization (3) Waste reuse and recycle (Table 2).

4.4 *Citation of Articles*

The most influential journal publications in the last twelve years were also investigated in VOSViewer, with the minimum citations set at 50. This resulted in a total of 9 out of 75 articles being selected as seen in Fig. 5 and Table 3. Cheng and Ma [15] is the most highly cited article in BIM for CDW Management. Cheng and Ma [15] was the first to develop a system that automatically estimates and plans for CDW using BIM. Akinade et al. [4] was the first to develop BIM based Deconstructibility Assessment Score (BIM-DAS) to aid in waste minimization through deconstruction. Akinade et al. [3] identified the limitation of existing CDW Management tools and gave factors to consider for the implementation of BIM for CDW Management. In terms of average normalized citation, Jalaei et al. [31] with an average normalized citation of 6.22, followed by Gharouni Jafari [21] with 5.02 and Guerra et al. [23] with 3.34 received the highest yearly based attention in BIM for CDW Management.

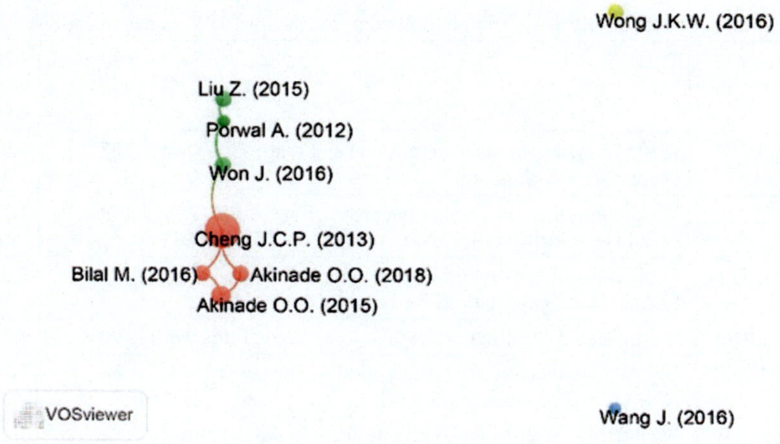

Fig. 5 Science mapping of most influential publications in BIM for CDW management research

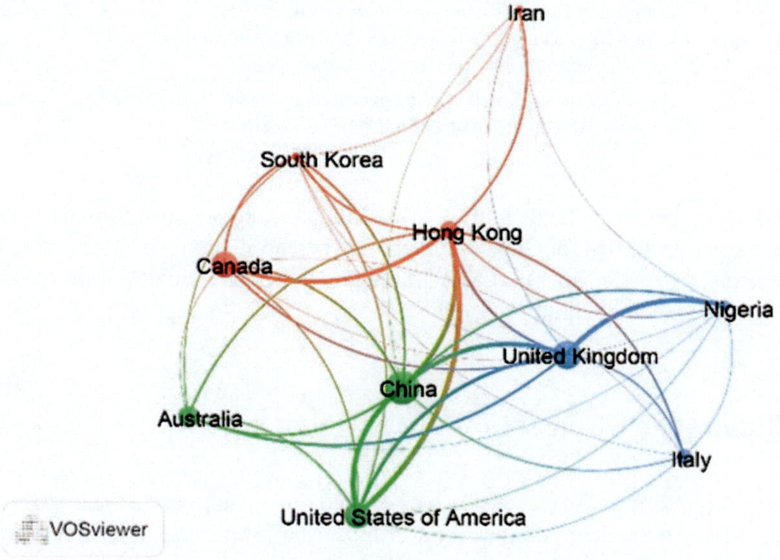

Fig. 6 Science mapping of countries active in CDW management research

4.5 *Countries Active in BIM for CDW Management Research*

Figure 6 and Table 4 show countries which are active in BIM for CDW Management research as generated by VOSViewer. Based on the total number of citations since 2011, countries like United Kingdom, Hong Kong and Nigeria have been the most active indicating high research interest in BIM for CDW Management. However,

Table 3 List of publications with the highest impact in BIM for CDW management research

Document	Title	Citations	Average norm citation
Cheng J.C.P. (2013)	A BIM-based system for demolition and renovation waste estimation and planning	197	1.99
Akinade O.O. (2015)	Waste minimization through deconstruction: A BIM based Deconstructability Assessment Score (BIM-DAS)	109	1.36
Akinade O.O. (2018)	Designing out construction waste using BIM technology: Stakeholders' expectations for industry deployment	91	2.37
Liu Z. (2015)	A BIM-aided construction waste minimization framework	90	1.12
Bilal M. (2016)	Big data architecture for construction waste analytics (CWA): A conceptual framework	88	1.24
Wong J.K.W. (2016)	Facilitating effective green procurement in construction projects: An empirical study of the enablers	87	1.22
Won J. (2016)	Quantification of construction waste prevented by BIM-based design validation: Case studies in South Korea	86	1.21
Wang J. (2016)	Combining life cycle assessment and Building Information Modelling to account for carbon emission of building demolition waste: A case study	68	0.96
Porwal A. (2012)	Building information modeling-based analysis to minimize waste rate of structural reinforcement	68	1.00

based on average normalized citation, Iran, Canada and USA have shown high recent research interest in BIM for CDW Management research. It is also observed that there is a growing interest in BIM for CDW Management research in developing countries like Nigeria, Iran and China.

5 Qualitative Discussion

In this section an in-depth discussion was carried out to summarize the main research themes, identify the research gaps and propose the future research direction in the field of BIM for Construction and demolition waste management.

Table 4 Countries active in BIM for CDW management research

Country	Documents	Citations	Average publication year	Average citation	Average norm citation
United Kingdom	10	540	2018	49	1.0
Hong Kong	7	377	2019	54	1.2
Nigeria	4	224	2017	56	1.0
United States of America	10	214	2019	21	1.4
Canada	10	185	2019	19	1.7
Australia	8	166	2019	21	0.7
South Korea	3	137	2016	46	1.1
China	13	127	2021	10	0.9
Italy	6	72	2020	12	0.6
Iran	3	64	2021	21	4.4

5.1 Current Research Areas

5.1.1 CDW Quantification and Estimation

Waste quantification and estimation is crucial for effective CDW management [15]. It can aid in pinpointing the main causes of waste, planning for waste control strategies, and developing benchmarking in the industry to aid in minimizing CDW and its Environmental impacts. Building Information Modelling (BIM) has been used in numerous CDW Management research to quantify and estimate Construction Waste (CW), Renovation Waste, and Demolition Waste (DW). BIM provides material and volume information that aid in the quantification of CDW. Guerra et al. [22] quantified construction waste (CW) by deducting the quantities of material purchased from the quantities of material needed for the structure. The materials needed are automatically extracted from BIM Bill of Quantities (BoQ) and the material purchased are obtained from general contractor's purchasing records. Similarly, other researchers employed the same method in their research [7, 23, 58]. The main drawback of this method is it considers all unused material in site to be waste, which isn't true in the case of materials in good condition which can be used in other sites. In other research, Construction Waste (CW) was quantified using CW quantification factors from Athena© [31] or [40] factors [55]. Depending on the research objective and the available information, the CW estimation method is selected. For instance, Won et al. [67], quantified the amount of CW prevented by BIM by using an experienced industry practitioner who estimated waste based on work scope to resolve each design error detected by BIM software. Most research in the quantification of waste is reported in the demolition stage. This is due to the massive amount of waste generated in the demolition stage compared to other stages. The most commonly employed method for Demolition waste (DW) quantification is obtaining material and volume information from BIM QTO, classifying the materials according to their type, then computing the amount of DW based on volume change factor and density [15, 35, 58, 62, 64, 71]. Other research however, didn't take account of volume change factor [31, 60, 61, 65]. In summary, current research is mainly focused on the estimation of Construction Waste and Demolition Waste only.

5.1.2 CDW Environmental Impact Assessment (EIA)

Environmental impact assessment of Construction and Demolition Waste (CDW) is a recent research theme in BIM for CDW Management field. With the recent rapid increase in the amount of CDW due to urbanization and population growth, it is imperative to identify waste management methods associated with the least Environmental Impacts. The environmental benefit of managing different kinds of CDW varies depending on the material type. For instance, the recycling of masonry waste has a far lower environmental benefit compared to the recycling of metal waste [65]. The most common and complete method to carry out EIA is Life Cycle Assessment (LCA). According to ISO 14040, carrying out LCA involves four steps: goal and scope definition, Life Cycle Inventory Analysis (LCI), Life Cycle Impact Assessment (LCIA), and Interpretation of results [30]. The goal and scope definition involves defining the goal of the LCA study (e.g., evaluating the EI of different CDW treatment method or evaluating the EI of different CDW materials) and from the study's goal, defining the scope. The scope mainly includes the definition of the system boundary, which involves selecting product/system life cycle stages, which will be included in the study. According to EN 15,978, the product or system stages include (1) The Product Stage, which is classified as A that includes; A1: Raw Material Extraction, A2: Transport and A3: Manufacturing; (2) Construction Stage is also classified under A and contains; A4: Transport, A5: Construction & Installation; (3) Use Stage, which is classified as B includes; B2: Maintenance, B4: Replacement, B6: Operational Energy construction, (4) End of Life Stages, which is classified as C and contains; C1: Demolition, C2: Transport, C3: Waste Processing, C4: Disposal [19]. In BIM for CDW Management fields research, the goals for conducting LCA are mainly:

1. To evaluate the EI of a waste treatment method [60, 61, 64, 65]. This is carried out before the demolition phase of the building to assess the best method to deal with demolition waste. The scope is the End-of-life stage (C).
2. To evaluate the EI of CDW. This is usually carried out in the design stage of the building to assess the full life cycle EI of CDW. The scope, in this case, is A1 to C4 [31].

In general, current research is mainly focused on End-of-Life EIA of CDW, rather than full life cycle EIA of CDW.

5.1.3 CDW Minimization

Waste minimization is the most extensively studied research theme in BIM for CDW Management field. Numerous tools, methodologies and technologies have been employed in minimizing CDW. According to Osmani and Villoria-Sáez [49], waste minimization involves dealing with waste before it has arisen by determining its root cause and avoiding its generation by implementing appropriate processes and practices. Waste minimization is also referred to as waste reduction, minimization at source, and waste prevention [49]. The root causes of waste, as identified by Pellegrini et al. [31] include, Poor design quality, Poor material handling, Improper procurement and planning, Making inappropriate decisions during design and unexpected design changes. Since 33% of CDW in the building is generated during the design and planning stage, BIM has been shown in literature to have beneficial effects in minimizing waste in this stage [45]. In design stage, Liu et al. [39] proposed an integrated BIM-aided CW Minimization process that addresses CW causes in the design stage using BIM-enhanced design activities (like detailing, clash detection, visualization and simulation, coordination, and communication) to aid architects and designers in decision making. Similarly, Haeusler et al. [26] developed a computational workflow to minimize construction waste by addressing poor decisions. Bilal et al. [11], proposed big data architecture for construction waste analytics due to the large size of data and computational complexity of designing out waste. Big data would aid in analysing waste since majority of waste is still classified as mixed waste which leads to inappropriate decisions. Bilal et al. [12] also developed a plug-in to minimize construction waste using big data. Banihashemi et al. [9] introduced a workflow where modular construction (MC) principles are simulated through parametric design (PD) to reduce construction waste by improving building flexibility. Cavalliere et al. [14] developed an assessment metric for functional flexibility of a building design to address CDW caused by functional change. Chidambaram [17], Khondoker [34], Porwal and Hewage [53] addressed the CW caused by reinforcement by improving design quality in BIM. Gharouni et al. [21] addressed CW caused by mismatches, likewise Porwal et al. [54] addressed CW caused by change order. Wu et al. [68, 69] developed a BIM based Parametric design platform which designed the layout of floor tiles to reduce waste caused by poor design quality. Similarly, Cheng et al. [16] and Malik et al. [42] minimized waste generated by cutting of materials on site. In planning stage, Pellegrini et al. [51, 52] integrated Waste Management (WM) strategies and Information management in the construction procurement process to minimize waste caused by procurement process Lastly, Liu et al. [38] developed a BIM-based automated design and planning for boarding of light-frame residential buildings to reduce waste caused by poor design and planning.

In summary, the current research is mainly focused on (1) Minimizing CDW due to Poor design quality, inappropriate decisions during design and unexpected design changes.(2) Minimizing CDW due Improper procurement and planning.

5.1.4 CDW Reuse, Recycle

Current research on the CDW reuse, recycle, and recovery is centered around the use of (1) planning and simulation, (2) information technology (i.e., material and component bank, material passport, GIS) (3) Radio frequency identification (RFID) technology. Guerra et al. [23] developed algorithms to estimate the recycle and reuse amount of concrete and drywall waste and planned for their on-site reuse and off-site recycle by developing 4D simulations to visually demonstrate the access of large equipment in the building's confined atrium and confirm clearances. Similarly, Kunieda and Codinhoto [36] developed a simulation tool which estimates and quantifies demolition waste using BIM enabled 4D-CAD combined with physics and game engines, and from the waste data developed a multi-factor waste recovery plan (e.g. machine transportation routes, collection order and timing) which increase purity levels and waste recovery rates. Moreover, Mei et al. [43] developed a framework to aid formwork planning for reuse.

Cai and Waldmann [13] proposed a material and component bank to effectively manage the recycling of materials and direct reuse of components. Similarly, Bertin et al. [10] developed a component database for the reuse of load bearing structural elements. Xing et al. [70] developed a cloud BIM platform which uses cyber-physical information for online identification, examination, and exchange of reusable components. Atta et al. [6] developed a digital material passport within BIM which provides information regarding the reusability and recyclability of building materials which can aid in improving reuse and recycle. Meng et al. [44] combined Building Information Modeling (BIM), Material Flow Analysis (MFA) and Geographic Information System (GIS) to analyze hotspots and evaluate the mining potential of a location to aid in urban mining thus improving material reuse and recycle.

Lastly, Iacovidou et al. [29] explored the potential pre-conditions for RFID to facilitate construction components reuse and developed a guidance for promoting their redistribution back to the supply chain. RFID tracks and archives the properties of structural construction components, thus aiding reuse.

5.2 Research Gap and Future Research Direction

Table 5 shows the research gaps and future research direction in each research area of BIM for CDW Management.

6 Conclusion

This review-based study analyzed a total of 75 papers published in BIM for CDW Management field since 2011 through a scientometric analysis and an in-depth qualitative discussion. The scientometric analysis revealed the following:

Table 5 Qualitative discussion summary

Research gaps	Future research direction
1. CDW quantification and estimation	
(1) Lack of renovation waste estimation research. Only Cheng and Ma [15] and Jalaei et al. [31] quantified waste due to renovation activities (2) Lack of regional dataset of waste volume adjustment factors and demolition waste recycling rate in most countries like Hong Kong (3) Lack of reliable regional dataset of CDW generation factors (4) Some waste streams like formwork are largely ignored in literature. Only Quiñones et al. [55] quantified waste from packing material and Bakchan et al. [8] developed a method to automatically quantify formwork waste using BIM (5) No clear framework for the use of Revit schedule for CDW estimation	(1) Estimation of renovation waste (2) Research on Volume change factor to aid in accurate waste estimation, Recycling rates to aid in planning for recycling (3) Development of reliable CDW database for CDW estimation and quantification (4) Estimation of formwork waste (5) Investigating universal methods that ensure highly accurate calculations when using Revit schedules
2. CDW Environmental Impact Assessment (EIA)	
(1) Lack of fully automated (dynamic) EIA tool (2) Lack of LCA modelling parameters like waste during transportation in construction phase, building material replacement fact in Use phase and type of treatment in EoL stage. Guimaraes et al. [24] developed a framework to improve LCA modelling parameter but there has been no further research after his work (3) Lack of trade-off analysis between environment and economy (4) Lack of research on environmental benefit of waste treatment method of CDW like recycling (5) Landfill charging based on mass/volume alone (6) Lack of full life cycle EIA of CDW	(1) Development of fully automated tools that can automatically evaluate EI of CDW (2) Refinement of each parameter with the use of statistical estimation and uncertainty analysis (3) Conducting trade-off analysis between environment and economy (4) EIA of various methods of treating CDW to ensure minimum EI (5) The current landfill charging fee system based on mass or volume should be revisited by taking other critical factors such as environmental impacts into consideration (6) Conducting full life cycle EIA of CDW
3. CDW minimization	
(1) Lack of research on quantification of CDW due to change orders and mismatches in design and as-built drawings	(1) The use of system dynamics model to comprehensively quantify CDW due to change orders (2) Use of machine learning to manage mismatches

(continued)

Table 5 (continued)

Research gaps	Future research direction
4. CDW reuse and recycle	
(1) Little research on formwork reuse since formwork make up a large portion of CW (2) Little research on concepts like Digital urban mining platform, which would enable a direct use of the demolished materials within new construction projects (3) RFID technical limitation. Technical operability and recycling of RFID tags seem to currently hinder its uptake for structural components' lifecycle management	(1) Further exploration on formwork layout planning, processing, and onsite storage is also encouraged for future work (2) Digital urban mining platform development research (3) Future technological innovations could provide solutions that would enable it to become a mainstream practice

1. The most influential journals in the field of BIM for CDW Management are Waste management and Journal of Clear production.
2. Keyword analysis revealed that current state of the art research areas include Environmental impact assessment of CDW using Life Cycle Assessment, and Reuse, and Recycle of CDW.
3. Article that have received the highest recent research attention Cheng and Ma [15], Gharouni Jafari et al. [21], Jalaei et al. [53] have been discussed.
4. Countries prominent in BIM for CDW Management research were identified and discussed. There is a growing interest of BIM for CDW Management research in developing countries like Nigeria, China, and Iran.

The qualitative analysis revealed the following:-

1. The BIM for CDW Management field is divided into four areas: CDW quantification and estimation, CDW Environmental Impact Assessment (EIA), CDW minimization and CDW reuse and recycle.
2. The main research gaps include, lack of reliable regional dataset of CDW generation factors, Lack of trade-off analysis between environment and economy, Lack of research on quantification of CDW due to change orders and mismatches in design and as-built drawings, Little research on concepts like Digital urban mining platform, which would enable a direct use of the demolished materials within new construction projects.
3. Future research directions include, Research on regional volume change factors and Recycling rates, Conducting trade-off analysis between environment and economy, the use of system dynamics model to comprehensively quantify CDW due to change orders and Digital urban mining platform development research.

Acknowledgements I would like to express my appreciation to the School of Research for granting me the scholarship for this research.

References

1. Ahankoob A, Khoshnava M, Rostami R, Preece C (2012) BIM perspectives on construction waste reduction. In: Management in Construction Research Association (MiCRA) Postgraduate Conference 195. Management in Construction Research Association, pp 195–199
2. Akinade OO, Oyedele LO (2019) Integrating construction supply chains within a circular economy: an ANFIS-based waste analytics system (A-WAS). J Clean Prod 229:863–873. https://doi.org/10.1016/j.jclepro.2019.04.232
3. Akinade OO, Oyedele LO, Ajayi SO, Bilal M, Alaka HA, Owolabi HA, Arawomo OO (2018) Designing out construction waste using BIM technology: stakeholders' expectations for industry deployment. J Clean Prod 180:375–385. https://doi.org/10.1016/j.jclepro.2018.01.022
4. Akinade OO, Oyedele LO, Bilal M, Ajayi SO, Owolabi HA, Alaka HA, Bello SA (2015) Waste minimisation through deconstruction: a BIM based Deconstructability Assessment Score (BIM-DAS). Resour Conserv Recycl 105:167–176. https://doi.org/10.1016/j.resconrec.2015.10.018
5. Akinade OO, Oyedele LO, Munir K, Bilal M, Ajayi SO, Owolabi HA, Alaka HA, Bello SA (2016) Evaluation criteria for construction waste management tools: towards a holistic BIM framework. Int J Sustain Build Technol Urban Dev 7(1):3–21. https://doi.org/10.1080/2093761X.2016.1152203
6. Atta I, Bakhoum ES, Marzouk MM (2021) Digitizing material passport for sustainable construction projects using BIM. J Build Eng 43. https://doi.org/10.1016/j.jobe.2021.103233
7. Bakchan A, Faust KM, Leite F (2019a) Seven-dimensional automated construction waste quantification and management framework: Integration with project and site planning. Resour Conserv Recycl 146:462–474. https://doi.org/10.1016/j.resconrec.2019.02.020
8. Bakchan A, Guerra BC, Faust KM, Leite F (2019b) BIM-based estimation of wood waste stream: the case of an institutional building project. In: Cho YK, Leite F, Behzadan A, Wang C (eds) ASCE international conference on computing in civil engineering 2019: visualization, information modeling, and simulation, i3CE 2019. American Society of Civil Engineers (ASCE), pp 185–192. https://doi.org/10.1061/9780784482421.024
9. Banihashemi S, Tabadkani A, Hosseini MR (2018) Integration of parametric design into modular coordination: a construction waste reduction workflow. Autom Constr 88:1–12. https://doi.org/10.1016/j.autcon.2017.12.026
10. Bertin I, Mesnil R, Jaeger JM, Feraille A, Le Roy R (2020) A BIM-based framework and databank for reusing load-bearing structural elements. Sustainability (Switzerland) 12(8). https://doi.org/10.3390/SU12083147
11. Bilal M, Oyedele LO, Akinade OO, Ajayi SO, Alaka HA, Owolabi HA, Qadir J, Pasha M, Bello SA (2016) Big data architecture for construction waste analytics (CWA): a conceptual framework. J Build Eng 6:144–156. https://doi.org/10.1016/j.jobe.2016.03.002
12. Bilal M, Oyedele LO, Qadir J, Munir K, Akinade OO, Ajayi SO, Alaka HA, Owolabi HA (2015) Analysis of critical features and evaluation of BIM software: towards a plug-in for construction waste minimization using big data. Int J Sustain Build Technol Urban Dev 6(4):211–228. https://doi.org/10.1080/2093761X.2015.1116415
13. Cai G, Waldmann D (2019) A material and component bank to facilitate material recycling and component reuse for a sustainable construction: concept and preliminary study. Clean Technol Environ Policy 21(10):2015–2032. https://doi.org/10.1007/s10098-019-01758-1
14. Cavalliere C, Dell'Osso GR, Favia F, Lovicario M (2019) BIM-based assessment metrics for the functional flexibility of building designs. Autom Constr 107. https://doi.org/10.1016/j.autcon.2019.102925
15. Cheng JCP, Ma LYH (2013) A BIM-based system for demolition and renovation waste estimation and planning. Waste Manage 33(6):1539–1551. https://doi.org/10.1016/j.wasman.2013.01.001

16. Cheng MY, Fang YC, Wang CY (2021) Auto-tuning SOS algorithm for two-dimensional orthogonal cutting optimization. KSCE J Civ Eng 25(10):3605–3619. https://doi.org/10.1007/s12205-021-0522-y
17. Chidambaram S (2019) Application of building information modelling for reinforcement waste minimisation. Proc Inst Civ Eng Waste Resour Manag 172(1):3–13. https://doi.org/10.1680/jwarm.17.00027
18. Eck NJV, Waltman L (2014) Visualizing bibliometric networks. In: Measuring scholarly impact. Springer, pp 285–320
19. En CEN (2011) 15978: 2011 Sustainability of Construction Works—assessment of environmental performance of buildings—calculation method. CEN, Brussels, Belgium
20. Ferdosi H, Abbasianjahromi H, Banihashemi S, Ravanshadnia M (2022) BIM applications in sustainable construction: scientometric and state-of-the-art review. Int J Constr Manag 1–13
21. Gharouni Jafari K, Ghazi Sharyatpanahi NS, Noorzai E (2021) BIM-based integrated solution for analysis and management of mismatches during construction. J Eng Des Technol 19(1):81–102. https://doi.org/10.1108/JEDT-02-2020-0044
22. Guerra BC, Bakchan A, Leite F, Faust KM (2019) BIM-based automated construction waste estimation algorithms: the case of concrete and drywall waste streams. Waste Manage 87:825–832. https://doi.org/10.1016/j.wasman.2019.03.010
23. Guerra BC, Leite F, Faust KM (2020) 4D-BIM to enhance construction waste reuse and recycle planning: case studies on concrete and drywall waste streams. Waste Manage 116:79–90. https://doi.org/10.1016/j.wasman.2020.07.035
24. Guimaraes GD, Gu N, Gomes V, Paniagua JO (2020) A conceptual framework for construction and demolition waste prevention in building's design phase. In: Ghaffarianhoseini A, Ghaffarianhoseini A, Naismith N et al (eds) 54th international conference of the architectural science association, ANZAScA 2020. Architectural Science Association, pp 31–40
25. Gupta S, Jha KN, Vyas G (2020) Proposing building information modeling-based theoretical framework for construction and demolition waste management: strategies and tools. Int J Constr Manag 1–11. https://doi.org/10.1080/15623599.2020.1786908
26. Haeusler MH, Gardner N, Yu DK, Oh C, Huang B (2021) (Computationally) designing out waste: developing a computational design workflow for minimising construction and demolition waste in early-stage architectural design. Int J Archit Comput 19(4):594–611. https://doi.org/10.1177/14780771211040169
27. Han D, Kalantari M, Rajabifard A (2021) Building Information Modeling (BIM) for construction and demolition waste management in Australia: a research agenda. Sustainability 13(23). https://doi.org/10.3390/su132312983
28. Hosseini MR, Martek I, Zavadskas EK, Aibinu AA, Arashpour M, Chileshe N (2018) Critical evaluation of off-site construction research: a scientometric analysis. Autom Constr 87:235–247
29. Iacovidou E, Purnell P, Lim MK (2018) The use of smart technologies in enabling construction components reuse: a viable method or a problem creating solution? J Environ Manage 216:214–223. https://doi.org/10.1016/j.jenvman.2017.04.093
30. ISO (2006) Environmental management—Life cycle assessment—Principals and framework. International Standard ISO 14040. International Organization for Standardization—ISO Geneva, Switzerland
31. Jalaei F, Zoghi M, Khoshand A (2021) Life cycle environmental impact assessment to manage and optimize construction waste using Building Information Modeling (BIM). Int J Constr Manag 21(8):784–801. https://doi.org/10.1080/15623599.2019.1583850
32. Jin R, Yuan H, Chen Q (2019) Science mapping approach to assisting the review of construction and demolition waste management research published between 2009 and 2018. Resour Conserv Recycl 140:175–188. https://doi.org/10.1016/j.resconrec.2018.09.029
33. Kaza S, Yao L, Bhada-Tata P, Van Woerden F (2018) What a waste 2.0: a global snapshot of solid waste management to 2050. World Bank Publications
34. Khondoker MTH (2021) Automated reinforcement trim waste optimization in RC frame structures using building information modeling and mixed-integer linear programming. Autom Constr 124. https://doi.org/10.1016/j.autcon.2021.103599

35. Kim YC, Hong WH, Park JW, Cha GW (2017) An estimation framework for building information modeling (BIM)-based demolition waste by type. Waste Manage Res 35(12):1285–1295. https://doi.org/10.1177/0734242X17736381
36. Kunieda Y, Codinhoto R (2018) Basic study of 4D-CAD application to demolition impact estimation. J Struct Constr Eng 83(748):773–779. https://doi.org/10.3130/aijs.83.773
37. Li CZ, Zhao Y, Xiao B, Yu B, Tam VWY, Chen Z, Ya Y (2020) Research trend of the application of information technologies in construction and demolition waste management. J Clean Prod 263:121458. https://doi.org/10.1016/j.jclepro.2020.121458
38. Liu H, Singh G, Lu M, Bouferguene A, Al-Hussein M (2018) BIM-based automated design and planning for boarding of light-frame residential buildings. Autom Constr 89:235–249. https://doi.org/10.1016/j.autcon.2018.02.001
39. Liu Z, Osmani M, Demian P, Baldwin A (2015) A BIM-aided construction waste minimisation framework. Autom Constr 59:1–23. https://doi.org/10.1016/j.autcon.2015.07.020
40. Llatas C (2011) A model for quantifying construction waste in projects according to the European waste list. Waste Manage 31(6):1261–1276
41. Lu W, Webster C, Chen K, Zhang X, Chen X (2017) Computational building information modelling for construction waste management: moving from rhetoric to reality. Renew Sustain Energy Rev 68:587–595
42. Malik N, Ahmad R, Chen Y, Altaf MS, Al-Hussein M (2021) Minimizing joist cutting waste through dynamic waste allocation in panelized floor manufacturing. Int J Constr Manag 21(10):1011–1023. https://doi.org/10.1080/15623599.2019.1602581
43. Mei Z, Xu M, Wu P, Luo S, Wang J, Tan Y (2022) BIM-based framework for formwork planning considering potential reuse. J Manage Eng 38(2). https://doi.org/10.1061/(ASCE)ME.1943-5479.0001004
44. Meng Z, Li M, Sun H (2020) Research on construction waste exploitation potential based on combination of BIM and GIS—a case study of residential buildings in Wuhan. In: You W, Cheng C, Zhang H (eds) 8th international symposium on project management, ISPM 2020. Aussino Academic Publishing House, pp 683–693
45. Mohammed M, Shafiq N, Al-Mekhlafi ABA, Al-Fakih A, Zawawi NA, Mohamed AM, Khallaf R, Abualrejal HM, Shehu AA, Al-Nini A (2022) Beneficial effects of 3D BIM for pre-empting waste during the planning and design stage of building and waste reduction strategies. Sustainability (Switzerland) 14(6). https://doi.org/10.3390/su14063410
46. Nikmehr B, Hosseini MR, Wang J, Chileshe N, Rameezdeen R (2021) BIM-based tools for managing construction and demolition waste (CDW): a scoping review. Sustainability 13(15). https://doi.org/10.3390/su13158427
47. Oraee M, Hosseini MR, Papadonikolaki E, Palliyaguru R, Arashpour M (2017) Collaboration in BIM-based construction networks: a bibliometric-qualitative literature review. Int J Project Manage 35(7):1288–1301
48. Osmani M (2011) Chapter 15—Construction waste. In: Letcher TM, Vallero DA (eds) Waste. Academic Press, Boston, pp 207–218. https://doi.org/10.1016/B978-0-12-381475-3.10015-4
49. Osmani M, Villoria-Sáez P (2019) Current and emerging construction waste management status, trends and approaches. Elsevier, pp 365–380
50. Pacheco-Torgal F, Labrincha JA (2013) The future of construction materials research and the seventh UN Millennium Development Goal: a few insights. Constr Build Mater 40:729–737
51. Pellegrini L, Campi S, Locatelli M, Pattini G, Di Giuda GM, Tagliabue LC (2020) Digital transition and waste management in architecture, engineering, construction, and operations industry. Front Energy Res 8. https://doi.org/10.3389/fenrg.2020.576462
52. Pellegrini L, Locatelli M, Meschini S, Pattini G, Seghezzi E, Tagliabue LC, Di Giuda GM (2021) Information modelling management and green public procurement for waste management and environmental renovation of brownfields. Sustainability (Switzerland) 13(15). https://doi.org/10.3390/su13158585
53. Porwal A, Hewage KN (2012) Building information modeling-based analysis to minimize waste rate of structural reinforcement. J Constr Eng Manage 138(8):943–954. https://doi.org/10.1061/(ASCE)CO.1943-7862.0000508

54. Porwal A, Parsamehr M, Szostopal D, Ruparathna R, Hewage K (2020) The integration of building information modeling (BIM) and system dynamic modeling to minimize construction waste generation from change orders. Int J Constr Manag. https://doi.org/10.1080/15623599.2020.1854930
55. Quiñones R, Llatas C, Montes MV, Cortés I (2021) A multiplatform bim-integrated construction waste quantification model during design phase. The case of the structural system in a spanish building. Recycl 6(3). https://doi.org/10.3390/recycling6030062
56. Rajendran P, Gomez CP (2012) Implementing BIM for waste minimisation in the construction industry: a literature review. In: 2nd international conference on management, no 2, pp 557–570
57. Saidu I, Shakantu W (2016) The contributions of construction material waste to project cost overruns in Abuja, Nigeria. Acta Structilia: J Phys Dev Sci 23(1):99–113
58. Shi Y, Xu J (2021) BIM-based information system for econo-enviro-friendly end-of-life disposal of construction and demolition waste. Autom Constr 125. https://doi.org/10.1016/j.autcon.2021.103611
59. Song J, Zhang H, Dong W (2016) A review of emerging trends in global PPP research: analysis and visualization. Scientometrics 107(3):1111–1147
60. Sözer H, Sözen H (2020) Waste capacity and its environmental impact of a residential district during its life cycle. Energy Rep 6:286–296. https://doi.org/10.1016/j.egyr.2020.01.008
61. Sözer H, Sözen H, Utkucu D (2020) Waste potential of a building through gate-to-grave approach based on life cycle assessment (LCA). Int J Sustain Dev Plann 15(2):165–171. https://doi.org/10.18280/ijsdp.150206
62. Su S, Li S, Ju J, Wang Q, Xu Z (2021) A building information modeling-based tool for estimating building demolition waste and evaluating its environmental impacts. Waste Manage 134:159–169. https://doi.org/10.1016/j.wasman.2021.07.025
63. Turkyilmaz A, Guney M, Karaca F, Bagdatkyzy Z, Sandybayeva A, Sirenova G (2019) A comprehensive construction and demolition waste management model using PESTEL and 3R for construction companies operating in Central Asia. Sustainability 11(6):1593
64. Wang J, Wei J, Liu Z, Huang C, Du X (2022) Life cycle assessment of building demolition waste based on building information modeling. Resour Conserv Recycl 178. https://doi.org/10.1016/j.resconrec.2021.106095
65. Wang J, Wu H, Duan H, Zillante G, Zuo J, Yuan H (2016) Combining life cycle assessment and building information modelling to account for carbon emission of building demolition waste: a case study. J Clean Prod 172:3154–3166. https://doi.org/10.1016/j.jclepro.2017.11.087
66. Won J, Cheng JCP (2017) Identifying potential opportunities of building information modeling for construction and demolition waste management and minimization. Autom Constr 79:3–18. https://doi.org/10.1016/j.autcon.2017.02.002
67. Won J, Cheng JCP, Lee G (2016) Quantification of construction waste prevented by BIM-based design validation: case studies in South Korea. Waste Manage 49:170–180. https://doi.org/10.1016/j.wasman.2015.12.026
68. Wu S, Zhang N, Luo X, Lu WZ (2021) Intelligent optimal design of floor tiles: a goal-oriented approach based on BIM and parametric design platform. J Clean Prod 299. https://doi.org/10.1016/j.jclepro.2021.126754
69. Wu S, Zhang N, Xiang Y, Wu D, Qiao D, Luo X, Lu WZ (2022) Automated layout design approach of floor tiles: based on building information modeling (BIM) via parametric design (PD) platform. Buildings 12(2). https://doi.org/10.3390/buildings12020250
70. Xing K, Kim KP, Ness D (2020) Cloud-BIM enabled cyber-physical data and service platforms for building component reuse. Sustainability (Switzerland) 12(24):1–22. https://doi.org/10.3390/su122410329

71. Xu J, Shi Y, Xie Y, Zhao S (2019) A BIM-based construction and demolition waste information management system for greenhouse gas quantification and reduction. J Clean Prod 229:308–324. https://doi.org/10.1016/j.jclepro.2019.04.158
72. Zoghi M, Kim S (2020) Dynamic modeling for life cycle cost analysis of BIM-based construction waste management. Sustainability (Switzerland) 12(6). https://doi.org/10.3390/su12062483

CFD Simulation of Smart Adaptive HVAC Vents for Efficient Cooling in Buildings

Phil S. T. Chan, Basil T. Wong, Victor N. S. Bong, Kok-Hing Chong, and Christopher J. Boniface

Abstract In recent years, various novel control methods have been shown to improve the energy efficiency of HVAC systems significantly, utilizing feedback from the cooled location to control the cooling effects. Despite this, the inflexibility of standard HVAC systems limits the potential energy savings. In this work, the performance of an easily retro-fitted smart adaptive vent system which was designed to overcome this inflexibility is studied. Computational fluid dynamics (CFD) was used to evaluate and compare the smart adaptive vent system against a standard diffuser. The directed airflow possible with the smart adaptive vent improved occupant thermal comfort under the same airflow rates at a cost of increased draft rates. The average PMV for all occupants were noted to reach a comfortable value at ~58% faster timeframe. The results demonstrate that the smart adaptive vent is capable of achieving greater thermal comfort at a similar or lower airflow rate, consequently improving the energy efficiency of the HVAC system.

Keywords Adaptive vent · Energy saving · HVAC · Directional airflow · Computational fluid dynamics

1 Introduction

An increase in the number of buildings particularly in developing countries has led to the widespread adoption of HVAC systems. Referring to Kubota [13], a survey in Johor Bahru, Malaysia in 2009 found that 65% of homes had air-conditioning. Projections also indicate that residential air-conditioning demand in the U.S. is expected to exceed electrical energy supply in the next decade due to the effects of climate change, and especially during extreme weather events [15].

P. S. T. Chan (✉) · B. T. Wong · V. N. S. Bong · K.-H. Chong · C. J. Boniface
Faculty of Engineering, Computing and Science, Swinburne University of Technology, Kuching, Sarawak, Malaysia
e-mail: pchan@swinburne.edu.my

C. S. Choo et al. (eds.), *Proceedings of ASEAN-Australian Engineering Congress (AAEC2022)*, Lecture Notes in Electrical Engineering 1072,
https://doi.org/10.1007/978-981-99-5547-3_16

Accordingly, a large quantity of research has been conducted with the goal of increasing the energy efficiency of HVAC systems, which will no doubt result in a significant decrease in overall energy consumption. A popular topic of research revolves around the use of advanced control techniques and strategies [4, 8]. One thing in common with most of these techniques is that they all require some form of feedback from the conditioned zone to function. Without feedback, the typical strategy has the HVAC system supplying conditioned air where it is diffused in all directions for uniform conditioning. Feedback opens up the option of using advanced control systems such as Model Predictive Control (MPC) which has been found to outperform other conventional controller types [5, 20]. However, even with feedback regarding the various parameters of the conditioned zone, there are still limitations on the potential improvements that can be implemented due to the inflexibility of standard HVAC systems and cost of retrofitting.

To that end, this work investigates a smart adaptive vent system comprising of wirelessly controlled actuators which are capable of active flow redirection. The aim of this system is to reduce the energy consumption of the HVAC system as well as improve the thermal comfort of occupants. Using computational fluid dynamics (CFD), the fluid behaviour can be simulated to determine: (1) the effects of the active adaptive vents on the airflow; (2) the effects of the directed airflow on the thermal comfort of occupants; (3) the change in the overall energy efficiency. Moving forward, Sect. 2 discusses various works related to adaptive vents. Section 3 covers the CFD simulation methodology and verification. Section 4 then presents and discusses the results of the simulations.

2 Adaptive Vents

There have been many methods for energy reduction that have been investigated in recent years [2, 6, 7, 9, 10, 18]. However, the use of adaptive vents has been a relatively recent development and the studies located by the authors have mostly been simulation-based. Below are several notable works.

One of the early works was by Watts et al. [19]. Watts experimentally investigated the use of wirelessly actuated vent registers which were capable of binary control of airflow volume to various zones and reduce the energy wasted in conditioning unoccupied spaces within a residential home. It was found that the concept was a viable method of energy reduction. A work by Abedi [1] describes an adaptive vent system autonomously controlled by a Reinforcement Learning algorithm to provide directional airflow. The system was found to reduce the required duration of airflow by 64% per cycle in an experimental test. Another work by Jazizadeh et al. [11] investigates the energy consumption of an adaptive vent system for five different scenarios in a CFD simulation with multiple occupants. The system was found to have an improved effect on occupant thermal comfort while reducing energy consumption by ~25%. Keskin and Mengüç [12] has also investigated an adaptive vent system as a cyber-physical-social system (CPSS), considering the many complex aspects of the design, development, and operation of the system. Experimental results were compared to numerical simulations.

In the simulation-based literature located by the authors, the works were very insightful. However, the effects of the flaps on the airflow were generally not included. The simulation of humidity, and the consequent effects on occupant thermal comfort were also not shown, with only temperature and/or air velocity being reported. This work therefore investigates the changes caused by the flaps on the airflow, including humidity and thermal comfort parameters to gain greater insight on the resulting effects on human occupants.

3 CFD Simulation

3.1 Simulation Model

SolidWorks Flow Simulation 2017 was used for all the CFD simulations performed in this work. SolidWorks Flow Simulation is a commercial software which embeds CFD into CAD software [16]. SolidWorks uses a Cartesian mesh and solves fluid flow using the well-known Navier–Stokes equations. For the solution of turbulent flow, it utilizes the Favre-averaged Navier–Stokes equations, and solves the Reynolds stresses with the modified k-e turbulence model. Solar radiation is solved using the discrete ordinates model with ray tracing.

The simulations were conducted in a room (Fig. 1) measuring 7.6 m wide by 7.6 m long by 2.5 m high, using transient flow analysis. The size of the room was arbitrary, made larger as to reduce the effects of the walls on the airflow.

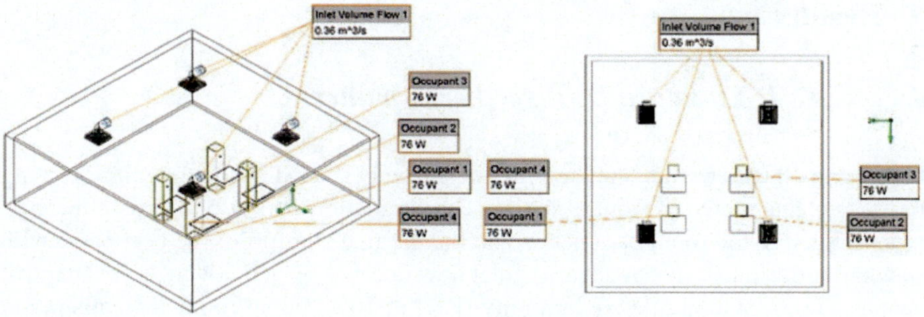

Fig. 1 Geometric model, isometric view (left) and top view (right)

3.2 Test Scenarios

In Case 1, four simulated occupants were placed within the test room along with four standard diffusers. The various parameters of temperature, humidity, and so forth were investigated to form a benchmark to compare against the performance of the adaptive vent. Case 2 replaces the standard diffusers with the adaptive vent to compare the changes in thermal comfort, energy efficiency and other parameters.

3.3 CFD Process Verification

For the verification of the CFD simulation process, the experimental work of Loomans [14] in a climate-controlled chamber was replicated. This experimental work has been replicated by several other works for verification of their respective CFD simulations [11, 17]. Figure 2a below shows the model of the experimental setup. The model setup is very similar in parameters, but the human occupant was simplified and modelled as a cuboid with dimensions of 350 mm × 150 mm × 1500 mm emitting 76 W, set at an initial temperature of 37 °C. The global mesh was set to the maximum level, resulting in 666,098 total cells with 507,410 fluid cells.

Figure 2b compares the simulated temperature data with the experimental temperature data at $z = 1.93$ m at $x = 1.5$ m, 2.25 m, 3.75 m. Figure 2c compares the simulated velocity data with the experimental velocity data at $z = 2.3$ m at $x = 1.5$ m, 2.4 m, 3.0 m. In general, the simulated results agree well with the experimental data from literature, validating the settings and simulation approach for this research.

4 Results

4.1 Case 1: Standard Diffuser Performance

In this case, there are four standard diffusers evenly spaced inside the room, serving four occupants seated in front of tables. The simulation ran for 600 s, with the goal of investigating the room parameters and the thermal comfort of the occupants with standard diffusers as a base scenario. Inlet flow rate was set at 0.009 m^3/s per adaptive vent at 13.0 °C with a relative humidity (RH) of 50%. Initial room air temperature was 30.0 °C with a relative humidity (RH) of 80%.

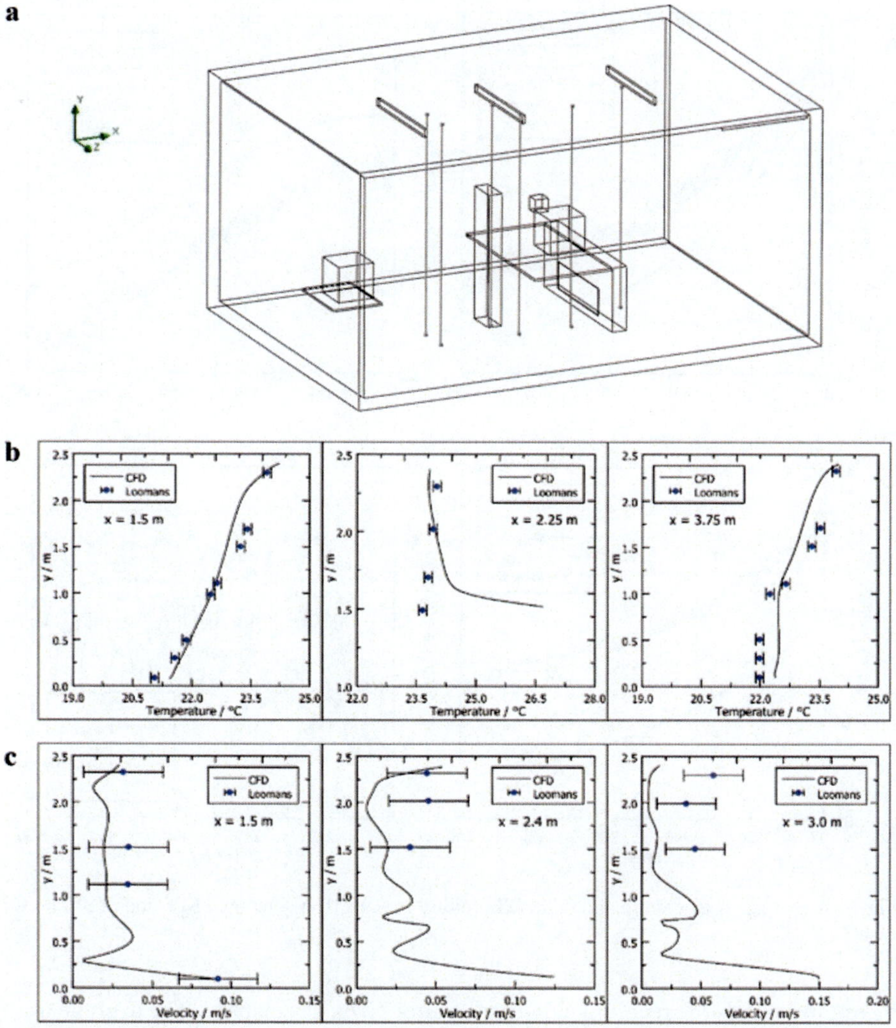

Fig. 2 **a** CFD verification setup model, **b** Simulated temperature data at z = 1.93 m, **c** Simulated velocity data at z = 2.3 m

Figure 3a shows the average temperature and relative humidity (RH) of the room and area surrounding the four occupants decreases with time as the inlet air is both cooler and drier than room conditions. It is observed that the average room temperature is higher than the average temperature surrounding the occupants as the occupants are also heat sources. From Fig. 3b, the PMV of occupants improves over time as the HVAC system conditions room air to more comfortable conditions. It is notable that the PMV of the occupants does not reach the recommended −0.5 to + 0.5 range by ASHRAE 55 (American Society of Heating [3]). The average PMV of

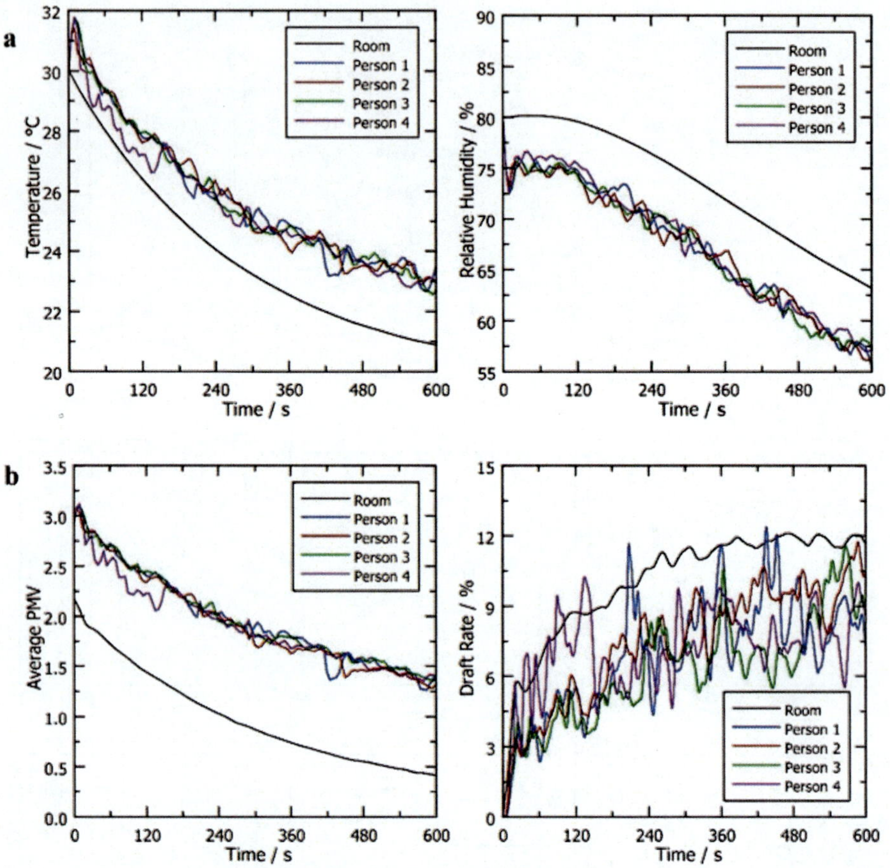

Fig. 3 **a** Average temperature and relative humidity for case 1, **b** Average PMV and draft rate for case 1

the room reaches the recommended range after 527 s has elapsed. The average draft rates of both the occupants and room do not exceed the ASHRAE recommended 20% throughout the simulation. This simulation will thus serve as a benchmark for the adaptive vent system to be compared against (Fig. 4).

4.2 Case 2: Adaptive Vent Performance

Case 2 is the same as Case 1, but it replaces the four standard diffusers with adaptive vents which have been set to redirect airflow towards the closest occupant (Figs. 5 and 6).

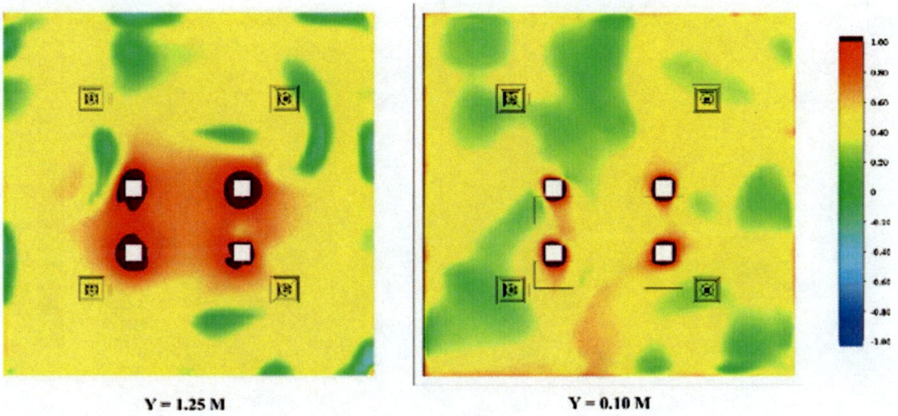

Fig. 4 Case 1 PMV at neck height (right) and ankle height (left)

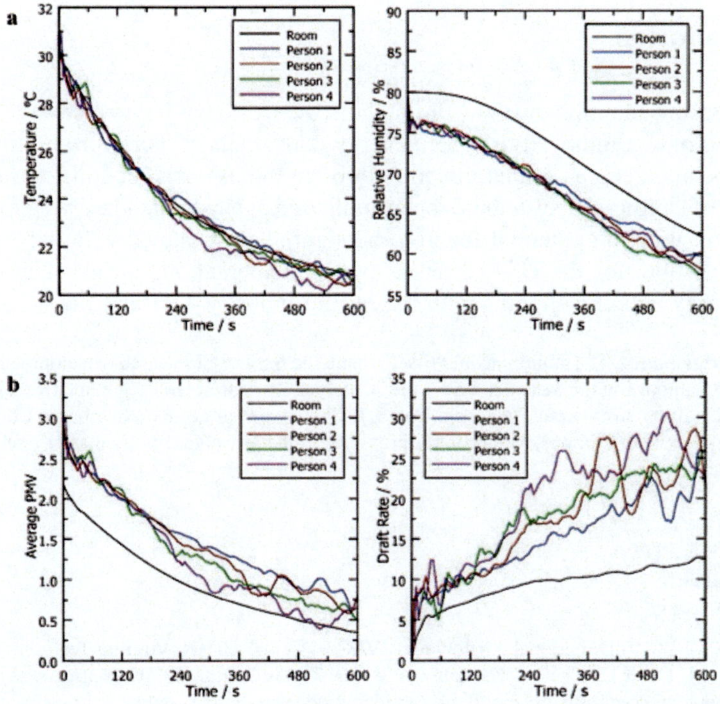

Fig. 5 **a** Average temperature and relative humidity for case 2, **b** Average PMV and draft rate for case 2

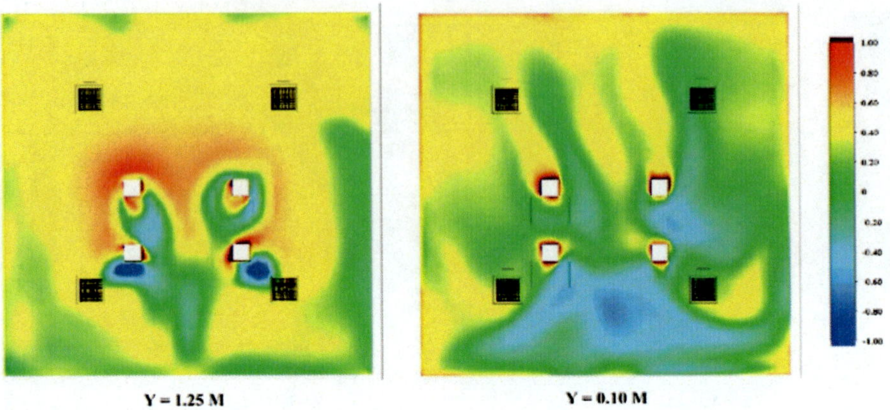

Fig. 6 Case 2 PMV at neck height (right) and ankle height (left)

5 Conclusion

The increasing adoption of HVAC throughout the world has led to many advances in the technologies employed to reduce energy consumption. The work outlined here aims to be another step in that direction through the use of directional airflow. The behavior of a prototype system has been compared against a standard system and the energy consumption potential found to be significant. Coupled with the AI-driven adaptive airflow for the HVAC vents, this implementation can assist in creating smart-energy building aligning with the current Industry 4.0 trend.

Acknowledgements The authors would like to thank the Sarawak Research Development Council (SRDC) for supporting this research under the "Catalyst" Research Grant Scheme (RDCRG/CAT/2019/09 (Catalyst)) in Sarawak, Malaysia. Phil S.T. Chan is also grateful to Swinburne University of Technology Sarawak Campus, Malaysia for providing full-fee-waiver for his master's scholarship.

References

1. Abedi M (2019) Directional Airflow for HVAC Systems. Thesis, Virginia Tech
2. Aftab M, Chen C, Chau C-K, Rahwan T (2017) Automatic HVAC control with real-time occupancy recognition and simulation-guided model predictive control in low-cost embedded system. Energy Buildings 154:141–156. https://doi.org/10.1016/j.enbuild.2017.07.077
3. American Society of Heating, Refrigerating and Air-Conditioning Engineers (2017) ASHRAE Standard 55-2017 Thermal Environmental Conditions for Human Occupancy. Atlanta, Ga.: The Society, 2017
4. Behrooz F, Mariun N, Marhaban M et al (2018) Review of control techniques for HVAC systems—nonlinearity approaches based on fuzzy cognitive maps. Energies 11:495. https://doi.org/10.3390/en11030495

5. Bianchini G, Casini M, Pepe D et al (2019) An integrated model predictive control approach for optimal HVAC and energy storage operation in large-scale buildings. Appl Energy 240:327–340. https://doi.org/10.1016/j.apenergy.2019.01.187
6. Comino F, Ruiz de Adana M, Peci F (2018) Energy saving potential of a hybrid HVAC system with a desiccant wheel activated at low temperatures and an indirect evaporative cooler in handling air in buildings with high latent loads. Appl Therm Eng 131:412–427. https://doi.org/10.1016/j.applthermaleng.2017.12.004
7. Díaz-Hernández HP, Macias-Melo EV, Aguilar-Castro KM et al (2020) Experimental study of an earth to air heat exchanger (EAHE) for warm humid climatic conditions. Geothermics 84:101741. https://doi.org/10.1016/j.geothermics.2019.101741
8. Esrafilian-Najafabadi M, Haghighat F (2021) Occupancy-based HVAC control systems in buildings: A state-of-the-art review. Build Environ 197:107810. https://doi.org/10.1016/j.buildenv.2021.107810
9. Hamanah W, Kassas M, Sahin A, El-Amin I (2019) Energy, exergy and economic analysis for ON/OFF cycle and VFD HVAC cooling systems. Int J Exergy. https://doi.org/10.1504/IJEX.2019.099293
10. Hou T, Chen Y (2020) Pressure drop and heat transfer performance of microchannel heat exchanger with different reentrant cavities. Chem Eng Process—Process Intensification 153:107931. https://doi.org/10.1016/j.cep.2020.107931
11. Jazizadeh F, Joshi V, Battaglia F (2020) Adaptive and distributed operation of HVAC systems: energy and comfort implications of active diffusers as new adaptation capacities. Build Environ 186:107089. https://doi.org/10.1016/j.buildenv.2020.107089
12. Keskin C, Mengüç MP (2020) An Adaptive Vent System for Localized and Customized Thermal Management in Buildings. J Heat Transfer 142. https://doi.org/10.1115/1.4045664
13. Kubota T (2010) Energy Consumption and Air-Conditioning Usage in Residential Buildings of Malaysia
14. Loomans MGLC (1998) The measurement and simulation of indoor air flow. Technovation
15. Obringer R, Nateghi R, Maia-Silva D, et al. (2022) Implications of increasing household air conditioning use across the united states under a warming climate. Earth's Future 10:e2021EF002434. https://doi.org/10.1029/2021EF002434
16. Sobachkin A, Dumnov G (2014) Numerical basis of CAD-embedded CFD. NAFEMS World Congress 2013:1–20
17. Stamou A, Katsiris I (2006) Verification of a CFD model for indoor airflow and heat transfer. Build Environ 41:1171–1181. https://doi.org/10.1016/j.buildenv.2005.06.029
18. Tewari P, Mathur S, Mathur J (2019) Thermal performance prediction of office buildings using direct evaporative cooling systems in the composite climate of India. Build Environ 157:64–78. https://doi.org/10.1016/j.buildenv.2019.04.044
19. Watts W, Koplow M, Redfern A, Wright P (2007) Application of multizone HVAC control using wireless sensor networks and actuating vent registers
20. Yao Y, Shekhar DK (2021) State of the art review on model predictive control (MPC) in Heating Ventilation and Air-conditioning (HVAC) field. Build Environ 200:107952. https://doi.org/10.1016/j.buildenv.2021.107952

Development of Smart HVAC Thermostat with AI Using Passive Infrared Sensors for Energy Saving

Basil T. Wong and Rui Li Sim

Abstract Global warming is causing climate change, disrupting the national economy, and affecting people's living habits. Numerous papers point out that human activity shows that large amounts of waste heat or energy are being continuously released into the climate system. This heat is thought to be responsible for the current global warming. In particular, there is strong evidence that HVAC systems consume more than 50% of total electricity consumption. Therefore, today's social products are environmentally friendly and energy efficient to reduce heat and energy waste. In addition, homeowners who use programmable thermostats correctly can save a lot of electricity. The reduction in energy consumption is due to the automatic use of energy-saving settings, i.e., shutting down the HVAC system when not needed. However, many programmable thermostats have user interfaces that are difficult to navigate and program. To overcome the shortcomings of programmed thermostats, smart thermostats can recognize occupancy and create schedules automatically without human involvement. The project aims to develop a system for sensing room occupants through passive infrared sensors. It then adjusts the thermostat temperature and fan speed accordingly to ensure thermal comfort. To some extent, energy savings can also be achieved through intelligent automatic control settings. Results show that using flow simulations, fluid motion in a room can be assessed, a database developed, and a smart thermostat can be used to automatically configure HVAC systems to reduce energy consumption while maintaining acceptable comfort levels for most indoor occupants. The results showed that the average energy consumption of an HVAC system controlled by a smart thermostat was 22.36% lower than that of a manually programmed thermostat.

Keywords Mart thermostat · Thermostat · Energy saving · Passive infrared sensor · Occupant detection

B. T. Wong (✉) · R. L. Sim
Faculty of Engineering, Computing and Science, Swinburne University of Technology, Kuching, Sarawak, Malaysia
e-mail: twong@swinburne.edu.my

C. S. Choo et al. (eds.), *Proceedings of ASEAN-Australian Engineering Congress (AAEC2022)*, Lecture Notes in Electrical Engineering 1072,
https://doi.org/10.1007/978-981-99-5547-3_17

1 Introduction

Global warming leads to climate change. This is an urgent issue that threatens sustainability. A careful examination of human activities shows that a large amount of waste energy is continuously emitted into the climate system. It is believed that this heat energy has caused the current global warming [3]. Several papers stated that the air-conditioning or heating, ventilation, and air conditioning (HVAC) system is one of the reasons that cause electricity consumption. Kuwait has summer temperatures around 50°C. Building cooling must consume 70% of peak and 40% of yearly electricity [11]. Residential structures emitted 149 tonnes of CO2 in 2008, 27% of the UK's total. These emissions come from space heating, which uses 60% of domestic energy [5]. In Canada, HVAC accounts for 50% of household energy use [18].

With the advancement of computing, networking, and sensing technologies, household appliances have become more automated, interconnected, and smart. Comfort, convenience, and a superior user experience are at the forefront of digital home technology. One of the purposes of household appliances is to save energy. Since the late 1920s, thermostats have been a mature part of the home. Originally, they were basic electromechanical devices with temperature controls and switches. Energy-saving methods, such as night frustration, are highly dependent on the occupants and need meticulous manual changes [19]. Besides, a paper has shown that intelligent thermostats can save energy, for example, for every 1 degree Celsius, the user increases the thermostat, and the user will reduce energy consumption by 6% (PIB [13]. Furthermore, a study has shown that households with programmable thermostats consume more energy than those that use manually operated thermostats, either because the user is misconfigured or because they are completely turned off [10]. To address heat waste and save energy, the research on the development of artificial intelligence smart HVAC thermostats that use PIR sensors for energy saving is of great significance for improving daily human life and reducing energy waste. This paper is primarily concerned with developing a system to sense the occupants of a room and then adjust the temperature on the thermostat accordingly to ensure thermal comfort. To a certain extent, energy can also be saved through intelligent automatic control settings. With the use of the flow simulation to evaluate fluid flow in the room based on various controlling parameters, a reference database is developed. It is then through the stored database that the system will automatically adjust the settings based upon, to provide the best possible comfortable and energy-saving environment for the occupants.

2 Literature Review

Programmable digital thermostats are called "energy-saving devices". Automatic energy-saving settings, such as turning off the heating and cooling system when not in use or "turning down" the thermostat in winter reduce energy use [15]. Most modern thermostats are programmable to overcome classic thermostat shortcomings. Programmable thermostats let users manually set the HVAC systems on time and set the point temperature. Users build a timeline depending on how much time they plan to spend at home and when they will be gone. If the user's expectations are met, the programmable thermostat will switch off (or change the setpoint) when they leave. Users can pre-cool or pre-heat the house (or room) before arriving to increase comfort [7]. Most households will not regulate their thermostats daily, and programmable thermostats are too complicated for most to operate. Recent studies show that households with programmable thermostats spend more energy than those with manually operated thermostats, either because they are misconfigured or turned off [10]. Many programmable thermostats have poor user interfaces, making configuration difficult. Even if users design a calendar, they must choose a static timetable that maximizes energy savings and comfort based on their dynamic occupancy patterns. Because daily activity and check-in habits change over time, for example, summer activities differ from winter activities, users must adjust the thermostat often to reflect changes in the optimal schedule, otherwise, energy-saving benefits are lost. Because of these obstacles, many individuals avoid configuring the thermostat, defeating the purpose. Intelligent thermostats can detect occupancy and set a schedule without human input. Numerous research has been undertaken on smart thermostats that monitor and anticipate occupancy trends using various technologies and sensors [7].

2.1 Related Works

- Ozgur created an IoT thermostat. Nodes are Arduinos with DHT22 temperature sensors and nRF24L01 modules. The sink node's online browsing uses ESP8266. IoT sensors and devices operate smart homes. Multiple sensors calculate the average home temperature. Smart thermostats can alter Combi boiler temperatures. Only the room temperature node communicates through ESP8266. The smart thermostat may then alter the boiler temperature based on the home's average temperature. This reduces energy usage and increases security over Internet-connected systems. The smart thermostat may turn off the Combi boiler when the user leaves and on when they return [12]. Apart from that, Nest offers aesthetically pleasing wall-mounted devices in addition to smartphone apps and web-based control. Along with access to schedules and real-time control, web and mobile apps provide energy history, which is a full historical record of the heating and cooling system's operation over time. Additionally, Nest offers a few intelligent capabilities that employ machine learning and motion detection, such as Auto-Schedule and Auto-Away [20]. The device is one of the most expensive consumer thermostats on the market at $249 [4]. While these methods are more precise in general, the increased cost and energy consumption of these devices make them unsuitable for many real-world applications.
- The authors created a smart thermostat that turns off the HVAC system while occupants are asleep or absent. Sensors detect movement and doors. Rapid response

systems employ a probabilistic model to detect if a tenant is awake, asleep, or absent. Without producing false vacancy detections, the device can react within minutes of a departure. Second, the system warms homes using historical occupancy trends and online sensor data. These ways are comfortable and energy efficient. Smart thermostats detect sleep using passive infrared (PIR) motion sensors and inlet reed switches. PIR and reed sensors are inexpensive and straightforward to install. They utilize a double-sided, wireless X10 sensor. Smart thermostats estimate the home's condition using Hidden Markov Models (HMMs) [10]. Last but not least, the ecobee3 thermostat's remote sensor also detects motion. Each remote sensor has a temperature sensor and a PIR sensor and communicates with the thermostat through a 915 MHz radio [6]. However, the list price of the Ecobee smart thermostat is $469, which is much higher than other thermostats [4]. These methods are capable of counting the number of people in a room, but it does not know where they are, therefore it may not provide adequate thermal comfort.

3 Methodology

The main objective of the research is to develop a smart thermostat that uses PIR sensors (Passive Infrared Sensors) to save energy by automatically shutting down the air conditioner when the room is unoccupied. The components for the smart thermostat are cheap and easy to obtain in the hardware and electronic accessories store. PIR (Passive Infrared Sensors) is one of the simplest and most cost-effective ways to detect occupancy. Furthermore, privacy concerns are avoided by avoiding the usage of cameras or films [8]. If the PIR sensor is activated, it indicates that there are still people in the room, and the room is marked as inhabited. The system assumes no inhabitants in the room if the PIR sensor detects no movement. The PIR sensor produces some noise, and the airflow of passing individuals may cause it to activate [1].

3.1 Occupancy Detection Algorithm

Energy-saving features the use of algorithms to check the occupants of the room. Figure 1 is the algorithm flow chart of the smart thermostat to save energy.

If the PIR sensor is activated, there are people in the room, and the room is marked as occupied. If no movement is detected by the PIR sensor, the system assumes that there are no occupants in the room. Therefore, the smart thermostat uses this occupancy detection algorithm against the four PIR sensors to identify the state of the room to save energy. Every sensor and the air conditioner have the same method for every situation. First, the air conditioner is turned on when the user enters the room. Next, check three cases of room area conditions. The first scenario is when the passive infrared sensor detects that there is no one in the area, the smart thermostat

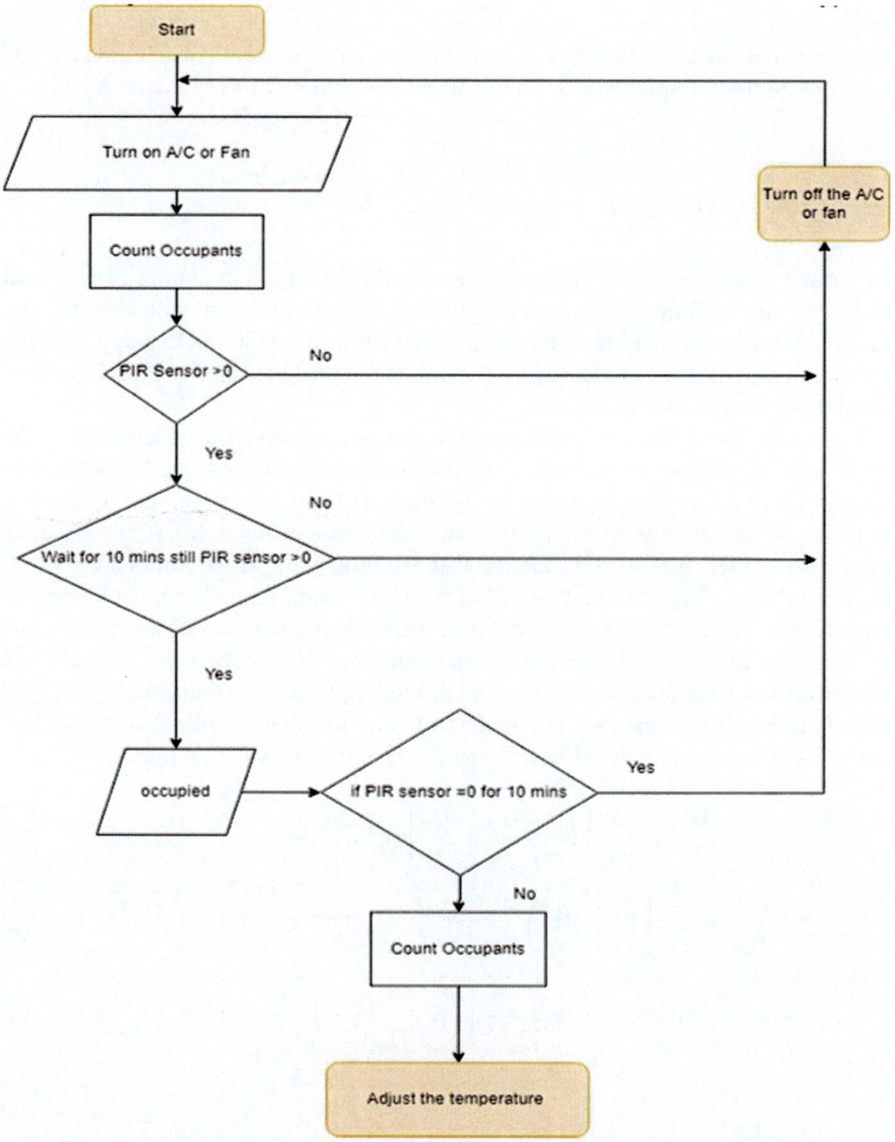

Fig. 1 Flow chart for smart thermostat system

shuts down the HVAC system and returns to the loop, waiting for the user to come in and turn on the air conditioner. Apart from that, Case 2 is when the output of the PIR sensor indicates high, which means the area is occupied and waits 10 min to ensure occupancy. Otherwise, the same as in Case 1, the HVAC system will shut down and return to the initial stage. Finally, Case 3 is when the area is occupied, and suddenly someone goes out. The passive infrared sensor does not detect an occupant for 10 min and then returns to the initial stage, waiting for user input. Otherwise, the smart thermostat will adjust the temperature of the air conditioner and switch it on and off according to the occupied area to maintain the thermal comfort of the room.

3.2 Flow Simulation

One of the objectives of the research is to simulate the fluid dynamics in the room to understand the fluid flow and accomplish the thermal comfort in the room. The fluid flow is complex and hard to predict. The flow simulation is to study how the airflow flows in the room and also how the number of occupants influences the room's thermal comfort and temperature.

The study stated the three fundamental equations of physics are derived from the three fundamental laws of conservation. The continuum equation, the Navier-Stokes equation, and the energy equation are all derived from the conservation of mass, momentum, and energy, respectively. The k-epsilon and Reynolds stress-viscosity models are chosen for investigation because the flow in the air-conditioned environment is turbulent. The standard k-epsilon model is a semi-empirical model based on the turbulent kinetic energy (k) model transmission equation and its dissipation rate (e). The transmission equation for k is derived from the precise equation, whereas the transmission equation for e is derived through physical reasoning and bears little resemblance to its mathematical counterpart. The following transmission equation can be used to calculate the turbulent energy, k, and its dissipation rate, ε:

$$\rho\frac{DK}{Dt} = \frac{\partial}{\partial x_i}\left[\left(u + \frac{\mu_t}{\sigma_k}\right)\frac{\partial k}{\partial x_i}\right] + \mathrm{G}_k + \mathrm{G}_b - \rho_\varepsilon - Y_M \quad (1)$$

$$\rho\frac{D\varepsilon}{Dt} = \frac{\partial}{\partial x_i}\left[\left(u + \frac{\mu_t}{\sigma_\varepsilon}\right)\frac{\partial \varepsilon}{\partial x_i}\right] + C_{1\varepsilon}\frac{\varepsilon}{k}(\mathrm{G}_k + C_{3\varepsilon}\mathrm{G}_b) - C_{3\varepsilon}\rho\frac{\varepsilon^2}{k} \quad (2)$$

$$\frac{\partial}{\partial t}(\rho E) + \frac{\partial}{\partial x_i}[u_i(\rho E + P)] = \frac{\partial}{\partial x_i}\left[\left(k + \frac{Cp\mu_t}{P}\right)\frac{\partial T}{\partial x_i} + u_j\left(\tau_{ij}\right)_{eff}\right] + S_h \quad (3)$$

3.3 Thermal Comfort

The Fanger comfort equation is used to calculate the indoor human thermal comfort index. The thermal balance of the entire human body is represented in the room by two metrics: Predicted Mean Voting (PMV) and Predicted Percent Dissatisfaction (PPD). According to ISO 7730:2005 (EN), the standard divides thermal comfort requirements into three levels, as shown in Table 1.

Table 1 Categories of thermal environment according to ISO/ FDIS7730:2005(E). [2]

Category	Thermal state of the body as a whole	
	PMV	PPD (%)
C	$-0.7 < PMV < 0.7$	<30
B	$-0.5 < PMV < 0.5$	<10
A	$-0.2 < PMV < 0.2$	<6

Additionally, there is another thermal comfort standard, ANSI/ASHRAE Standard 55-2020, which has similar thermal comfort requirements for computer simulation software. Acceptable thermal comfort requirements are that the PPD must be less than 10% and the PMV range should be less than 0.5 and greater than −0.5 [9]. The method entails comparing field test results of thermal acceptability values to the requirements given in ISO 7730:2005 and ANSI/ASHRAE Standard 55-2020 while analyzing comprehensive measurements of indoor climate.

3.4 Room Model

The selected SolidWorks Flow Simulation location is Tutorial Room 17, 3rd Floor, the Swinburne University of Technology Sarawak Campus A Block. The placement of chairs and tables is also included in the modeling. Overall, the size of the tutoring room is 9.2 m × 7.1 m × 2.75 m (L × W × H). In addition, the model has 28 tables and 28 chairs. In addition, four air conditioners and two exhaust fans are installed at the side and top of the room. The sensitivity of the PIR sensor is up to 5 meters, and the horizontal and vertical viewing angles are 82 degrees and 100 degrees, respectively. Four PIR sensors will be installed in the room connected to each air conditioner. The location of the PIR sensor will be placed on the ceiling of the room facing the table and chairs, as the height of the room is 2.75 meters, which is still within the sensitivity range of the sensor.

3.5 Boundary Conditions

A room's temperature distribution and comfort factors are studied using SolidWorks Flow Simulation. The temperature will first be set to 25 °C. The inlet velocity is 1.38 m/s, according to the manufacturer's specifications. In addition, the exhaust fan is represented by the pressure outlet. Each exhaust fan has a 0.11 m2 surface area and can remove 838 m3/h of indoor air. All of the exhaust fans are on the ceiling. Set occupants or humans as heat flux and release heat through natural convection for thermal boundary conditions. Humans emit an average of 100W per person, according to scientific literature. Thermal radiation is taken into account in the simulation model. The glass and wall heat fluxes are set at 81.8063 W/m2 and 130 W/m2, respectively.

4 Results and Discussion

4.1 Simulation Results

Four scenarios are depicted in Fig. 2a, where different numbers of air conditioners are activated to explore the fluid temperature in the room. Meanwhile, 28 people were seated in the room. When the thermostat is set to 20 °C (i.e., a typical standard condition), the coldest temperature in the room is about 21.3 °C. As shown in Fig. 2a, the temperature stabilized for about 800 seconds, indicating that the smart thermostat can pre-cool the room 14 min before the occupant enters. Also, the room is warmest when only one air conditioner is running. When the air conditioner was turned on, the temperature in the room was 25.4 °C and started to stabilize at 800 seconds. In addition, the number of air conditioners used increased by 3 and 4, with little change from 800 seconds. Also, a room with three air conditioners was slightly cooler than a room with two, at 22.3 °C and 22.8 °C, respectively. In addition, to facilitate further discussion, specific points were selected to study and observe the transient temperature change over time, as shown in Table 2.

The results showed that people sitting on the right side of the room felt warm. This result from solar radiation near the wall and the windows on the right. Therefore, the thermostat can determine from the database that the left air conditioner must be 1°C

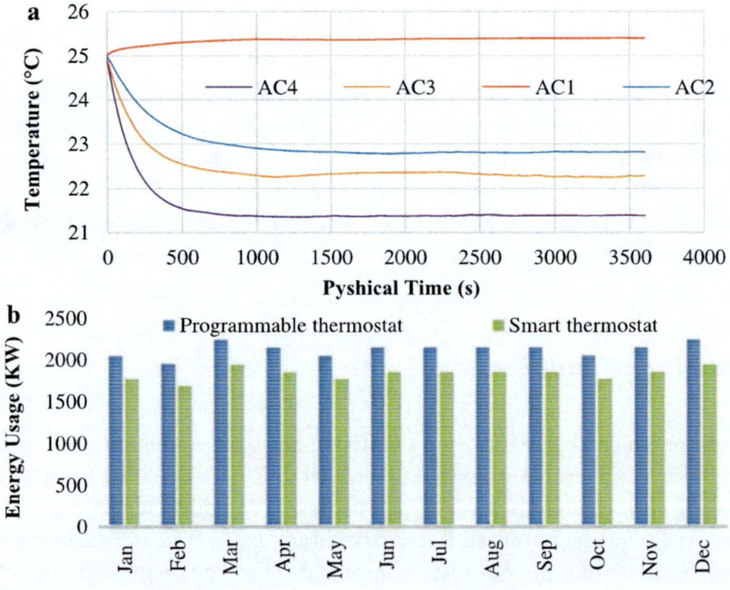

Fig. 2 **a** This figure shows the different number of air-conditioners and the average temperature of the simulated room, **b** This figure shows the monthly energy of the HVAC required for cooling

Table 2 Summary of the temperature at different specific points in the simulated room.

	Scenario			
Location	20AC4	20AC1	20AC2	20AC3
Right front Occupant Average Temperature (Fluid) [°C]	28.74	32.80	30.26	29.76
Right back Occupant Average Temperature (Fluid) [°C]	27.63	31.69	29.10	28.74
Left front Occupant Average Temperature (Fluid) [°C]	30.22	30.22	28.07	28.05
Left back Occupant Average Temperature (Fluid) [°C]	26.23	29.87	27.18	26.79

cooler than the right air conditioner to cool the occupant sitting on the right side of the room and keep the room temperature balanced.

4.2 Thermal Comfort

Figures 3a and b result from PDD and PMV for AC3, respectively, where the room uses three air conditioners from the SolidWorks Flow simulation. As a result, the PMV of the occupants in the room is in the 0.13 to -0.20 region. Figure 3a shows that the occupants are green and dark blue in the PDD results and are in the 5.0 to 5.8 region Hence, the room has three air conditioners classified as Class "A" in ISO 7730: 2005 and meets the thermal comfort requirements of ANSI/ASHRAE Standard 55-2017 with a PPD of less than 10% and an occupant PMV range of -0.5 to 0.5. The room with three air conditioners is the most comfortable scenario because it is the only scenario that is classified as "Class A", and the rest of the scenarios are "Class B" in ISO 7730: 2005, although all scenarios fulfill the thermal comfort requirement for the ANSI/ASHRAE Standard 55-2017. Hence, the AC 3 scenario is comfortable for the occupant sitting at the front of the room, and the temperature difference is slight from AC 4, thus encouraging the room to turn on 3 air conditioners when there are not many occupants.

4.3 Energy Saving Strategy

Usually, the air conditioner turns on when someone enters and off after the last person leaves or at a set interval. When no one is in the area, the smart thermostat switches off the AC. A static timetable was contrasted with a smart thermostat technique that made occupants comfortable. For static energy usage, refer to *Public Holidays in Kuala Lumpur for 2021 (2021)*. Use Kuala Lumpur's five working days in 2021 to calculate monthly HVAC energy use. 20 °C is used to determine energy usage and bills. A typical programmable thermostat assumes 20°C. The smart thermostat's temperature setting will remain at 20 °C, except in rooms with four air conditioners, where it will be 22 °C.

a

b

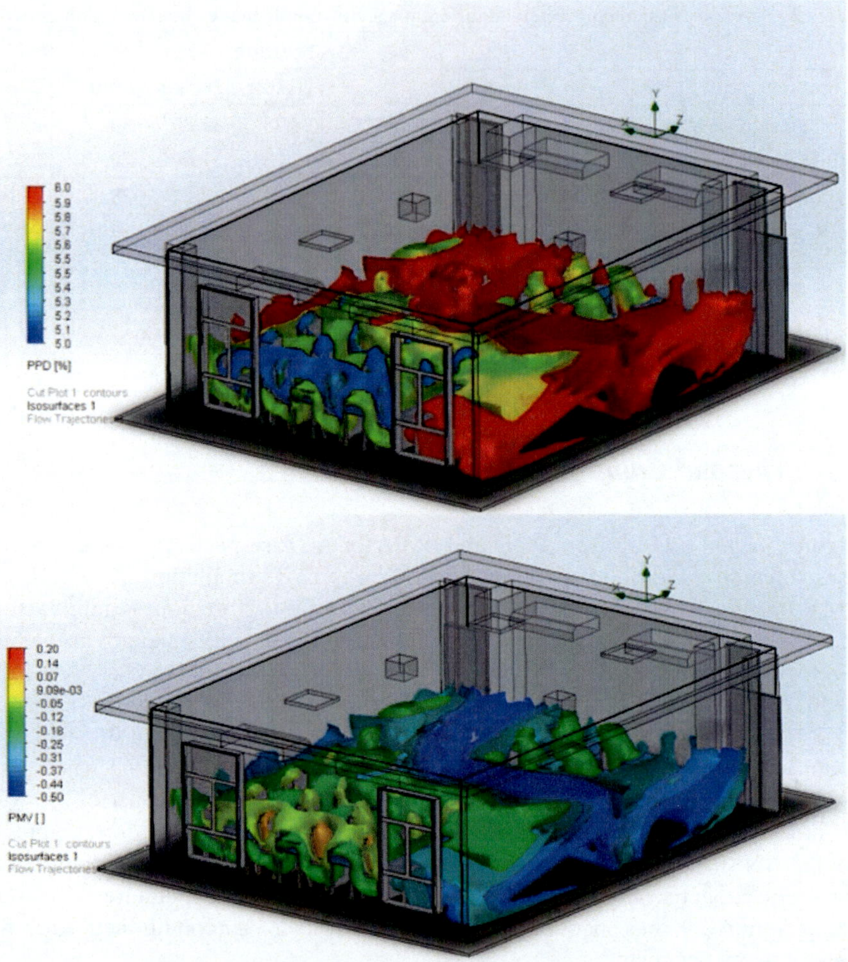

Fig. 3 **a** Three-dimensional view of PPD distribution in the simulated room. All values are in percentage, **b** Three-dimensional view of PMV distribution in the simulated room

Assume the space is used 11 hours a day from 8 am to 6 pm. The energy-saving strategy of the smart thermostat is where the area of the room is occupied, and the result is the number of air conditioners. The scenario is a normal tutoring room suppose 8 people are sitting at the front of the room after lunchtime (1:00 pm), there are 28 people, and the room is full and finally, there were only three people at the front of the classroom in the evening. Therefore, Fig. 2b is a bar graph comparing the expected HVAC energy usage in the Scenario between a programmable thermostat and a smart thermostat. It can be seen that the programmable thermostat uses more energy for the HVAC system. Overall, smart thermostats use 22.36 %

less energy than manual programmable thermostats for air-conditioners. Additionally, the expected total electricity bill for the HVAC system. For HVAC systems, programmable thermostat control consumes more power. The annual electricity bill for a manually programmable thermostat-controlled HVAC system is RM8,030.76 On the other hand, smart thermostats can help users save around RM 1800 in HVAC electricity bills in a year, totaling RM 6,234.79.

5 Conclusion

To sum up, the research aims to develop a smart thermostat that senses the occupants of a room and then adjusts the temperature on the thermostat accordingly to ensure thermal comfort. Use SolidWorks Flow Simulation to explore the room's temperature distribution and comfort parameters, then automatically adjust the thermostat from the analyzed database. Additionally, these strategies can save a room 22.36 % in energy annually, while still maintaining the HVAC system in compliance with ASHRAE and ISO standards. In the future, there are several plans to be done, including forecasting strategies, generating schedules based on histograms and historical data, and deploying new occupancy estimation methods to collect occupancy distribution data.

Acknowledgements The authors would like to thank the Swinburne University of Technology Sarawak for providing the necessary resources for this research.

References

1. Agarwal Y, Balaji B, Gupta R, Lyles J, Wei M, Weng T (2010) Occupancy-driven energy management for smart building automation. BuildSys'10—Proceedings of the 2nd ACM Workshop on Embedded Sensing Systems for Energy-Efficiency in Buildings, (November):1–6
2. Andreasi WA, Lamberts R, Cândido C (2010) Thermal acceptability assessment in buildings located in hot and humid regions in Brazil. Building Environ 45(5):1225–1232
3. Bian Q (2019) The Nature of Climate Change- equivalent Climate Change Model's Application in Decoding the Root Cause of Global Warming. Int J Environ Climate Change. (December):801–822
4. Burrough M, Gill J (2015) Smart thermostat security: turning up the heat. Univ, Illinois Urbana-Champaign, Urbana, IL
5. Combe N, Harrison D, Craig S, Young MS (2012) An investigation into usability and exclusivity issues of digital programmable thermostats. J Eng Design 23(5):401–417
6. Huchuk B, Sanner S, O'Brien, W (2019) Comparison of machine learning models for occupancy prediction in residential buildings using connected thermostat data. Building Environ 160(March):106177. <https://doi.org/10.1016/j.buildenv.2019.106177>
7. Iyengar S, Kalra S, Ghosh A, Irwin D, Shenoy P, Marlin B (2015) IProgram: Inferring smart schedules for dumb thermostats. BuildSys 2015 - Proceedings of the 2nd ACM International Conference on Embedded Systems for Energy-Efficient Built 211–220

8. Kim SH, Moon HJ, Yoon YR (2017) Improved occupancy detection accuracy using PIR and door sensors for a smart thermostat. In: Abstracts of Proceedings of the 15th IBPSA Conference San Francisco, CA, USA, 7-9 Aug 2017, pp. 2753–2758
9. Kolková Z, Hrabovský P, Matušov J, Nosek R (2018) Analysis of thermodynamic parameters and their influence on the thermal comfort in the working environment MATEC Web of Conferences 168
10. Lu J, Sookoor T, Srinivasan V, Gao G, Holben B, Stankovic J, Field E, Whitehouse K (2010) The smart thermostat: Using occupancy sensors to save energy in homes. SenSys 2010 - Proceedings of the 8th ACM Conference on Embedded Networked Sensor Systems, pp. 211–224
11. Maheshwari GP, Al-Taqi H, Al-Murad R, Suri RK (2001) Programmable thermostat for energy saving. Energy Build 33(7):667–672
12. Özgür L, Akram VK, Challenger M, Dağdeviren O (2018) An IoT based smart thermostat. 2018 5th International Conference on Electrical and Electronics Engineering, pp. 252–256
13. PIB Delhi (2018) Frequently Asked Questions on BEE recommendations on temperature setting of Air Conditioners. https://pib.gov.in/PressReleaseIframePage.aspx?PRID=1537124#:~:text=24 degree Celsius%3F-,Answer%3A It is estimated that by increase in temperature of,is 24-25 degree Celsius. Accessed 11 April 2022
15. Pritoni M, Meier AK, Aragon C, Perry D, Peffer T (2015) Energy efficiency and the misuse of programmable thermostats: The effectiveness of crowdsourcing for understanding household behavior. Energy Res Social Sci 8:190–197. https://doi.org/10.1016/j.erss.2015.06.002
16. 'Public Holidays in Kuala Lumpur for 2021' (n.d.) 2021. https://www.3ecpa.com.my/sme/kuala-lumpur-public-holidays-2021/ Accessed 26 May 2022
18. Stopps H, Touchie MF (2019) Reduction of HVAC system runtime due to occupancy-controlled smart thermostats in contemporary multi-unit residential building suite. *I*OP Conference Series: Materials Science and Engineering 609(6)
19. Tamas R, O'Brien W, Quintero MS (2021) Residential thermostat usability: comparing manual, programmable, and smart devices. Building Environ 203(June):108104. https://doi.org/10.1016/j.buildenv.2021.108104
20. Yang R, Newman MW (2013) Learning from a learning thermostat 93–102
21. Yongson O, Badruddin IA, Zainal ZA, Aswatha Narayana PA (2007) Airflow analysis in an air conditioning room. Building Environ 42(3):1531–1537

Improving Construction Project Management Via 4D and 5D BIM in Sarawak: An Overview

San-Chuin Liew, Victor Nee-Shin Bong, and Raymond Mark Yean-Yong Fong

Abstract The fourth industry revolution or Industry 4.0 is set to spur growth in the Architecture, Engineering and Construction (AEC) industry in line with the modern digital age. Building Information Modelling (BIM), at the forefront of this digital revolution, is a well-known solution to enhance the productivity of the construction industry through digitalization of building information which enables close collaboration amongst all parties involved. It is, therefore, not only useful for AEC professionals in terms of geometric modelling, but also provides the platform for project managers to improve management efficiencies. To ensure the effective deployment of BIM in any project, the time (4D) and cost (5D) elements must be integrated with the geometric models (3D). This is essential to enable project managers to gain a full insight into the overall project planning and progress to facilitate effective decision-making. This paper explores 4D and 5D BIM implementation to promote BIM usage amongst the local AEC professionals in Sarawak to improve construction project management. The 4D and 5D BIM simulations were performed on 3D BIM models to confirm the benefits. It is also noted that at the stage of the current study, BIM implementation in Sarawak (a region in Malaysia) was at a relatively infancy stage. Thus, a questionnaire was also setup to solicit feedback from local AEC professionals. It was discovered that the respondents were generally aware and agreeable to BIM's huge potential to aid construction project management by increasing both efficiency and productivity.

Keywords Building Information Modelling · Construction project management · 4D BIM · 5D BIM

S.-C. Liew · V. N.-S. Bong (✉)
Faculty of Engineering, Computing and Science, Swinburne University of Technology Sarawak Campus, Kuching, Sarawak, Malaysia
e-mail: vnsbong@swinburne.edu.my

R. M. Y.-Y. Fong
Project and Programme Matters Sdn Bhd, Kuching, Sarawak, Malaysia

C. S. Choo et al. (eds.), *Proceedings of ASEAN-Australian Engineering Congress (AAEC2022)*, Lecture Notes in Electrical Engineering 1072,
https://doi.org/10.1007/978-981-99-5547-3_18

1 Introduction

The global rise of populations has seen an ever-increasing demand for new infrastructures and buildings. This represents a challenge to the Architecture, Engineering and Construction (AEC) industry, which is constantly evolving to find new ways to be more productive and efficient. Construction projects often involve multiple parties of different disciplines including architects, engineers and contractors to complete a common task [21]. Often, communication between the different parties involved can be complex due to little overlapping of expertise between parties. A construction project manager is, therefore, appointed to bridge the communication gap and manage/coordinate the overall project. The larger the scale of the project, the more stakeholders would be expected and thus more multi-disciplinary collaboration between parties will be required, thus increasing the overall complexity of the project management [4]. The issues may be complicated further when not all the information involving the project is made readily available to the relevant stakeholders, and communication channels have not been well-established. To overcome this, multiple parties have identified Building Information Modelling (BIM) as a platform for effective collaboration, which can be done through the sharing of single 3D model-based data amongst the project stakeholders. In fact, BIM is widely regarded as a transforming force in the construction industry, and as the "enabler" for digital transformation and the "thrust" to improve productivity, all of which are consistent with modern-day industry trends. The multi-dimension of BIM also ensures that it can be properly integrated with scheduling (4D), costing (5D), sustainability analysis (6D) as well as facility management (7D), thus effectively encompassing the entire project lifecycle. The scheduling and costing aspects are amongst the most critical for any construction project manager with the pressing demand of the modern day—to complete the project on schedule and to stay within the budget allocated. The implementation of 4D BIM and 5D BIM has been known to save time and cost in a construction project with proper scheduling and cost estimations [6]. Such information also enables construction project managers to gain a full insight into the overall project planning and progress to facilitate effective decision-making. This paper explores 4D and 5D BIM implementation to promote BIM usage amongst the local AEC professionals in Sarawak to improve construction project management. This is particularly of interest as most localized studies have been focusing on the barriers to BIM implementation in general [3, 17]. The study was done via questionnaires, sample 3D and 4D BIM modelling and secondary data collection from readily available resources. It provides a snapshot of a localized context on 4D and 5D BIM implementation in Sarawak (a region in Malaysia), which would be useful for future BIM framework and implementation plans here.

2 Literature Review

2.1 *BIM and Project Management*

BIM has been identified in various literature as the key solution for effective management and control of any construction project regardless of the project scale. It is an intelligent 3D model-based process that allows for a collaboration process where all project stakeholders can gather around the same model throughout the design and construction process [1]. The key principle of BIM is that it is based on a single model over all stages of a construction project. A major difference between BIM and conventional Computer Aided Engineering (CAD) design is its three-dimensional nature and information content. A change in component information will update all plans, sections, and elevations automatically across the BIM model. In addition, the BIM model also offers "smart objects" which store physical and semantic data (functions and properties of an element), which contain necessary data for information management throughout the project life cycle [19]. As priorly mentioned, the major strength of BIM is its ability to hold all necessary information and to provide a platform for information management. Through the implementation of a new project delivery approach known as the Integrated Project Delivery (IPD) [18], BIM's potential can be better exploited to realize the various benefits which will lead to cost and time savings. The construction project managers will be able to consolidate project information and resolve the common fragmentation issues which have been plaguing the industry. BIM enables the formation of a cohesive information flow and sharing process with all the project stakeholders, which are essential in the formation of a single 3D model to be used for project coordination and management. Updates can be done swiftly by relevant project participants and the outcome can be communicated with more immediacy through the single shared model, thus facilitating an effective and efficient decision-making process. To reap the maximum benefits from BIM implementation, it is crucial that the BIM process is implemented starting from the initial phase of the project (presumably the conceptual design phase), and to be carried on throughout all the different phases of the project.

2.2 *4D BIM: Scheduling*

BIM in the 4th dimension relates to scheduling, or more specifically the planning of the construction site by adding a new element "time" to the geometrical information of a 3D BIM model. This new element outlines the timeline of the construction project from initiation to completion and details how the project will progress or evolve within that timeframe. This will help to elaborate the schedule of the full project, including the mobilization, installation/construction sequences, etc. This will allow the project team to better plan the resources needed and avoid any conflicts during the construction process. It also provides visualization aid for the site status management and the evaluation of impacts from any changes planned or undertaken throughout the entire project. 4D BIM allows the related parties to be on the same page in terms of duration, timeline and creating or changing the schedule for the actual project. Each task duration, starting date and completion date can be outlined and elaborated to specific building model components and the sequencing of the activities related to these components. 4D BIM allows the schedule to be updated when there are changes in the schedule to detect clashes that might happen during the actual construction [10, 20]. Another benefit of the 4D model often highlighted involves safety planning in which BIM is seen to be able to automate the safety checking process by exploring its effectiveness and efficiency in detecting and preventing fall-related hazards. By investigating the schedule for the construction, potential hazards can be identified and resolved in the planning phase to prevent casualties during the actual operation [15]. Through the integration of 4D BIM, the analysis of the overall project could generate more data for the reduction or prevention of errors through validation of the construction plan and improving communications between the project team. The issue of constructability can be resolved more efficiently, while new practices and workflows to reduce environmental impacts and improve on-site safety can be better visualized, planned and implemented [9, 11].

2.3 5D BIM: Costing

BIM in the 5th dimension is a critical activity to compose a complete BIM model where the cost of each element is inserted into a BIM authoring or compatible software to conclude the total cost of the project. This method is incredibly explicit and is crucial for projects under allocated budget to reduce waste and prevent cost overruns. Its benefits as well as implementation challenges have been discussed extensively [8]. 5D BIM implementation gives rise to the opportunity for cost savings as it provides proper allocation of expenses for activities and contingency plans as well as preventing duplication of activities [7]. As a form of digital tool, 5D BIM allows for the automation of the construction quantity take-off, which will help to resolve the cost estimation dilemma frequently encountered especially by the quantity surveyors. By simply implementing the BIM approach to quantity take-off, necessary information on the construction items and components can be included in the BIM model which in turn allows more accurate determination of the total project sum [16]. The often-cited labour-intensive work of having to manually calculate the amount of each and every material and component required for building work can now be simplified, therefore, enabling the emphasis on value engineering to achieve optimal output [22]. Quantity surveyors, for instance, can focus on costing and factoring risk mitigations and defining assemblies of construction instead. 5D BIM can assist in precisely quantifying modelled elements and materials by reducing assumptions made [5]. Early automation enabled by 5D BIM during quantity take-offs leads to the enhancement of the decision-making process by exploring different cost plan that fits the budget of the project. With the addition of construction time schedules (4D BIM) throughout the project lifecycle, quick and precise budget tracking is made possible. It has also been highlighted in some studies how the BIM 4D and 5D processes can be integrated with modern business intelligence to enhance the digitalization of the construction industry [2].

3 Methodology

The research framework is summarized in Fig. 1. Primary data was collected through questionnaires and BIM sample modelling. The questionnaires were designed to collect the necessary data regarding the perception towards BIM amongst the AEC professionals in Sarawak. A sample size of 30 respondents was considered based on their expertise, proficiency and experiences in the local construction industry was acquired. The results provide an insight into the attitude of the local AEC professionals towards BIM implementation, which to the best of the authors' knowledge, has yet to be well documented. In addition, a simple structural sample model was taken from the samples in Autodesk Revit and put through Autodesk Navisworks for assessment of BIM capabilities. To evaluate the effectiveness of BIM in project management, 4D and 5D BIM simulations were performed using that model. The simulations provide a visualization of the construction project timeline and the project cost. In terms of secondary data, past studies, research papers, questionnaires and experiments which have been conducted by other researchers were referenced for information concerning BIM-based project management models and frameworks, their capabilities, benefits, challenges and barriers. It provides a macro-view of BIM implementation and is used to compare with the results from the primary data collection. The following sub-sections discussed the 3D, 4D and 5D sample modelling done in the preliminary study of this work.

3.1 3D BIM Modelling

A 3D BIM model which contains all the geometrical information and other specified information can be generated through any BIM authoring software. One of the commonly used software is Autodesk Revit. Autodesk Revit can be used to develop the architectural, structural, mechanical, plumbing and electrical models of a building. These models are generated independently by the respective BIM modellers and will contribute to the single file model for communication and coordination purposes. To ensure a smooth transition from the 3D geometrical modelling to 5D cost modelling, it is inherent that the software inter-operability and file compatibility are to be emphasized. While there are multiple BIM-ready or BIM-compatible software around, it is inherent that the inter-operability and file compatibility issues are checked and confirmed prior to the start of the project. The 3D BIM models used for the 4D and 5D BIM simulations are shown in Figs. 2 and 3, respectively. For the

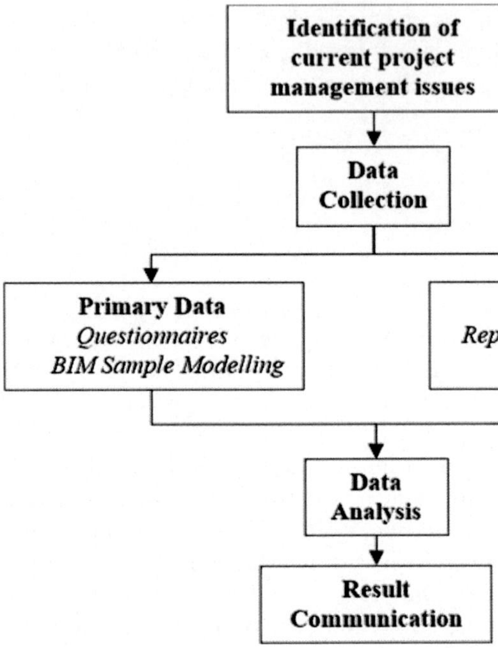

Fig. 1 Research framework

preliminary study, Autodesk Revit can be used for timeline visualization as well as to quantify the project cost from the 3D geometrical model. For advanced studies involving coordination and multi-disciplinary collaboration, Autodesk Navisworks is used owing to its advanced timeline, quantification and optimized collaboration models.

3.2 4D BIM Modelling

The 3D model for a double-story house was modelled. Both the architectural and structural part of the building model was included to obtain the full design of the building. The scheduling phase involved the actual against-planned schedule for the construction activities. The planned schedule for the construction activities was allocated manually in Microsoft Project. Both the actual and planned schedules were integrated in Autodesk Navisworks for the 4D simulation of the house model. To complete the 4D simulation, the Time Liner function in Autodesk Navisworks was deployed and the schedules were allocated to the construction processes to visualize the construction sequences.

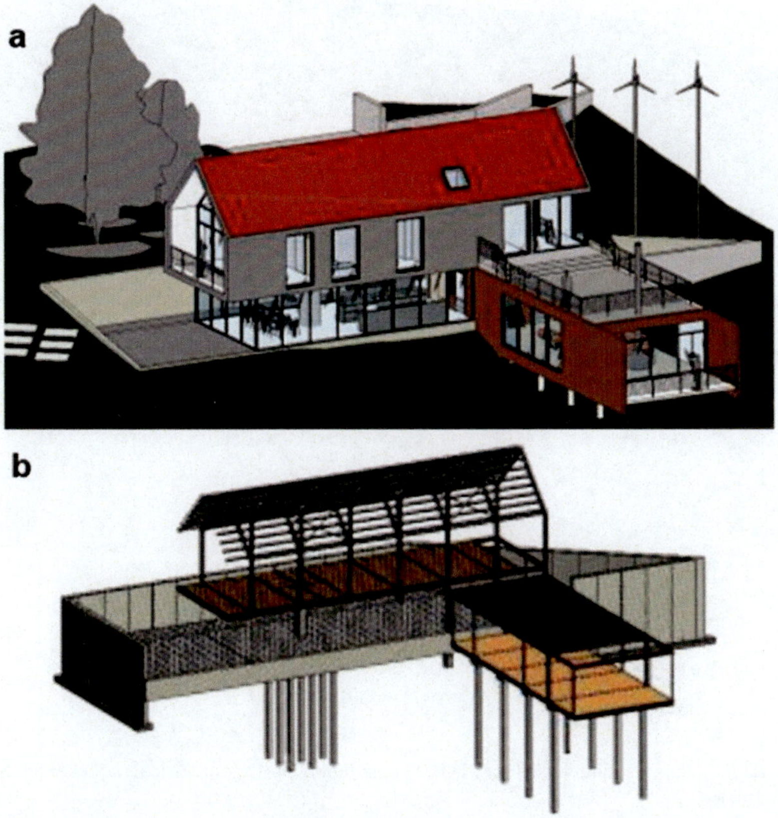

Fig. 2 3D BIM sample model used for 4D BIM simulation: **a** architectural model and **b** structural model

3.3 5D BIM Modelling

To ensure structural stability, the 3D BIM model generated in Autodesk Revit was imported into SPACE GASS as shown in Fig. 3. To ensure an accurate quantification, the Revit schedules were created for all the elements and materials involved with updated cost based on the JKR schedule of rates 2019. This schedule of rates is commonly referred to by the local AEC industry for costing purposes. A total of eight schedules were created: Roofs, Walls, Windows, Doors, Purlins, Floors, Columns and Beams. For this study, project scenarios were assumed to alter the grand total cost to simulate the quantification and costing process.

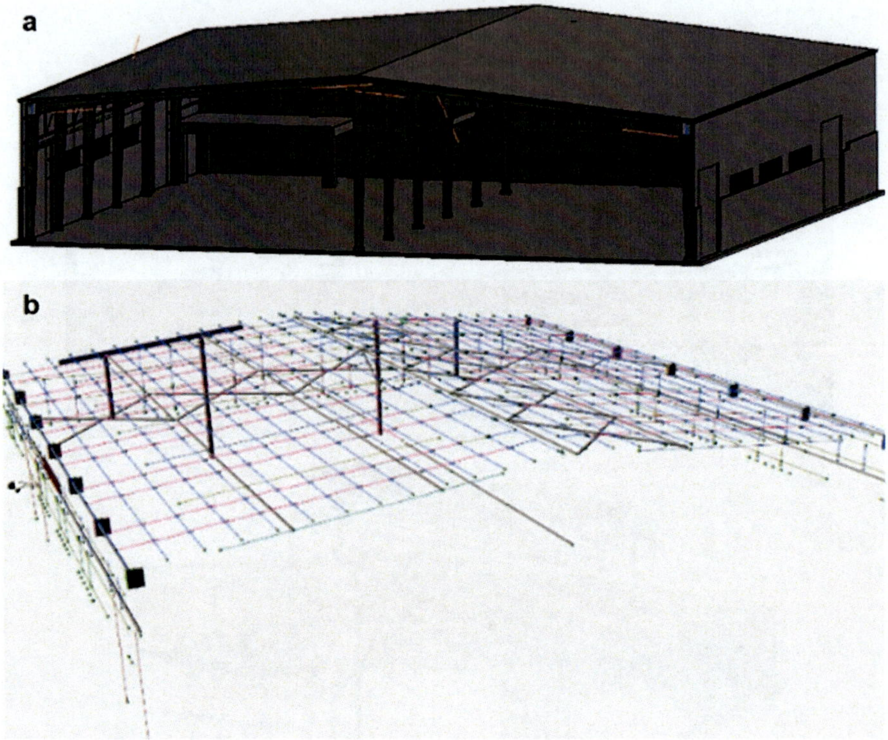

Fig. 3 3D BIM sample model used for 5D BIM simulation: **a** 3D geometrical model and **b** SPACE GASS model

4 Results and Discussions

4.1 Demographics of the Respondents

Table 1 summarizes the role and background of the respondents, who confirmed to have considerable knowledge regarding BIM. The demographics also showed that there was only one participation from the architects, while most of the respondents were from engineering consultancies. This is mainly contributed to the fact that BIM is still relatively new in the Sarawak construction industry. Its utilization is mostly within large-scale construction projects in which BIM has been mandated. Hence, there are still many parties who lack exposure to BIM knowledge and experience. Out of the 30 respondents, however, both experienced and inexperienced BIM users have agreed that BIM has a huge potential to aid construction project management by increasing both efficiency and productivity. The results are summarized in Fig. 4. The current sample size acquired for the questionnaire study may be insufficient to provide representative data of the local AEC industry in Sarawak, and a more detailed

Table 1 Demographics of the respondents

	Consultancy	Contractor	Authority	Architectural practice
Project manager	7	2	0	0
Architect	0	0	0	1
Civil and structural engineer	13	2	1	0
Mechanical and electrical engineer	1	1	0	0
Others	2	0	0	0

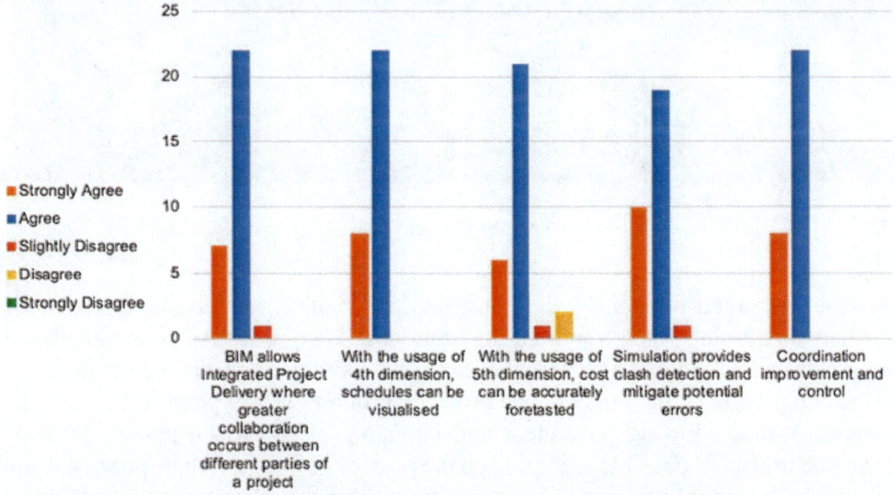

Fig. 4 Respondents' perception towards BIM-based project management

and in-depth study will be conducted in the future to further validate the current data. However, it shows that the general awareness of BIM is already present in the industry. More studies should also be conducted on the implementation framework which can be better suited to the local AEC industry.

4.2 *Advantages of 4D BIM*

Most of the respondents believed that 4D BIM will provide the project manager with the ability to visualize construction activities and schedules. This claim is further reinforced with the implementation of a platform for coordination. The coordination platform used in this study is Autodesk Navisworks. A simulation of construction activities was performed to provide clear visualization of the project progress

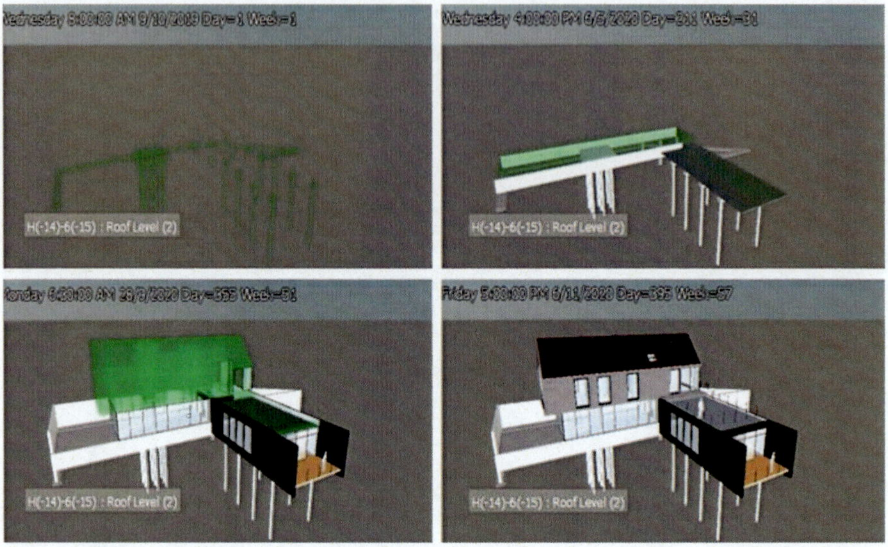

Fig. 5 4D BIM simulation of a sample construction project

timeline. As noted previously, the construction schedule was originally produced in Microsoft Project, then imported into Autodesk Navisworks to be assigned to a sample 3D model. Figure 5 shows snapshots of the simulation from the substructure to the completion of the project. The project's progress at any point in time can be visualized, thus allowing for a clear understanding of all project phases. Furthermore, the impact of possible scenarios on the project's progress can be predicted and assessed. This could be a crucial element to aid the project manager to produce a clash-free schedule for the construction project. The questionnaire and 4D BIM simulation implicated that 4D BIM greatly enhances the project manager's knowledge of the project progress and schedule, which can help them to make more informed decisions and improve the productivity of the construction project. In the local context, this is an encouraging sign that the practitioners are aware of the benefits and have a generally positive attitude towards 4D BIM.

4.3 Advantages of 5D BIM

Just like 4D BIM simulation, 5D simulation can provide visualization of the project at any progress along with its cost of activities. After the model is completed and attached to cost databases, quantification can be done instantly. Furthermore, the quantification will change to match any alteration of the model simultaneously. Figure 6 shows the sample results obtained from the 5D BIM study based on the

JKR schedule of rates 2019. Both the cost of an original roof design and an alternative design was computed. Manual calculations were done to validate the simulated results. The numbers obtained were identical. Apart from visualization, BIM allows different practices such as architect and mechanical to collaborate in the singular information-rich project model. The model can be integrated with the information provided by each practice. Thus, every detail of a building can be modelled in BIM. Here again, both the primary and secondary data indicate the similar benefits of BIM towards project management. The availability of visualizations helps the project manager to clearly communicate with the project team. Moreover, visualization of the project provides stakeholders to understand what the building will look like via virtual construction. The integration of information-rich 3D BIM models with added dimensions of schedules (4D) and costs (5D) ensures better management of project costs and scheduling through various changes throughout the project.

a

<Roof Schedule>			
A	B	C	D
Family and Typ	Area	Cost	Total Roof Cost
Basic Roof:Generic - 80mm:627045: Basic Roof:Generic - 80mm:627045			
Basic Roof:Ge	1423.204 m²	$112.10	159,541.14
Roofs 1:Roofs 1:1458577:1: Roofs 1:Roofs 1:1458577:1			
Roofs 1:Roofs	0.003 m²	$112.10	0.30
Roofs 2:Roofs 1:1458676:1: Roofs 2:Roofs 1:1458676:1			
Roofs 2:Roofs	0.003 m²	$112.10	0.30
Grand total: 3			159,541.73

Original design

<Roof Schedule>			
A	B	C	D
Family and Type	Area	Cost	Total Roof Cost
Basic Roof:Generic - 80mm:627045: Basic Roof:Generic - 80mm:627045			
Basic Roof:Gener	1423.204 m²	$82.20	116,987.35
Roofs 1:Roofs 1:1458577:1: Roofs 1:Roofs 1:1458577:1			
Roofs 1:Roofs 1:	0.003 m²	$82.20	0.22
Roofs 2:Roofs 1:1458676:1: Roofs 2:Roofs 1:1458676:1			
Roofs 2:Roofs 1:	0.003 m²	$82.20	0.22
Grand total: 3			116,987.78

Alternative design

b

	Area (m^2)	Cost (RM)	Cost of each element (RM)	Total (RM)
Model "O"	1423.204	112.1	$Area \times Cost = 1423.2 \times 112.1 = 159540.72$	159541.3
	0.003	112.1	$Area \times Cost = 0.003 \times 112.1 = 0.3363$	
	0.003	112.1	$Area \times Cost = 0.003 \times 112.1 = 0.3363$	
Model "ARCH"	1423.204	82.2	$Area \times Cost = 1423.2 \times 82.2 = 116987.04$	116987.5
	0.003	82.2	$Area \times Cost = 0.003 \times 82.2 = 0.2466$	
	0.003	82.2	$Area \times Cost = 0.003 \times 82.2 = 0.2466$	

Fig. 6 5D BIM simulation of a sample construction project: **a** Extract shows a comparison between the cost of the original design and the alternative design. **b** Extract shows the validation using manual calculation

5 Conclusions

The current research works are still very much a work in progress, with ongoing implementation in a real-life project. The current results confirm BIM's huge potential in the project management aspect of the construction industry. The additional dimensions: time (4D) and cost estimation (5D) provide useful insight to project managers to make informed decisions to improve the project efficiency, thus alleviating the overall productivity of the project team. Without a doubt, BIM will revolutionize the way AEC professionals, especially project managers, approach construction projects. It must also be noted that BIM must be implemented throughout the entire project lifecycle to fully realize its potential. While the current sample size may not be a good representation of the AEC industry in Sarawak, it provides insights into the awareness and local attitude towards BIM. Further studies will be required to validate the data, while more importantly, studies on a feasible framework that suits the local context would be crucial to ensure the success of BIM implementation. The current work can also be extended to include other dimensions of BIM, such as sustainability analysis (6D) [12–14] and facility management (7D). With further emphasis on sustainable construction these days, it is inherent that BIM, for its whole-of-project-lifecycle capability to also support the operations and maintenance of a building, should be integrated into future construction projects. This will not only ensure minimal wastage during the construction phase, but also an efficient and effective management of the building throughout its entire lifecycle.

Acknowledgements The current research work is a collaboration between Swinburne University of Technology Sarawak Campus and Project and Programme Matters Sdn Bhd.

References

1. Abdirad H (2015) Advancing in building information modeling (BIM) contracting: TRENDS in the AEC/FM industry. In: AEI 2015, pp 1–12
2. Çelik U (2020) 4D and 5D BIM: a system for automation of planning and integrated cost management. Advances in building information modeling: first Eurasian BIM forum, EBF 2019, Istanbul, Turkey, May 31, 2019, revised selected papers 1. Springer, pp 57–69
3. Chan JR, Sadeghifam AN, Joo MS (2022) The barrier factors to the implementation of building information modelling for construction industry in Sarawak. Civ Sustain Urban Eng 2(1):44–55
4. Dai J, Goodrum PM, Maloney WF (2009) Construction craft workers' perceptions of the factors affecting their productivity. J Constr Eng Manage 135(3):217–226
5. Diaz P (2016) Analysis of benefits, advantages and challenges of building information modelling in construction industry. J Adv Civ Eng 2(2):1–11
6. Fazli A, Fathi S, Enferadi MH, Fazli M, Fathi B (2014) Appraising effectiveness of building information management (BIM) in project management. Procedia Technol 16:1116–1125
7. Goucher D, Thurairajah N (2012) Usability and impact of BIM on early estimation practices: cost consultant's perspective. In: Proc CIB MCrp, management of construction: research to practice, vol 2, pp 555–569
8. Hasan AN, Rasheed SM (2019) The benefits of and challenges to implement 5D BIM in construction industry. Civ Eng J 5(2):412

9. Honnappa D, Padala SS (2022) BIM-based framework to quantify delays and cost overruns due to changes in construction projects. Asian J Civ Eng 23(5):707–725
10. Hu Z, Zhang J (2011) BIM-and 4D-based integrated solution of analysis and management for conflicts and structural safety problems during construction: 2. Development and site trials. Autom Constr 20(2):167–180
11. Jupp J (2017) 4D BIM for environmental planning and management. Procedia Eng 180:190–201
12. Khoshdelnezamiha G, Liew S, Bong V, Ong D (2019) BIM-based approach for green buildings in Malaysia. In: IOP conference series: earth and environmental science, vol 1. IOP Publishing, p 012052
13. Khoshdelnezamiha G, Liew S, Bong V, Ong D (2020a) A BIM-based automated assessment tool for green building index. In: IOP conference series: materials science and engineering, vol 1. IOP Publishing, p 012059
14. Khoshdelnezamiha G, Liew SC, Bong VNS, Ong DEL (2020) Evaluation of BIM application for water efficiency assessment. J Green Build 15(4):91–115
15. Koo B, Fischer M (2000) Feasibility study of 4D CAD in commercial construction. J Constr Eng Manage 126(4):251–260
16. Li X, Xu J, Zhang Q (2017) Research on construction schedule management based on BIM technology. Procedia Eng 174:657–667
17. Naing TM, Sadeghifam AN, Joo MS (2022) Identifying the critical barriers factors to the implementation of building information modelling (BIM) in the Sarawak's construction industry. Civ Sustain Urban Eng 2(1):21–32
18. Piroozfar P, Farr ER, Zadeh AH, Inacio ST, Kilgallon S, Jin R (2019) Facilitating building information modelling (BIM) using integrated project delivery (IPD): a UK perspective. J Buil Eng 26:100907
19. Politi RR (2018) Project planning and management using building information modeling (BIM). Izmir Institute of Technology, Turkey
20. Sampaio AZ, Gomes NR, Gomes AM, Sanchez-Lite A (2022) Multi-project in an integrated BIM model: clash detection and construction planning. J Softw Eng Appl 15(10):345–358
21. Yates JK, Eskander A (2002) Construction total project management planning issues. Proj Manag J 33(1):37–48
22. Zainon N, Mohd-Rahim FA, Aziz NM, Kamaruzzaman SN, Puidin S (2018) Catching up with building information modeling: challenges and opportunities for quantity surveyors. J Surv Constr Prop 9(1):19–31

Investigation of Voltage Profile Variation and Stability of PV-Grid Interconnected System

Chong Keat Yeoh and Yun Ii Go

Abstract Large-scale solar projects in Malaysia had grown from an export capacity of 451 MW_{ac} to 563 MW_{ac}, indicating an increase of 25% from 2018 to 2020. The power quality of the distribution network is important to be monitored via an optimal power system management scheme. Voltage fluctuation could occur due to overloading, harmonic distortion, output power variation, intermittency of renewable resources, etc. This paper investigates the voltage profile variation and develops a reactive power compensation scheme by implementing an optimal bus topology into the LSS system. In Stage 1, a preliminary resources assessment has been carried out on the LSS sites from Suruhanjaya Tenaga Malaysia. In Stage 2, load flow analysis and contingency analysis have been carried out to identify optimum voltage profile using different bus topologies. Integration of a flexible alternating current transmission system demonstrated an increase of 5% in voltage profile performance. The outcome exhibits minimal loss in the system using the in-electricity standards in Malaysia Grid Code. This research serves as a reference to the LSS stakeholders, national policy-decision makers, and non-profit organizations who planned similar initiatives in meeting regional energy targets.

Keywords Intermittency · LSS · Load-flow · 9-Bus

1 Introduction

The voltage profile variation and stability of a PV-grid interconnected system at all buses in the system had been widely researched by numerous experimental studies. The increased renewable energy installed capacity and high penetration of photovoltaic plants to power grid system affects power grid voltage stability. Stabilize voltage stability is one of the proven research [1–5] to ensure that the solar generator

C. K. Yeoh · Y. I. Go (✉)
School of Engineering and Physical Science, Heriot-Watt University Malaysia, No. 1, Jalan Venna P5/2, Precinct 5, 62200 Putrajaya, Malaysia
e-mail: y.go@hw.ac.uk

C. S. Choo et al. (eds.), *Proceedings of ASEAN-Australian Engineering Congress (AAEC2022)*, Lecture Notes in Electrical Engineering 1072,
https://doi.org/10.1007/978-981-99-5547-3_19

operates in parallel to conventional synchronous generator to maintain the stability of the power system. The development of an optimum power system management scheme and load flow analysis has been carried out to have a further understanding of potential issues and influence on the power system quality. By having this configuration, the researcher is able to identify the optimum bus configuration or topology for the power management scheme and achieve minimal losses by integrating FACTS into the bus topology design.

A reactive power compensation scheme will be introduced to have an optimized system to stabilize the voltage fluctuations [6]. The aim of this paper is to research the resource assessment of selected LSS sites from Suruhanjaya Tenaga Malaysia and generate an economical assessment of the cost of energy of the selected LSS sites [7]. This is followed by the development of the load flow analysis of the transmission line and integration of flexible alternating current transmission system (FACTS) into IEEE 9-bus topology. The study was conducted by identifying the optimal placement of FACTS and LSS in a 9-bus topology to achieve an optimum voltage profile variation and stability of the PV-Grid interconnected system.

2 Methodology

2.1 Resources Assessment

Suruhanjaya Tenaga Malaysia has released a selection of locations for a LSSPV. The location and export capacity are stimulated at the resource assessment to collect data for irradiance, solar path, land availability, energy consumption, nominal power, and energy cost [8]. The collected data is transferred to PSS SINCAL for load flow analysis to determine the maximum energy yield annually [9]. The resource assessment is able to pinpoint the location needed for design so that it can calculate the solar path at the design location. It also provides the latitude and longitude of the location so that the research location's accuracy is ensured, as shown in the parameters. The resource assessment provides the data on the designated location's solar path. Figure 1a and b show the solar path at Marang and Kuala Muda, whereas **c** indicates the overall workflow. Marang, Terengganu, experiences a reduction in irradiance at the end of the year. Due to the last quarter of the year, Malaysia has its Monsoon season, which is the country's rainy season. Marang, Terengganu, is located near the South China seacoast. Therefore, the solar farm during the monsoon season would be less efficient due to the weather variation. Weather conditioning has to be taken into consideration when designing an optimized solar farm.

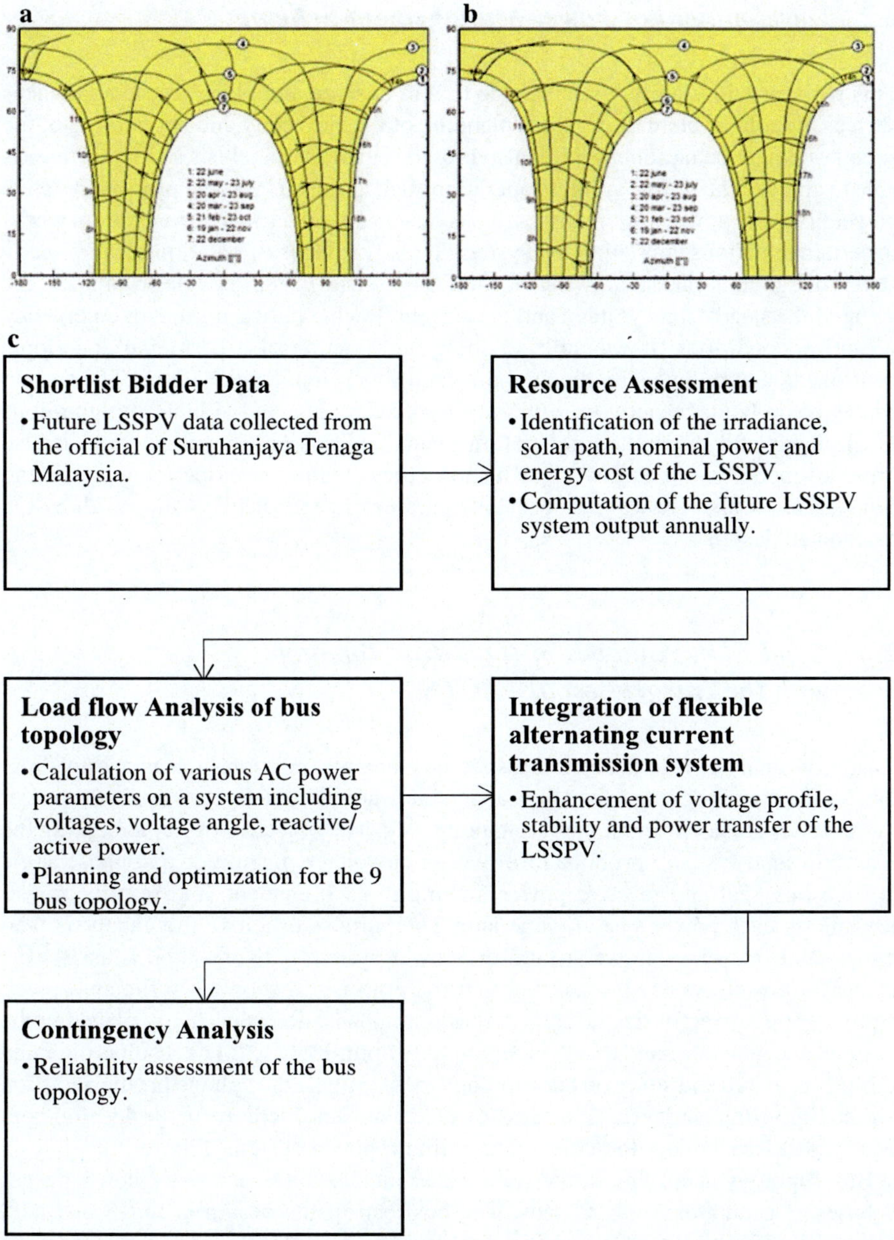

Fig. 1 **a** Sunpath at Marang, **b** Kuala Muda at legal time, and **c** Methodological stages

2.2 Development of Power Management Scheme

This work provides an industrial-grade tool to analyze and plan electric appliances such as the solar system. It provides planning of the electricity and networking of the solar system to the national grid. It also provides load flow analysis so that the user is able to analyze the power systems operation such as the bus voltage profiles, reactive power flow, and power system losses. It ensures that the electrical power system works in certain situations applied in the system. It also provides calculations of the steady state voltage and current flowing of the design. The system is able to provide the value of the steady state voltage and the current flowing during normal or emergency operating conditions. It was able to show the function of a harmonic and ripple control calculation by using the load flow and short circuit calculations. The system shows a low-voltage network with various feeders and individual configurations of the loads and distributed generation plants and charging infrastructure. It also provides modelling of a network as it shows the calculation of operating points and time series. It also shows the necessary parameters and displays the result of the designated design.

2.3 Load Flow Analysis of the 9-Bus Topology with the Integration of FACTS

Load flow analysis is to determine a specific element condition such as overloading. During the operating stage, load flow analysis ensures that each generator in the power system runs at the optimum operating point [10]. The objective of implementing the load flow analysis is to produce information on voltage magnitude and phase angle at each bus, real and reactive power flowing in each element, and reactive power loading on each power system generator. The purpose of a load flow analysis is to understand the power flows around the power system network [11]. It assists the electrical power system in ensuring that the power grid supplies sufficient power, equipment is correctly sized, reactive power compensation is perfectly placed in the system, and generator and transformer are fully optimized [12]. The result is obtained through research and study on bus topology to investigate the voltage profile variation and stability. Precise data and analysis are taken into consideration in the development of this research project to achieve an optimum power system. Figure 2 displays a 9-Bus Topology according to the IEEE rules and Malaysia standards, consisting of 9 buses, 3 generators, and 3 loads. The bus load of the design is LO13 at Bus 8 (100 MW, 35MVAR), LO14 at Bus 5 (125 MW, 50 MVAR), and LO15 at Bus 6 (90 MW, 30MVAR). The figure displays the voltage profile of each node (bus) of the research topology with minimal voltage loss.

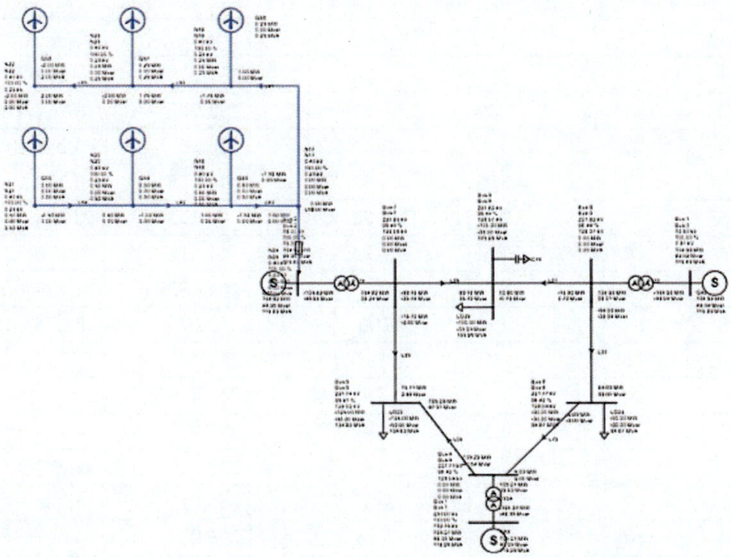

Fig. 2 9-Bus topology power system

2.4 Contingency Analysis

Contingency analysis in this research allows the power system to predict any future fault or failure to prevent future crises [13]. It provides reliability checks and corrective actions for the user to have future planning and operation. Contingency analysis is performed to assess the reliability of system. The contingency analysis is implemented to detect any outage of the power system. Table 1 displays a contingency analysis of the research project result. The purpose of this contingency analysis is to assess the new value (V_w, V_{piTV}, V_{piTI}) to withstand the corresponding limit of the 9-bus topology power system.

3 Results and Discussion

3.1 The Implementation of FACTS into the Research Design

The simulation results are presented in the form of charts for voltage profile and distance. The results from the simulation have been compared with the corresponding experimental results to have a better understanding of the voltage profile of the research. From the result achieved in Figs. 3 and 4, the performance of a flexible alternating current transmission system improves the overall voltage profile. The

Table 1 Contingency analysis of the 9-bus topology

Name	Key	No. All	Pos	V_w	V_{piTV}	V_{piTI}
2T34	1	7	1	−100	233.276	14.274
2T33	2	7	2	−85.694	232.74	14.267
L35	3	7	3	−42.778	46.59	10.258
L26	4	7	4	−42.773	46.235	10.272
L30	5	7	5	−28.607	46.635	10.264
L27	6	7	6	−14.306	47.912	10.276
G68	7	8	7	0	266.132	14.274
L68	8	7	8	100	48.076	10.275
2T33	9	6	1	−100	247.449	20.797
2T34	10	6	2	−85.694	247.449	20.797
2T33	11	7	3	−42.778	232.936	14.269
2T34	12	7	3	−42.778	233.477	14.277
2T33	13	7	4	−42.773	235.715	14.363
2T34	14	7	4	−42.773	236.069	14.361
2T33	15	7	5	−28.607	237.049	14.307
2T34	16	7	5	−28.607	254.401	14.476
2T33	17	7	6	−14.306	236.159	14.307
2T34	18	7	6	−14.306	239.592	14.337
2T33	19	7	7	0	288.769	20.797
2T33	20	7	8	100	239.955	14.344
2T34	21	7	8	100	237.329	14.323

performance of FACTS allows the voltage profile to increase in the result achieved. On the other hand, without FACTS implementation, the voltage profile decreases drastically. FACTS has minimized the voltage profile average loss of 5% at Bus 2, 8, and 9. FACTS prevents significant deviation of bus voltage due to the reactive drawn by the 9-bus topology generators. It provides acting voltage support to the power system to avoid any possibility of voltage losses. It also improves the power system's AC grid controllability and power transfer capability. It is able to confine power flow to a designated route in the power system to ensure the right amount of power is flowing through the entire power system. Lastly, FACTS is able to prevent any incoming cascading outages by providing emergency control to the power system.

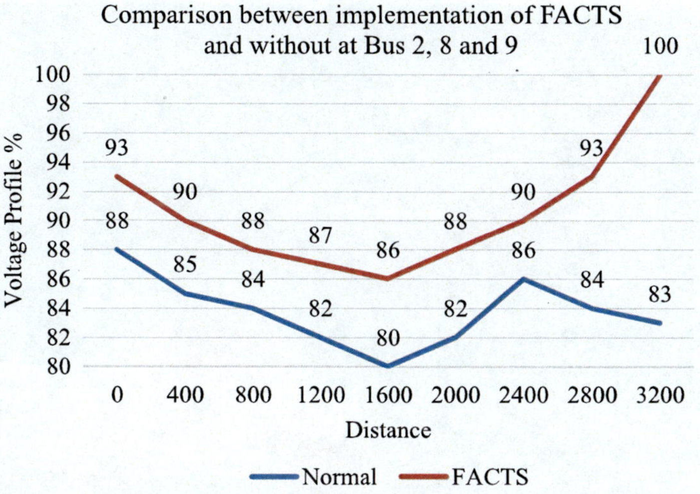

Fig. 3 Comparison between the implementation of FACTS and without at Bus 2, 8, and 9

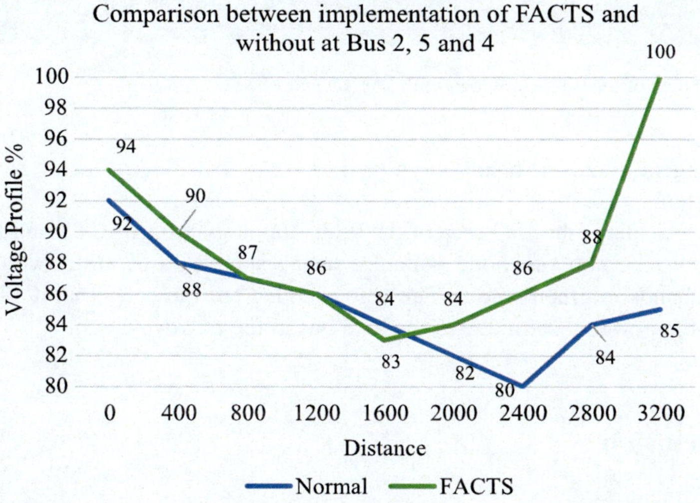

Fig. 4 Comparison between the implementation of FACTS and without at Bus 2, 5, and 4

3.2 *The Implementation of Contingency Analysis into the Research Design*

The value achieved through this research can be used to analyze the system's power flow and understand the network behaviour of the system. The value also provided a corrective action when a fault or outage occurs in the 9-bus topology power system.

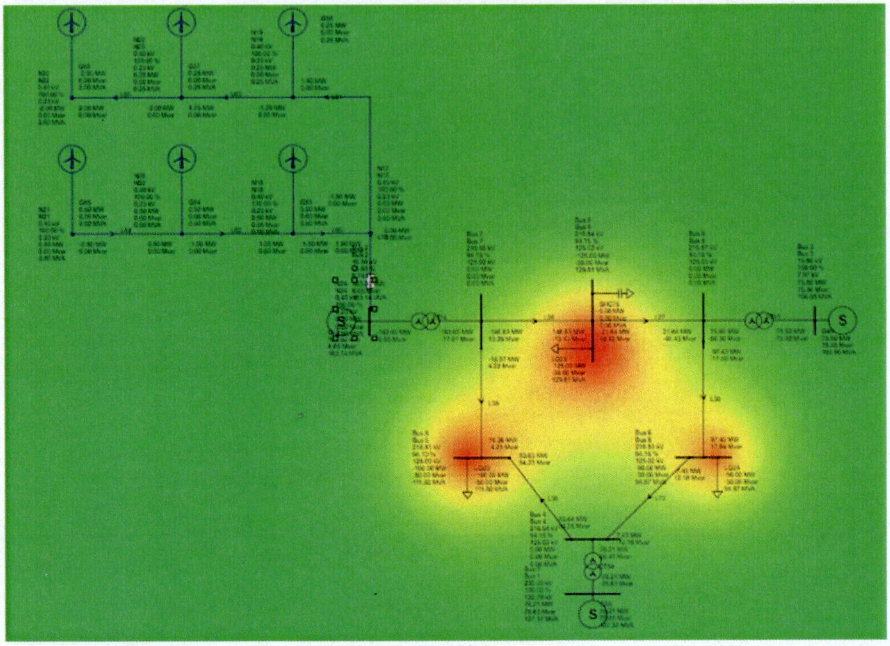

Fig. 5 ISO area of the 9-bus topology with FACTS and LSS

The contingency analysis result provided the system to achieve a secure zone for the power system.

Figure 5 displays the ISO area of the 9-bus topology with FACTS and LSS. The figure shows the network node voltages during standard operations voltages and operating bands in green colour. The colour-coded red indicates that it's in critical ranges in the power system due to the load bus in the power system.

4 Conclusion

The investigation of voltage profile variation and stability of a PV-Grid-Energy interconnected system is a concept that implements a power management scheme for the system, with a successful reactive power compensation scheme adding more efficiency, reliability, and power quality to the established concept of harvesting of the LSSPV. The research also has identified the economic assessment to evaluate the energy cost of the LSSPV which the research has identified the total investment of the Cycle-3 LSSPV at Marang, Terengganu and Kuala Muda, Kedah. The research project has identified an optimum bus configuration approved by the IEEE, showing minimal losses in the designed system in accordance with the Malaysia grid code standards. Based on the research project result shown, the voltage profile variation and the stability of photovoltaic grid energy interconnected system design can introduce to the large-scale solar system in Malaysia. At the same time, they are proving a higher power quality, voltage variation, and voltage profile. The best performance voltage of implementing a flexible alternating current transmission system into the bus topology has resulted in an increase in voltage profile performance of at least 5% on average against the underperforming design. The finding proved that the proposed work had provided a power management scheme that provides a minimal loss in the system. The project also has performed a contingency analysis to enable the detection of component outages of the power system to have a corrective action for future planning and operation. Future LSSPV projects in Malaysia could take this research project as a stepping stone for an optimum power management scheme. The implementation of solar and renewable energy related projects [14–16] addressed the rising issue of climate change. It is aligned with Malaysia's energy target in meeting world energy challenges by 2030.

References

1. Fernandez MI, Go YI (223) Power management scheme development for large-scale solar grid integration. J Electr Syst Inf Tech 10(15). https://doi.org/10.1186/s43067-023-00080-7
2. Mohanan M, Go YI (2020), Optimized power system management scheme for LSS PV grid integration in Malaysia using reactive power compensation technique. Global Challenge, Wiley. https://doi.org/10.1002/gch2.201900093
3. Khan R, Go YI (2019) Assessment of Malaysia's large-scale solar projects: power system analysis for solar PV grid integration. Global Challenge, Wiley. https://doi.org/10.1002/gch2.201900060
4. Solution ET (2021) Load flow analysis. Electricla Testing Solution [Online]. Available: http://electricaltestingsolutions.com/index.php/engineering/load-flow-analysis. [Accessed 2 Nov 2021]
5. R. E. A. (IRENA) (2017) Renewable energy statistics, Abu Dhabi, UAE
6. Vanishree J, Ramesh V (2014) Voltage profile improvement in power systems-a review. In: 2014 International conference on advances in electrical engineering (ICAEE)
7. Tenaga S (2020) Request for proposal (RFP) for the development of large scale solar photovoltaic (LSSPV) plants in peninsular Malaysia for commercial operation in 2021. Suruhanjaya Tenaga
8. PVsyst (2020) PVsyst PVsyst. https://www.pvsyst.com/. Last Accessed 19 Oct 2020. (Online)
9. Energy S (2021) Most efficient way of controlling reactive power. Siemens, [Online]. Available: https://www.siemens-energy.com/global/en/offerings/power-transmission/facts/portfolio/mscmscdn.html#:~:text=Mechanically%20switched%20capacitors%20(MSC)%20or%20filter%20circuits%20(FC),at%20the%20point%20of%20connection. [Accessed 22 Jan 2021]
10. Said MS (2015) Comparison of load flow methods on IEEE 9-bus system transmission network using Gauss-Seidel and Newton Raphson
11. Sharma S, Velgapudi NS, Pandey K (2017) Performance analysis of IEEE 9 Bus system using TCSC. In: 2017 Recent developments in control, automation & power engineering RDCAPE, pp 251–256
12. Zhao W, Guo Q, Sun H, Ge H, Li H (2018) Practical short-term voltage stability index based on voltage curves: definition, verification and case studies. IET Gener Transm Distrib 12:4292–4300

13. Meng X, Pian Z (2016) Derivation of distribution network vulnerability indicators based on voltage stability smart power distribution system, vol 19
14. Thadani HL, Go YI, Large scale solar system design, optimal sizing and techno-economic-environmental assessment. Sustain Energ Res 10(11) (2023). https://doi.org/10.1186/s40807-023-00081-0
15. Thadani HL, Go Yl (2021) Integration of solar energy into low-cost housing for sustainable development: case study in developing countries. Heliyon 7:e08513. https://doi.org/10.1016/j.heliyon.2021.e08513
16. Faruhaan A, Go YI (2021) Energy storage sizing and enhanced dispatch strategy with temperature and safety considerations: a techno-economic analysis. Energ Storage e260:1–33. https://doi.org/10.1002/est2.260

Solar Photovoltaics Thermal Profiles Development and Visualization: Experimental, Numerical Analysis and Validation

Muhammad Danish Arman Zairin and Yun Ii Go

Abstract The current energy market is moving towards digitalization for energy yield monitoring. As an intermittency resource, solar energy production needs to be monitored closely which is highly dependent on environmental factors. Various research had been carried out to improve the safety, reliability and life cycle of solar PV systems. Thermal flow has been identified as one of the measurements that will significantly affect the system's capability in terms of output and efficiency. Thermal flow studies received increased attention in sensing and visualization mechanisms among the solar industry's stakeholders. The aim of this research paper is to design and implement a visualizing tool for solar PV thermal flow. The development of the graphical user interface (GUI) was also discussed in the paper to display and visualize the temperature condition. GUI has been developed to display images that determine the temperature condition in selected resolutions. The implication of the project is discovering the most optimal temperature measurement technique that can be utilized for upcoming real-time condition monitoring systems of solar PV systems. The significance of the output from this research would bring great benefits to future solar panel researchers regarding the importance of applying such visualizing techniques to PV systems by providing better temperature data analysis, identifying thermal flow patterns and allowing manufacturers to have quick action towards common faults. Results show that nearest-neighbour interpolation offers a slightly more accurate temperature reading compared to the Bicubic interpolation with an approximation error difference of 0.00596. Results have been validated using a thermal gun where bicubic interpolation produces a higher range of temperature reading compared to nearest-neighbour interpolation, with a difference in percentage accuracy reading of 0.5462%.

Keywords Thermal flow · Sensing · Interpolation · GUI · Monitoring

M. D. A. Zairin · Y. I. Go (✉)
School of Engineering and Physical Science, Heriot-Watt University, Malaysia. No. 1, Jalan Venna P5/2, Precinct 5, 62200 Putrajaya, Malaysia
e-mail: y.go@hw.ac.uk

C. S. Choo et al. (eds.), *Proceedings of ASEAN-Australian Engineering Congress (AAEC2022)*, Lecture Notes in Electrical Engineering 1072,
https://doi.org/10.1007/978-981-99-5547-3_20

1 Introduction

Renewable energy sources make up approximately 26% of the world's electricity. The latest report from International Energy Agency (IEA) has shown that in the year 2024, it is expected that renewable energy will take up to 30% of world energy consumption. The IEA predicts that solar energy will be constantly increased in its share of the global renewable generation as the retail electricity prices in most countries are much higher compared to solar electricity. Solar energy represents an alternative way to generate green renewable sources of energy. Solar power is among the most effective way to reduce carbon footprint [1– 3]. According to statistics from the British Petroleum, BP (2019–20) on World's Energy [4], the solar installed capacity was 586 GW, which was a net increase of 98 GW compared to the year 2018. British Petroleum (2019–20) also stated that solar energy has constantly increased its share and represents more than 26% of the world's renewable generation by source, which is an increase of 14% compared to the year 2015. The global demand for solar array installations is exponentially increasing and is expected to expand from 45–50% of energy by the year 2050, according to International Energy Agency IEA and IRENA [5, 6].

2 Objectives

The aim of the research is to design and develop a suitable sensing system to measure the temperature of solar PV. These data will be processed to produce thermal profiles via various interpolation techniques. Thermal images will be reconstructed using a suitable visualization tool. Experimental work will be carried out to identify the optimal position of thermistors and the interpolation algorithm. The results will be validated using direct measurement of thermal gun. By creating a graphical user interface (GUI) for thermal reconstruction using MATLAB, the thermal profile was generated and visualized in different resolutions under a multi-node sensing mechanism. Temperature profiles can be reconstructed in enhanced resolutions under multi-node sensing mechanisms. This study will determine the algorithm that offers the highest accuracy for predicting the thermal profile of a PV module via different interpolation techniques other than cubic spline and bilinear.

3 Literature Review and Research Gap

Various research [7–13] had been carried out on solar thermal effect and its implication on system performance. When the PV system is operating, thermal heat is continuously flowing across the solar cell which cannot be observed via naked eye. Hot spots occur in a PV module due to mismatch in solar cells, physical defects, or cells are partially shaded. Hot spot heating in PV module occurs when its operating current exceeds the reduced short-circuit current of a shadowed or faulty cell. When such a condition occurs, the affected cell is forced into reverse bias and dissipates power, which may cause local overheating. It will also create a partial blockage or short-circuiting caused by junction box error. Similarly, the higher the number of microcracks that occurs on PV cells, the higher the percentage of output power that will be lost. The concept of this research originated from the statistical report on the short-term and long-term defects of PV system in the United States was evaluated by Zhen et al. [14]. It was observed that among the 115 defective PV modules, 25% were due to cell hot spots and microcracks. These common faults have led to the utilization of a visualization tool that mitigates such issues. Visualization tool for monitoring can minimize or prevent such defects to achieve higher operational efficiency. According to Ogbomo et al. [15] majority of the solar panels are around 15–25% efficient due to mechanical issues caused by the presence of these external resistances or tropical climate. An experiment conducted by Daliento et al. [10] includes producing a thermographic image of a monocrystalline solar panel integrated with a traditional bypass diode being tested for visualizing hot spots to maximize power dissipation. Zhen et al. [14] investigated PV modules of about 200 MWp that had been running in the United States for 1–3 years, tested and analysed the performance of failure modules, and summed up the reasons for the degradation and failure of these modules. According to an experiment of hot spot testing conducted [11], hot spot heating occurs in a PV module when its operating current exceeds the reduced short-circuit current (Isc) of a shadowed or faulty cell or group of cells. When such a condition occurs, the affected cell or group of cells is forced into reverse bias and dissipates power, which can cause local overheating. The consequences of the absence of prior monitoring and visualizing system for thermal flow through a PV system may lead to electrical or potential fire hazards if the thermal flow is not properly monitored or maintained. Recent research has shown that hot spots can be visualized via thermal image processing with degradation analysis by Chaudhary [9].

4 Research Methodology

NTC thermistor and thermal gun to collect temperature data for locations on the PV module. Interpolation techniques predict and display the temperature data of the PV module through a graphical user interface (GUI). Arduino IDE is used as analogue-to-digital converter to compute and process the measured temperature data. In this case, a 10-bit analogue-to-digital converter (ADC) is selected to convert the analogue input from the NTC thermistor into readable temperature values. Visualization tool will be used to develop and display thermal images that can to closely predict and monitor the condition of solar PV. Different interpolation techniques and display of thermal images on GUI using MATLAB. The solar panels utilized in this experimental setup are tabulated below, including the illustration and specification. The solar panels were labelled as 'Solar Panel 1', 'Solar Panel 2', 'Solar Panel 3' and 'Solar Panel 4'. Bicubic interpolation is cubic interpolation in two dimensions consisting of multiple nearest-neighbour algorithms. The two-dimensional interpolation form that will be used is the bicubic interpolation data fit. The bicubic technique uses the values of 16 nearest-neighbour pixels instead of the 4 pixels used by the bilinear approach to determine the value of the unknown pixel. Since the bicubic technique uses the nearest 16 pixels, the quality of the generated output image is theoretically more optimal compared to the bilinear and nearest neighbour. The triangular bicubic interpolation is a simple bicubic interpolation function and returns lower values for a high value and input for a low input that causes blurriness. Smooth edges can be achieved with the stated conditions when the nearer pixel values are high, and the further pixel values are set to very low. This will reduce the blurred effect, and compared to the cubic interpolation algorithm, this will generate smooth edges [16–19]. The conditions indicate that the value computed by the b-spline bicubic function is very low for the high value of variable x. For example, a very low weight is assigned to the distant intermediate pixels. This approach is an enhanced variant of the two-dimensional normal grid over the cubic interpolation. By taking the nearest 4×4 surrounding pixels for a total of 16 pixels into account, the bicubic interpolation goes one step beyond the limits of bilinear. The interpolated surfaces are smoother than the equivalent surfaces obtained by the bilinear interpolation and the nearest-neighbour methods stated earlier in the review of the literature.

The pre-processing phase of the workflow involves the data collection of sample data for different sensor arrangements. The temperature data were measured using a thermal gun and Arduino. For each type of solar panel, the temperature data were sampled once every 1 h with a total of 10 samples for different sensor arrangements. After sampling the temperature data, the next phase is the processing phase to convert the data from analogue-to-digital value. After MATLAB computed the digital data according to the manufacturing data sheet of the NTC thermistor, the interpolation algorithm was implemented to generate a better resolution of the thermal profile. Since the sensor placement was evenly distributed, the interpolation algorithm allows us to predict the temperature of the remaining area at the centre of the panel. The last phase of the thermal reconstruction is the post-processing phase, where MATLAB generates the thermal profile image based on the scaled colour bar. The thermal profile images were reconstructed in a GUI including a scaled colour bar that displays and determine the condition of the temperature for each point shown in the GUI. This phase also discusses the comparative analysis between the implementation of the interpolation technique and the use thermal gun. Figure 1 summarizes the workflow and the flowchart of the MATLAB programming code for reconstructing the thermal profile image [7, 13, 15–19].

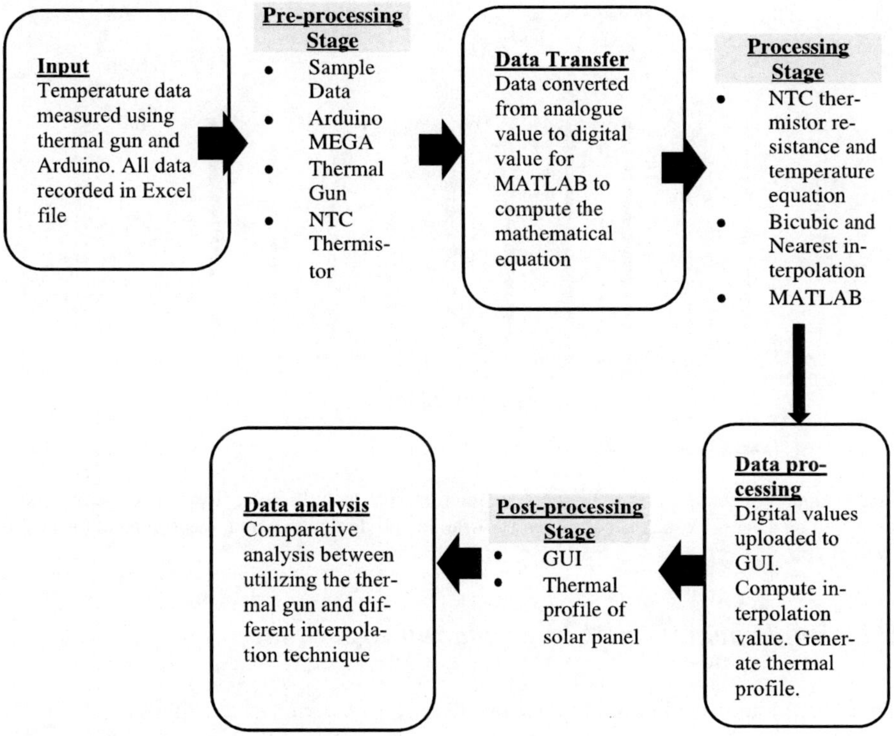

Fig. 1 Workflow diagram of the project

5 Results and Discussion

The monitored temperature measurement data collection for this type of solar panel is slightly different compared to the other panels introduced in this study (Fig. 2). The tabulated data collection for Solar Panel 3 were split into two different sections, as this type of solar panel consisted of two small panels integrated with each other using a hinged. Based on these numerical findings, all the averaged sample data values will be implemented with the bicubic and nearest-neighbour interpolation algorithm to further study and analyse the thermal flow pattern. In real practice, it is rather difficult and inconvenient to solely rely on tabulated data to monitor the conditions of a solar PV system. For faster and more reliable data analysis, the computation of GUI using MATLAB to monitor the temperature condition of the solar PV for the study of thermal flow pattern was introduced.

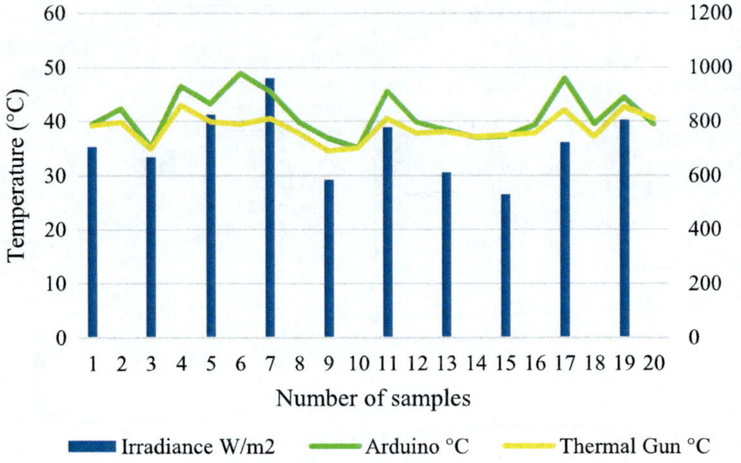

Fig. 2 Data collected during the experimental stages include solar irradiance, temperature measured by a sensor which is connected to Arduino and direct temperate measurement through a thermal gun

5.1 Implementation of Interpolation Algorithm

This part of the results and discussion illustrates the generated thermal profile of the averaged sample data in the GUI for the 2 × 2 sensor arrangement for solar panels 1, 2, 3 and 4. The proposed nearest and bicubic interpolation algorithm was able to interpolate the measured data values from the 2 × 2 matrix to a higher resolution of the 3 × 3 matrix, which allows operators and users to clearly visualize the direction of the thermal flow. As shown in the experimental setup earlier, the four NTC thermistors were placed in each corner of the back of the solar area. The reason behind this is to allow the interpolation algorithm to compute, predict and add the missing temperature value data other than the measured solar cell area, for example, the centre cell area of the panel. The interpolation algorithm that was implemented on the 2 × 2 sensor arrangement for this research were the bicubic and nearest-neighbour interpolation. The figures below show the visualization of the thermal flow for each type of solar panel.

Nearest-neighbour interpolation technique results in a slightly lower resolution compared to using the bicubic interpolation technique. The computation time for utilizing bicubic interpolation compared to the nearest neighbour is relatively lower. In addition to that, the bicubic interpolation algorithm was able to compute a higher range of temperature value data compared to the nearest-neighbour algorithm. Figure 3 (left) offered a total of nine different temperature values while, Fig. 3 (right) offered a total of four different temperature values. Bicubic interpolation was able to approximately distinguish more direction of thermal flow across the solar panel based on temperature differences, as for nearest-neighbour technique displayed a lower number of thermal profile directions and temperature data values due to the

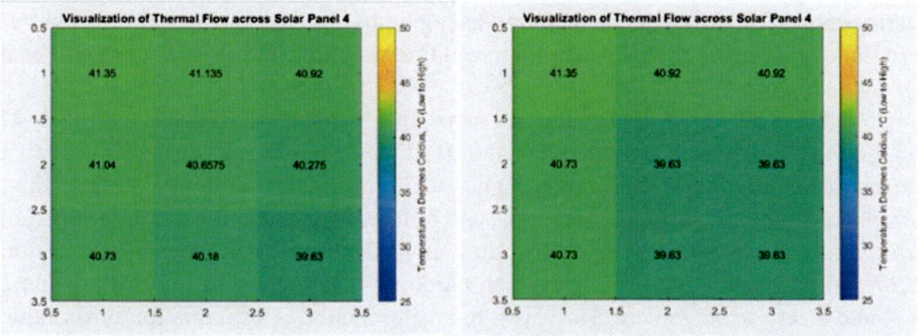

Fig. 3 Visualization of thermal flow using bicubic interpolation (left) and nearest-neighbour interpolation (right)

simplicity and limitation of the mathematical computation in the algorithm. Although this deduction does show that the temperature differences are exceedingly low, these findings can bring great impact and differences in the bigger solar energy application as it would be difficult to determine the exact direction of the thermal flow. These findings can be deduced that the bicubic interpolation tends to be a more reliable technique as it can approximately pinpoint the direction of the thermal profile across the solar panel. In terms of accuracy for temperature readings, a comparative analysis for which interpolation technique offers a better temperature performance was carried out, which will be further discussed in the next sub-topic. The experimental results show that the temperature difference between the front and back part of the solar panels are relatively low. When the temperature difference between the back and front part of the solar panel is high, this results in the potential occurrence of the hot spot phenomenon on the surface of the solar panel, which results in a high thermal profile flowing across the panel. This monitoring technique can be implemented in bigger applications for solar energy farms as it can pinpoint or predict possible faultiness of solar cell areas to prevent inefficiency of solar energy output and fire or electricity hazards.

5.2 Validation

The comparative analysis between the utilization of measuring the temperature of the solar panel using a thermal gun and the interpolation technique was discussed here. The approximate error and average error percentage calculation determines which measuring technique offers a more accurate reading. The average approximation error calculation is the difference between the interpolated temperature data value computed in the GUI and the average measured temperature data value using the thermal gun. The difference is then divided again by the average measured temperature data value using the thermal gun to obtain the final average error (Tables 1 and 2).

Calculated averaged approximation error values for using bicubic and nearest-neighbour interpolation are 0.02473 and 0.01877, respectively. It can be deduced that the nearest neighbour offers slightly higher accuracy in terms of temperature reading compared to the bicubic interpolation, with a difference in averaged approximation error value of 0.00596. During the conduction of the experiment, the approximation error and uncertainty can be clearly compared with lower resolution, such as the 1 × 2 and 2 × 1 arrangement. However, the difference in approximation error value between utilizing the bicubic and nearest-neighbour interpolation was exponentially getting smaller when testing out with higher resolution, such as the 2 × 2 and 3 × 3 arrangement. This is due to bicubic interpolation would be able to compute higher range of temperature values in higher dimension for a set of known points. Nearest-neighbour offers better accuracy for temperature reading, whereas bicubic interpolation would theoretically offer lower approximation error for temperature reading and become a much more reliable interpolation technique in the bigger solar panel application, as it is able to compute and produce a higher range of temperature reading compared to the nearest-neighbour interpolation.

Table 1 Approximate error calculation of averaged sample data with sensor arrangement using bicubic interpolation

Types	Approximate error value of sample data for 2 × 2 sensor matrix (Bicubic interpolation)			Averaged approximation error
Solar panel 1	0.04258	0.02617	0.00918	0.025980
Solar panel 2	0.00707	0.00828	0.00949	0.008280
Solar panel 3	0.03358	0.02720	0.02074	0.027170
Solar panel 4	0.05239	0.04347	0.03439	0.043416

Table 2 Approximate error calculation of averaged sample data with sensor arrangement using nearest interpolation

Types	Approximate error value of sample data for 2 × 2 sensor matrix (Nearest interpolation)			Averaged approximation error
Solar panel 1	0.04077	0.00467	0.00467	0.016700
Solar panel 2	0.00562	0.01069	0.01069	0.009000
Solar panel 3	0.03969	0.02247	0.02247	0.028210
Solar panel 4	0.06473	0.02921	0.02921	0.021631

6 Conclusion

The outcome of this work complements related research projects [21–25] in large-scale solar integration and storage for national grid in meeting world energy challenge 2023. With the computation of the approximation error for comparative results analysis, it has been deduced that the bicubic interpolation offers a better accurate temperature reading of the surface of the panel compared to other temperature measurement techniques. Collection of data for temperature measurement of solar panels under tropical weather conditions in Malaysia using a thermal gun and different interpolation techniques in enhanced resolutions were also experimentally demonstrated to generate the thermal profile of solar PV system in the GUI. Recommendation and future work that can be addressed for this research include testing with different sensors that can be connected to the Arduino platform such as the radiometric sensor. The sensor values may not be highly influenced by the atmospheric conditions as it can consider incident and reflected radiation. This project can also be integrated with the Internet of Things (IoT) platform, such as ThingSpeak can be considered to efficiently record, store and visualize data obtained from the sensor online.

References

1. Thadani HL, Go YI (2023) Large scale solar system design, optimal sizing and techno-economic-environmental assessment. Sustain Energy Res 10:11, Springer Nature, 2023, ISSN: 2731–9237. https://doi.org/10.1186/s40807-023-00081-0
2. Hamzah AH, Go YI (2023) Design and assessment of building integrated PV (BIPV) system towards net zero energy building for tropical climate. Adv Electr Eng Electronics Energy 3:100105, ISSN 2772–6711. https://doi.org/10.1016/j.prime.2022.100105
3. Thadani HL, Go YI (2021) Integration of solar energy into low-cost housing for sustainable development: case study in developing countries, Heliyon 7, e08513. https://doi.org/10.1016/j.heliyon.2021.e08513
4. BP Statistical Review of World Energy (2020)
5. IEA International Energy Agency (2011) Trends in photovoltaic applications. Survey report of selected IEA countries between 1992 and 2010
6. IRENA (2019) Renewable energy technologies
7. Sheikh N, Kocaoglu DF (2011) A comprehensive assessment of solar photovoltaic technologies: literature review. Proc PICMET Portland, OR
8. Wang G, Wang B, Yao Y, Lin J, Chen Z, Hu P (2020) Parametric study on thermodynamic performance of a novel PV panel and thermal hybrid solar system. Appl Therm Eng 180:115807
9. Chaudhary AS, Chaturvedi DK (2023) Live hotspots visualization and degradation analysis of solar photovoltaic panels using QR code and thermal image processing. In: Tiwari S, Trivedi MC, Kolhe ML, Singh BK (eds) Advances in data and information sciences. Lecture Notes in Networks and Systems 522. Springer, Singapore
10. Daliento S, Di Napoli F, Guerriero P, d' Alessandro V, (2016) A modified bypass circuit for improved hot spot reliability of solar panels subject to partial shading. Sol Energy 134:211–218
11. Deng S, Zhang Z, Ju C (2017) Research on hot spot risk for high-efficiency solar module. Energy Procedia 130:77–86
12. Eder G, Voronko Y, Hirschl C, Ebner R, Újvári G, Mühleisen W (2018) Non-destructive failure detection and visualization of artificially and naturally aged PV modules. Energies 11:1053

13. Kretzmann JE (2016) Thermographic image processing toolbox for pulsed and transient thermography manual, 2nd ed. Wiley Pub, Indianapolis
14. Zhen (2016) Study of bypass diode reliability under non-uniform irradiance distribution on PV module surface. Acta Energiae Solaris Sinica. 37(2):355–360
15. Osarumen OO, Emeka HA, NN E, (2016) A review of photovoltaic module technologies for increased performance in tropical climate. Renew Sustain Energy Rev 75:1225–1238
16. Mrityunjay Patil M (2018) Interpolation techniques in image resampling
17. Zhou et al (2010) An edge-directed bicubic interpolation algorithm
18. Hafiz FA (2020) Monitoring and visualization of solar pv thermal flow via interpolation techniques. JEST J Energy Saf Technol 1(2):01–09
19. Han D (2020) Comparison of commonly used image interpolation methods. In: Proceedings of the 2nd international conference on computer science and electronics engineering (ICCSEE 2013)
20. Adrian Cotfas P (2018) Solar cell capacitance determination based on an RLC resonant circuit. Energies 2018(11):672
21. Fernandez MI, Go YI (2023) Power management scheme development for large-scale solar grid integration. J Elect Syst Info Tech 10(15) Springer Nature. https://doi.org/10.1186/s43067-023-00080-7
22. Alias ND, Go YI (2023) Decommissioning platforms to offshore solar system: road to green hydrogen production from seawater. Renew Energy Focus. ISSN 1755-0084. https://doi.org/10.1016/j.ref.2023.05.003
23. Citalingam K, Go YI (2022) Hybrid energy storage design and dispatch strategy evaluation with sensitivity analysis: techno-economi-cenvironmental assessment. Energy Storage 4:e353. https://doi.org/10.1002/est2.353
24. Faruhaan A, Go YI (2021) Energy storage sizing and enhanced dispatch strategy with temperature and safety considerations: a techno-economic analysis Energy Storage 3(6):1–33, e260. https://doi.org/10.1002/est2.260
25. Mohanan M, Go YI (2020) Optimized power system management scheme for LSS PV grid integration in Malaysia using reactive power compensation technique. Global Challenge. https://doi.org/10.1002/gch2.201900093

The Influence of Weather Regimes on Off-Grid Peer-To-Peer Energy Trading

A. S. Y. Ling and W. A. W. Z. Abidin

Abstract Standalone solar PV power system is being used as an option for electrification in remote areas around the world providing basic electricity needs. However, the approach suffers from power mismatch and energy efficiency issues. The peer-to-peer energy trading (P2PET) concept enables to encounter the issue of combining multiple standalone power systems to form an energy community. Energy transaction between peers is permitted and even sell energy to business for economic activities which maximizes the community's welfare from electrification. However, the generation of the solar PV system is intermittent due to its fluctuating nature which depends on solar power. This paper conducts an analysis of the influence of weather on P2PET, focusing on multiple standalone solar power system design specifications in rural Sarawak, Malaysia. The analysis is based on an off-grid community-based energy trading with a priority-based market clearing model under a combined multiple standalone solar PV system. Simulation studies are applied to perform the analysis of the research.

Keywords Rural electrification · Peer-to-peer energy trading · Solar PV · Energy transaction · Market clearing model

1 Introduction

Based on the report of the International Energy Agency (IEA), 993 million people around the world are still unable to access electricity and 73% of them live in rural areas [6]. Financial and geographical issues have restricted the potential of rural electrification with grid propagation for these countries [18]. A large investment is essential for grid extension to reach the rural area which is considered an uneconomical option as the low energy demand from a particular area [3]. To accomplish

A. S. Y. Ling (✉) · W. A. W. Z. Abidin
Department of Electrical and Electronic Engineering, Faculty of Engineering, Universiti Malaysia Sarawak, Kota Samarahan, Malaysia
e-mail: 19020164@siswa.unimas.my

C. S. Choo et al. (eds.), *Proceedings of ASEAN-Australian Engineering Congress (AAEC2022)*, Lecture Notes in Electrical Engineering 1072,
https://doi.org/10.1007/978-981-99-5547-3_21

Fig. 1 Installed solar PV systems at Longhouse in Sarawak

full electrification globally, renewable energy sources is the cheapest way to achieve, especially in rural area [6]. The renewable energy system is an outstanding solution for remote areas with low power demand areas due to the scalability of input power sources. The maturity of PV technologies and continuing decrease in the cost of solar PV over the years led to the usage of the grid and off-grid electricity increasing [1, 12].

For rural electrification of a remote area in Sarawak, the diesel generator is the common solution although it is costly and has difficulty with fuel transportation. The utilization of local energy resources such as solar, wind, and hydro brings significant benefits such as energy cost saving and greener energy [14]. In Sarawak, several rural electrification schemes utilizing renewable energy as primary sources have been introduced such as Sarawak Alternative Rural Electrification Scheme (SARES), standalone solar hydro hybrid station, Sarawak Energy's CSR Project [11, 19]. Those electrification schemes is aiming to supply electricity on a daily basis in rural communities with standalone solar PVs or integrated with micro-hydro. Figure 1 shows an example of a standalone solar PV system in rural Sarawak.

However, energy efficiency is still low in rural areas although implemented with renewable energy resources. Current electrification schemes either use energy-limiting meters to distribute energy from the standalone system or implement a mini-scale of solar PV system for each household. The energy allocation mechanism in the existing system leaves some surplus energy unused, preventing the system from been fully utilized [8]. Electrification does grow the economy in the long term by increasing the productivity and income of beneficiaries [13]. The existing schemes cannot support rural communities to run micro-businesses that required higher energy demand.

A community-based distributed generation paradigm should be identified as the future power systems paradigm in rural areas. The concept of community-based distributed generation is sharing energy among partners which brings more advantages than traditional distributed generation systems. One of the decent models for this concept is Peer-to-peer energy trading (P2PET) [10].

Due to the advancement of Information Communication Technologies (ICT) and Blockchain technologies, Decentralized P2PET architecture is the potential for future local energy trading in local markets and communities [4]. The consumer who owns a power generation system such as a solar PV system is considered a prosumer. P2PET is an effective market-driven bi-directional energy transactions system [17]. In P2PET

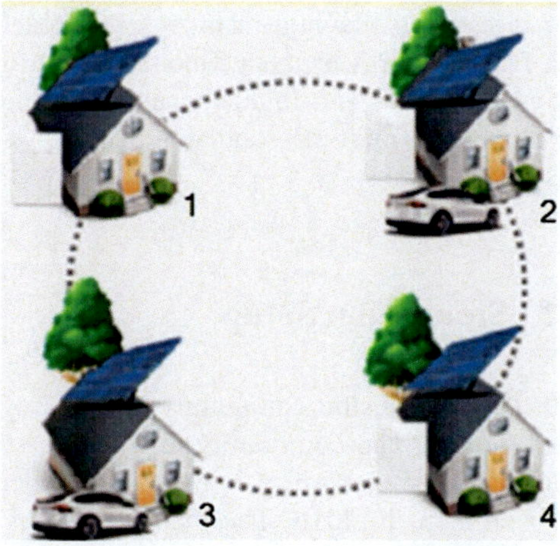

Fig. 2 P2P energy trading where prosumers are linked together to form a network

energy trading, the peers (Household 1, Household 2, Household 3, Household 4) are represented as energy traders who sell and buy energy directly without third parties such as electric utility companies shown in Fig. 2. For example, Household 1 can directly sell surplus energy to Household 4 through an energy network since Household 1 has surplus energy and Household 4 is willing to buy at an acceptable price. Thus, P2PET does reduce costs for energy consumers and brings profit for energy producers through energy transactions.

The architecture of P2PET energy trading can combine solar PV or microgrid as a peer of the network. Since P2PET energy trading is a community-based distributed power system, the energy generation system is belonging to the peers themselves. Besides, P2PET energy trading might be stimulating the growth of the economy in rural areas since P2PET energy trading aggregate the energy resources from peers and improve standards of power for higher energy loads. Thus, socio-economic activities which have high power loads can be carried out by trading the energy with the community members. The involvement of socio-economic activities such as food processing, packaging and storage, school facilities, and agriculture activities in energy trading creates economic value for the community and brings revenues to the community through P2PET and business [9].

Many studies have been published on the application of P2PET system for the urban, grid-connected microgrid. In [2], industrial P2PET energy trading among charging stations of electric vehicles and business buildings with PV solar generation is proposed with a dynamic pricing mechanism. Game theory is frequently used by the researcher for interaction between buyers and sellers in energy trading. The author [15] combines multiple microgrids to form a Stackelberg game-based energy trading with power network constraint consideration. In [7, 16], the power network usage fee is considered in-game theory-based energy trading. The author [16] also focuses on social welfare maximization and fairness of profit allocation in the pricing mechanism. Community-based energy trading introduced by Paudel and Beng [10] focuses on cost saving and proves the impact of the implementation of energy trading. A P2PET energy exchange model of microgrid for rural electrification is introduced by Harish et al. [5]. In this paper, the studies on the influence of weather on P2PET is based on multiple standalone solar power system design specification in rural Sarawak, Malaysia.

2 Simulation Setup

In the studies, the community consists of 18 members: three consumers and 15 prosumers. The community in the case study as shown in Fig. 3 is simulated at Rumah Manggat which is electrified under a rural electrification scheme from the government in 2016. The main source of income of the community members is agriculture, fishing, and forestry. Under the scheme, each of the members has been provided with a small-scale solar PV system. Thus, excess energy cannot be utilized since those systems are standalone.

By applying the P2PET concept, multiple solar PV systems are linked together. We proposed that the energy can support any productive activity from agriculture and rural industries such as post-harvest processing in fruit and vegetable, irrigation, and milling. This feature of P2PET enables the improvement of rural welfare and poverty reduction. A range of micro-businesses such as kiosks, bakeries, and halls can also be achievable by setting them as consumers in energy trading. In the case study, we simulated two small workshops and one food processing center as the consumer in the community. The load consumption profile of the community members is generated according to the electrical appliances assumed in the design specification as shown in Table 1. The different pattern of load consumption is generated based on the member's nature of job and business and the wealth of the family.

Table 2 explains the assumption of different load profiles used in the simulation. The load consumption pattern from the community members is shown in Fig. 6. Energy for the basic needs of the community members is mostly consumed for cooking, lighting, and other electric appliances. In the case study, the load consumption behavior of the community members is based on the "normal" type of load. For the workshop load profile, the consumption pattern assumes machinery operation. The power is assumed to supply two machines at once. Besides, a food processing

Fig. 3 Solar home project at Rumah Manggat, Batang Ai

Table 1 Assumption of electrical appliances in design specification

No	Electrical appliances	Unit	Watt
1	Bulbs	5	11
2	Television with video player	1	105
3	Fan	1	38
4	Rice cooker	1	500
5	Freezer	1	100

center with freezers is simulated as a consumer in the case. The freezers need to operate the whole day.

The prosumers in the community are assumed to own an individual 1 kW solar PV generator with two days of autonomy which is a 12 kWh of battery storage with a limit discharge SOC of 10%. The solar PV generation graph is stated in Fig. 4. The efficiency of the charging and discharging of the battery is set to 95%. The prosumers are set to have the capability to control the battery charging and discharging and PV generation. Besides, the power capacity of the system is limited to 700 W to prevent system overloading. Since there are 18 families in the community, the simulated power system is an 18-bus radial distribution system which is in Fig. 7.

In Table 3, each community member's data is displayed. The table depicts the strategy of members, locations in buses, and load profile. The load profile depends on the trading strategy selected by the member. Figure 5 shows that all community members are prosumers who participate in P2PET with a selected strategy based on Table 3. Members 9 and 10 are the light workshop that operates some machinery in the community so the load profile is set as a workshop while member 11 is the central cold storage to store meat for business.

The completion of peer-to-peer energy trading depends on the strategy chosen by the peers involved. There are four types of strategy options:

- DM for peers who are concerned about the availability of electricity supply and need more energy.
- SS is for those interested in self-sufficiency and selling surplus energy for profit.

Table 2 Load profile assumption explanation

Type of load	Operation hour	Energy usage	Description
Normal	Whole day	Cooking, lighting, electric appliances	Assume the family goes out to work in the early morning, rests in the afternoon (at home), and go work until evening. Electric mainly used for cooking, TV, and lighting
Workshop	Working hours (8–5)	For machinery tools	Use to operate the machine when working hours but a rest period at (12–13)
Freezer	Whole day	For freezer	Freezer required to operate whole day long, increased power is used for cooling after opening the freezer

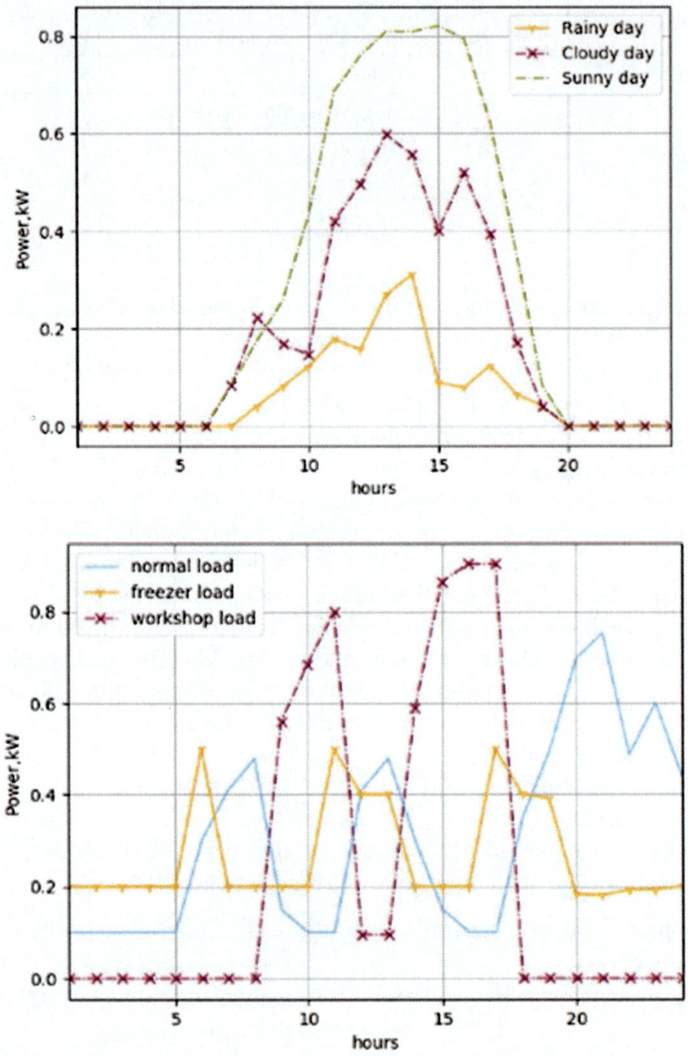

Fig. 4 PV generation and different load profile patterns according to the strategy selected

- AM for peers who prioritize profit earning, even if it means sacrificing personal usage.
- BC for buying consumers who need to purchase energy from the trading community.

Each strategy reflects a different preference and willingness of the peers, and the choice of strategy can determine the success of the energy trading venture.

Table 3 Community member data

Member	Information			Member	Information		
	Strategy	Bus	Load profile		Strategy	Bus	Load profile
1(P)	SS	1	Normal	11(C)	BC	11	Freezer
2(P)	AM	2	Normal	12(P)	AM	12	Normal
3(P)	SS	3	Normal	13(P)	SS	13	Normal
4(P)	AM	4	Normal	14(P)	SS	14	Normal
5(P)	AM	5	Normal	15(P)	AM	15	Normal
6(P)	DM	6	Normal	16(P)	AM	16	Normal
7(P)	AM	7	Normal	17(P)	AM	17	Normal
8(P)	SS	8	Normal	18(P)	DM	18	Normal
9(C)	BC	9	Workshop				
10(C)	BC	10	Workshop				

Note (P) Prosumer, (C) Consumer

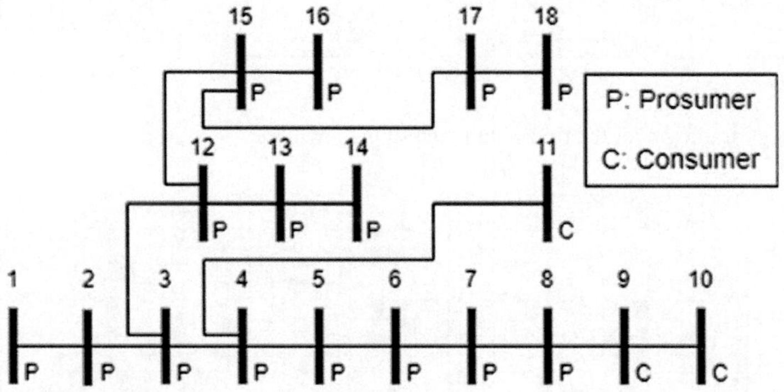

Fig. 5 18-bus radial distribution test system [7]

3 Results and Discussion

The implementation of the proposed trading system is based on standalone solar PV power systems in rural to improve energy utilization. As solar power is considered the primary energy source, weather becomes a critical factor that affects the stability of operation. Because solar PV power systems convert sunlight directly into electricity which means the system is high dependence on solar radiation. The change of weather is one of the factors that varies the solar radiation from the sun to earth due to the cloudiness of the sky. Therefore, the analysis of energy trading between different weather is done. In this case, the energy trading is simulated on sunny days, cloudy days, and rainy days. The difference in the weather within simulation affects the solar energy generation efficiency. This variability manifests in the generation profile of solar PV systems under different weather. The simulation results of three different

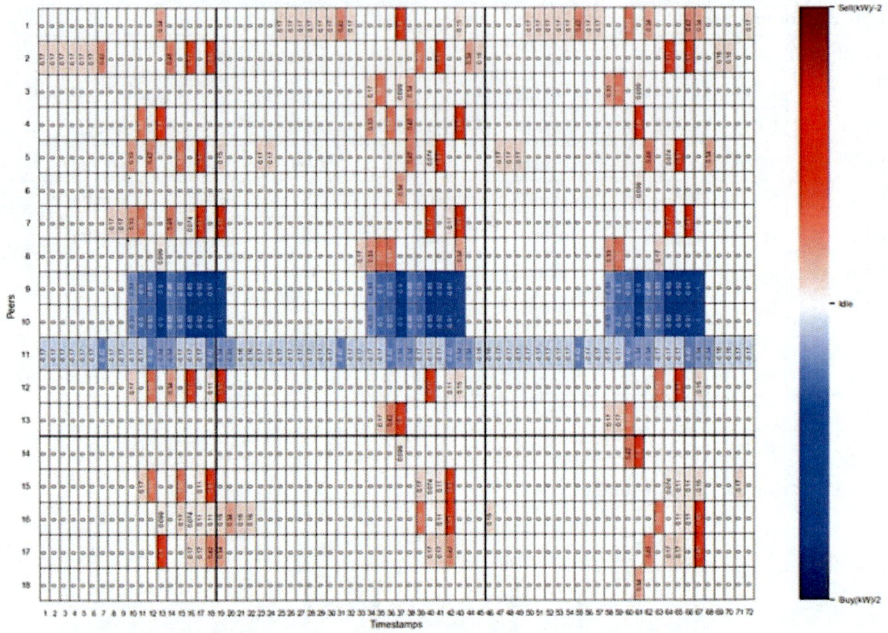

Fig. 6 Energy trading over the timestamp on a sunny day

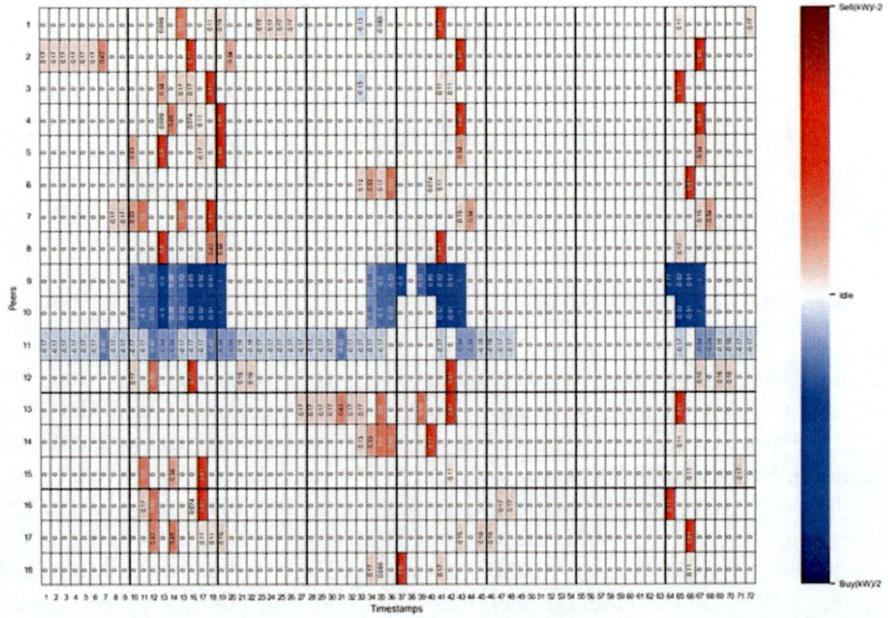

Fig. 7 3 days of energy trading over the timestamp in cloudy days

types of weather are compared in terms of profit and trading frequency. Energy trading under sunny days and cloudy day based on the proposed system model with 72 h (3 days) is shown in Figs. 6 and 7, respectively. The trading is not formed during the 40th to 63rd hour. In Fig. 8, energy trading under a rainy day based on the proposed system model with 48 h (2 days) is shown. After the 21st hour of trading on a rainy day, the energy trading is suspended due to the absence of the seller due to the weather conditions.

The total frequency of trade among the peers in different weather is compared in Fig. 9. From Fig. 9, consumers (peers 9, 10, and 11) trade the most among the peers in the simulation because they only can consume electricity by buying from prosumers. Based on the simulated results, the prosumers buy energy several times only. They mostly sell energy when involved in energy trading due to the effect of selected strategies and priority-based market clearing mechanisms.

In Fig. 10, The total trading frequency throughout the timestamp is compared with the total generated energy by the system in different weather. Based on Figs. 9 and 10, the sequence of trading frequency, from highest to lowest is arranged as follows: sunny days, cloudy days, and rainy days. This can be explained by the facts that the trading frequency has a strong relationship with the total generated energy from the solar PV system. In the proposed system, the forming of the trading is depending on the availability of the energy stored in the battery. Therefore, the reduction of the energy production of solar PV systems will reduce the energy in

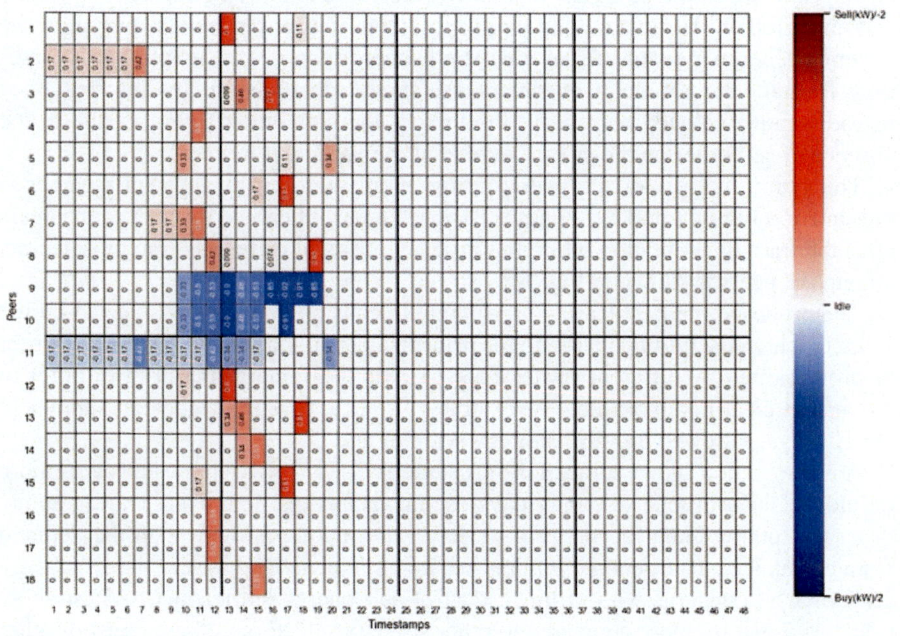

Fig. 8 2 days of energy trading over the timestamp on rainy days

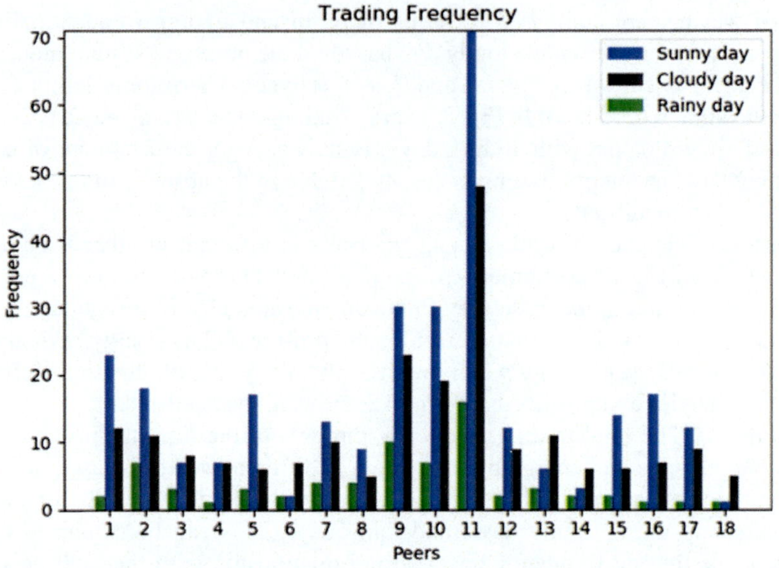

Fig. 9 Comparision of trading frequency of community members in different weather

a battery that can be traded after personal energy consumption. So the amount of energy that can be traded on the market will decrease and even go to zero. During a certain period on cloudy days and rainy days, there is an absence of trading since the generated energy is enough to sustain the prosumer's energy consumption and conditions for trade are not met. Consumers also show a decrement in trading frequency in bad weather (cloudy and rainy) throughout the simulation in Fig. 9 due to the absence of sell order on the energy market at the particular moment.

The profit analysis of the peers during energy trading in different weather is presented in Figs. 11 and 12. Figure 11 shows the profit earned by each of the peers after the trading simulation while the profit analysis of different strategy selectors among the peers is shown in Fig. 12.

The proposed market clearing algorithm depends on the strategy selected by the seller to match the trading pair with the prioritized energy buyer. In Fig. 12, the order of profit earned by different strategy selectors is the same as the priority order in the market clearing algorithm. "AM" strategy selector has the highest in sunny and cloudy days, followed by "SS" strategy selector and DM.

A phenomenon shows some of the prosumers (3,6,13,14,18) trade and earns more on cloudy days instead of sunny days as shown in Figs. 9 and 11. During sunny days, the total amount of energy for sale on the market is more than the demand from consumers. Thus, energy trading will be more beneficial for the "AM" strategy selector who concerns with profit instead of self-energy sufficient. "AM" strategy selectors will earn more compared to other strategy selectors as the interest of other strategy selectors is not concerned with profit earning. However, less energy is generated on cloudy or rainy days compared to sunny days. The viability of the surplus

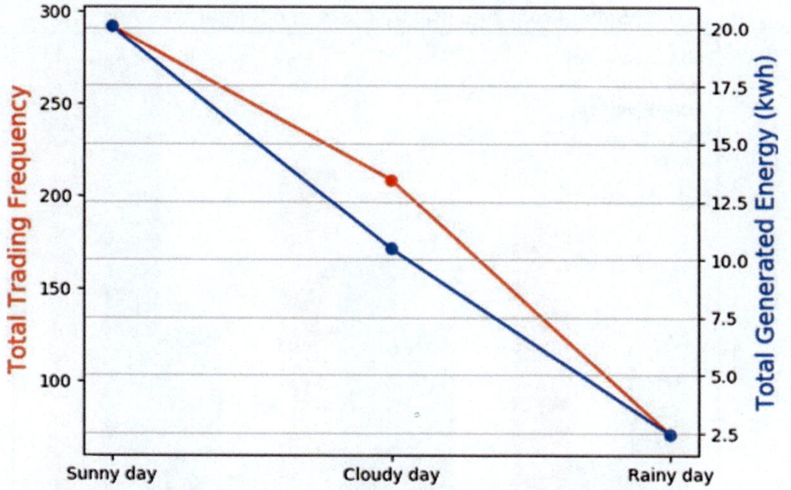

Fig. 10 Total generated energy versus trading frequency in different weather

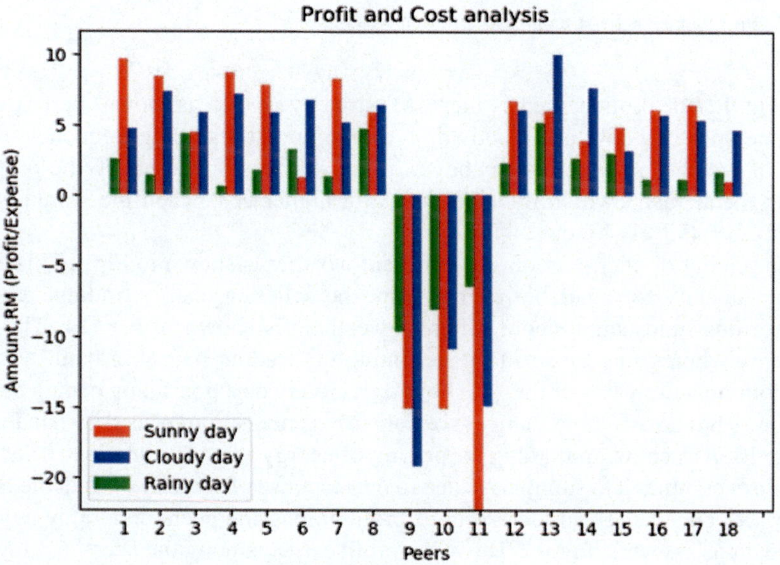

Fig. 11 Profit analysis of peers in different weather

energy of each peer will be less. This causes other strategy selectors to have a greater chance to form the trading pair with the seller as the "AM" strategy selector already sold their surplus. So "SS" and "DM" strategy selectors sell more frequently on cloudy days because they have surplus energy to sell and have the chance to sell.

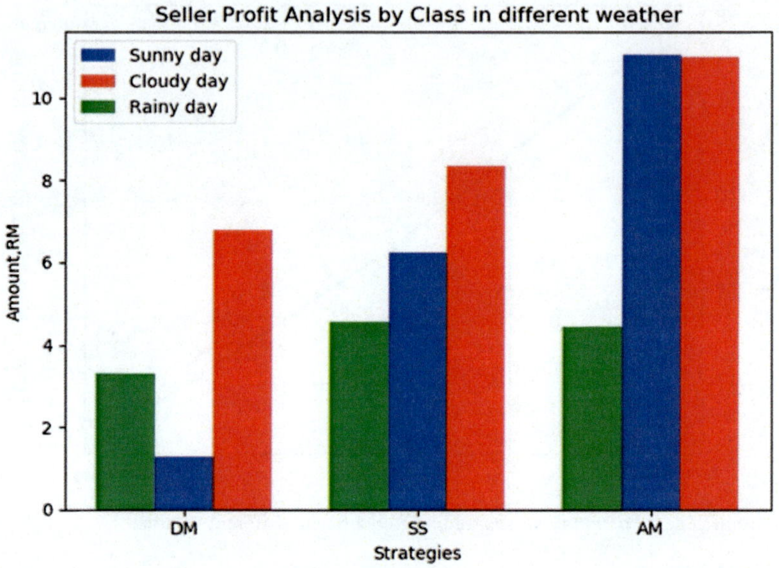

Fig. 12 Profit analysis by strategies in different weather

In Fig. 12, the profit earned by the "AM" strategy selector is shown approximately the same on sunny days and cloudy days even though the trading frequency in both kinds of weather is different as shown in Fig. 9. This is caused by the proposed pricing mechanism, which trades energy at a higher cost when the stored energy capacity is relatively low.

The pricing of energy trading in different weather is shown in Fig. 13. The total battery capacity for available energy from the seller in energy trading, recorded over various timestamps under different weathers is shown in Fig. 14. The price of energy when going to zero to representing not trading pair is formed. After the 20th hour on a rainy day or the 48th hour on a cloudy day, no trading occurs because the battery has no available energy to be sold in the energy market as shown in Fig. 14. In Fig. 13, it is showing that higher pricing of energy is traded on cloudy days and rainy days compared to sunny days due to the response of scarcity of tradable energy in Fig. 14. This causes an increment of profit among the peers on cloudy days and rainy days as shown in Fig. 12. However, profit earned among the peers in rainy days is relatively low than in other weathers due to low frequency in energy trading and low surplus energy.

Besides, the tradable energy available in the energy market is the surplus energy that is not consumed by prosumers when generated by the solar PV system. Its performance will vary depending on the weather. In analyzing different weather conditions, we analyzed that solar radiation that varies from different weather affects energy trading. When energy generated from the solar panel is decreased until the

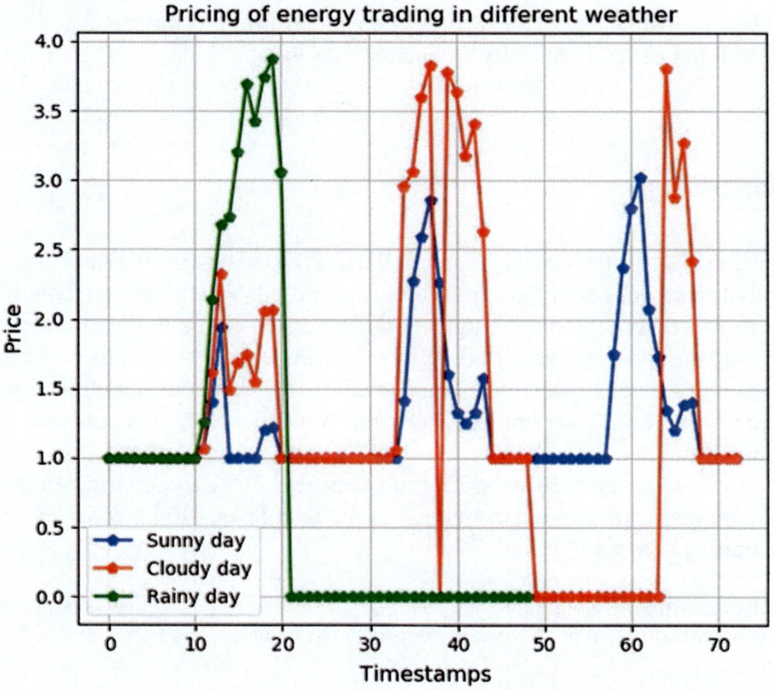

Fig. 13 Pricing over the timestamp of the energy trading in different weather

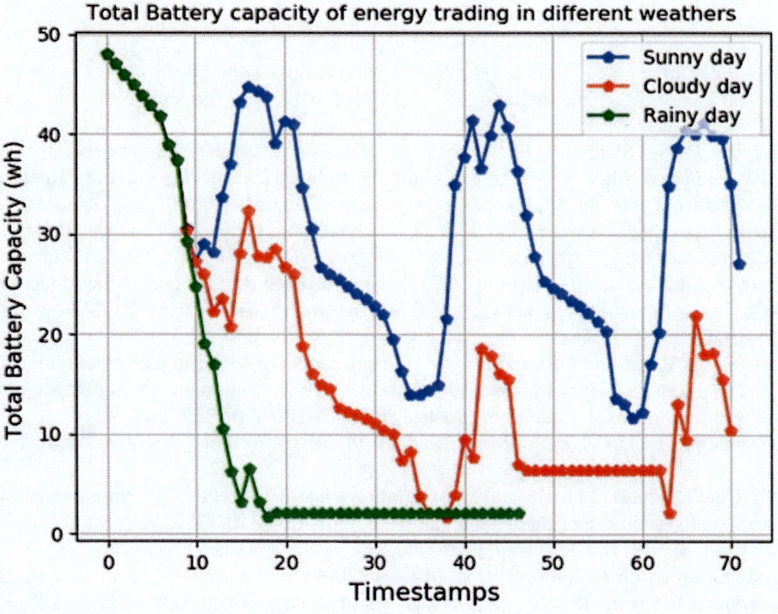

Fig. 14 Total battery capacity of energy trading over the timestamp in different weathers

system cannot sustain the personal energy consumption of prosumers fully, the energy trading will not be held until surplus energy is available.

4 Conclusion

In this paper, peer-to-peer energy trading (P2PET) based on existing rural electrification with a standalone solar power system is proposed. The result based on the case studies proves that the proposed algorithm ensures the community has better social welfare and profit from energy trading when selling energy to the business entity in the community. The proposed market clearing model that considers limited energy allocation based on ESP operation is proven to maximize the benefits received by the community.

For future work, power forecasting of renewable DER and distributed methods of data communication and management in the data layer will be considered in the energy trading solution.

Acknowledgements All contributing research grants and organizations to the work presented in the full paper must be appropriately acknowledged by the authors.

References

1. Ahmad Z, Halyani MY, Norain I et al (2015) Stand alone solar photovoltaic for rural electrification: a case study in Kampung SG LAH Tapah Malaysia. J Teknol 76:65–69. https://doi.org/10.11113/jt.v76.5487
2. Aznavi S, Fajri P, Shadmand MB, Khoshkbar-Sadigh A (2020) Peer-to-peer operation strategy of PV equipped office buildings and charging stations considering electric vehicle energy pricing. IEEE Trans Ind Appl 56:5848–5857. https://doi.org/10.1109/TIA.2020.2990585
3. Díaz P, Arias CA, Peña R, Sandoval D (2010) FAR from the grid: a rural electrification field study. Renew Energy 35:2829–2834. https://doi.org/10.1016/j.renene.2010.05.005
4. Guerrero J, Chapman AC, Verbic G (2019) Decentralized P2P energy trading under network constraints in a low-voltage network. IEEE Trans Smart Grid 10:5163–5173. https://doi.org/10.1109/TSG.2018.2878445
5. Harish VSK V, Anwer N, Kumar A (2019) Development of a Peer to peer electricity exchange model in micro grids for rural electrification. In: 2019 2nd international conference on power energy, environment and intelligent control (PEEIC). IEEE, pp 259–263
6. International Energy Agency (2018) 2018 world energy outlook: executive summary. Oecd/Iea
7. Jin Y, Choi J, Won D (2020) Pricing and operation strategy for peer-to-peer energy trading using distribution system usage charge and game theoretic model. IEEE Access 8:137720–137730. https://doi.org/10.1109/ACCESS.2020.3011400
8. Kirubi C, Jacobson A, Kammen DM, Mills A (2009) Community-based electric micro-grids can contribute to rural development: evidence from Kenya. World Dev 37:1208–1221. https://doi.org/10.1016/j.worlddev.2008.11.005

9. Mandelli S, Barbieri J, Mereu R, Colombo E (2016) Off-grid systems for rural electrification in developing countries: definitions, classification and a comprehensive literature review. Renew Sustain Energy Rev 58:1621–1646. https://doi.org/10.1016/j.rser.2015.12.338
10. Paudel A, Beng GH (2018) A hierarchical peer-to-peer energy trading in community microgrid distribution systems. In: 2018 IEEE power & energy society general meeting (PESGM). IEEE, pp 1–5
11. Sarawak Energy Berhad (2019) Power to grow—lighting up rural Sarawak
12. Singh GK (2013) Solar power generation by PV (photovoltaic) technology: a review. Energy 53:1–13. https://doi.org/10.1016/j.energy.2013.02.057
13. Torero M (2015) The impact of rural electrification: challenges and ways forward. Rev Econ Dev 23:49. https://doi.org/10.3917/edd.hs03.0049
14. Balaji V, Sekar K, Duraisamy V et al (2015) Performance analysis of energy management controller for stand alone solar power generation system using soft computing techniques. J Teknol 76:111–117. https://doi.org/10.11113/jt.v76.5889
15. Yan M, Shahidehpour M, Paaso A et al (2021) Distribution network-constrained optimization of peer-to-peer transactive energy trading among multi-microgrids. IEEE Trans Smart Grid 12:1033–1047. https://doi.org/10.1109/TSG.2020.3032889
16. Zhong W, Xie S, Xie K et al (2020) Cooperative P2P energy trading in active distribution networks: an MILP-based nash bargaining solution. IEEE Trans Smart Grid 1–1:https://doi.org/10.1109/TSG.2020.3031013
17. Zia MF, Benbouzid M, Elbouchikhi E et al (2020) Microgrid transactive energy: review, architectures, distributed ledger technologies, and market analysis. IEEE Access 8:19410–19432. https://doi.org/10.1109/ACCESS.2020.2968402
18. Exploring Rural Electrification Challenges in ASEAN I Energy Watch : Global & Regional Energy Insights, Thought leadership, & Conversations. https://www.energywatch.com.my/blog/2018/04/24/exploring-rural-electrification-challenges-in-asean/. Accessed 14 Nov 2019a
19. CSR Project in Batang Ai: Solar Home System for Rumah Kino and Rumah Manggat I Power to Grow. https://sarawakenergy.wordpress.com/2015/01/05/csr-project-in-batang-ai-solar-home-system-for-rumah-kino-and-rumah-manggat-2/. Accessed 14 Apr 2021b

UAV-Based Lane-Changing Reduction Using Traffic Video Analyser

Jit Boon Bong, Elizabeth Eu Mee Chong, and Chung Siung Choo

Abstract Multi-lane roundabouts allow drivers to change lanes by merging and diverging among circulating lanes, yielding weaving in the roundabout. Weaving yields traffic conflicts as crossing traffic slows down the circulating flow. Speed reduction creates significant gaps that reduce capacity. Thus, weaving is crucial due to its significant effect on roundabout capacity. However, little information is available on weaving in multi-lane roundabouts with inscribed circle diameters (ICD) of more than 100 m. A possible reason for the lack of weaving data is the difficulty of collecting lane change data due to the traffic flow complexity, which involves several lane changes and the required resources. Typical data collection methods include using ground-level recorders and surveyors. Ground-level recorders could only provide limited data points due to view coverage constraints. Increasing the workforce and equipment can overcome this limitation, but it may prolong the post-recording analysis due to more views for analysis. Conversely, Unmanned Aerial Vehicles (UAVs) can capture complex traffic interaction data in a single session. Extraction of UAV-based recording by traffic video analyser could provide a wide range of outputs in a shorter time than manual analysis. This paper proposes a novel methodology to extract UAV-based lane change data using a traffic video analyser. The fieldwork involves flying a UAV at a multi-lane roundabout with an ICD of 151 m in Kuching, Malaysia. Turning movement and lane change were obtained using a traffic video analyser. Virtual gates are specified at 10 m intervals to get lane change location. A lane change can be detected when a vehicle crosses two consecutive virtual gates over two adjacent circulating lanes. Weaving occurs when two lane change swap paths along circulating lanes. The findings show that 86% of lane change happens in the weaving section. This paper demonstrated that using UAV and traffic video analyser is feasible to investigate lane change in a multi-lane roundabout with ICDs more than 100 m. Since using a traffic video analyser on UAV-based data gives a promising outcome, it is worth considering this methodology in similar lane change studies in the traffic field.

J. B. Bong (✉) · E. E. M. Chong · C. S. Choo
Faculty of Engineering, Computing and Science, Swinburne University of Technology, Kuching, Sarawak, Malaysia
e-mail: jbbong@swinburne.edu.my

C. S. Choo et al. (eds.), *Proceedings of ASEAN-Australian Engineering Congress (AAEC2022)*, Lecture Notes in Electrical Engineering 1072,
https://doi.org/10.1007/978-981-99-5547-3_22

Keywords Lane change · Weaving · Large multi-lane roundabouts · UAV · Traffic video analyser

1 Introduction

Weaving is crossing two or more traffic streams (in the same direction) without the aid of a traffic control device [4–6]. Weaving occurs in multi-lane roundabouts due to more movement freedom, allowing vehicles to swap paths [2, 11]. Crossing traffic slows circulating flow, creating more significant gaps between successive vehicles in the weaving section [12]. The gap enlargement causes fewer vehicles to pass through the weaving section gaps, leading to a failure to achieve roundabout design capacity [13]. Therefore, weaving is crucial in the capacity analysis due to its significant effect.

However, little information is available on collecting weaving data in multi-lane roundabouts with ICDs more than 100 m [9] set up a Portable Vision-Based Traffic Analyzer (PVBTA) to collect weaving section flow for two hours. The researchers then adopted Semi-Automatic Video Analyser (SAVA) as an image processing tool to identify conflicting (i.e., crossing of two vehicles) and non-conflicting flows (i.e., vehicles remain car-following states). The SAVA tools also help extract merging gaps in the weaving section. The limitation is that the work adopted only one weaving section from a conventional roundabout with an inscribed circle diameter (ICD) of 89 m. Weaving in a multi-lane roundabout with ICD exceeding 100 m remains unclear.

Wang and Yang [12] collected data from eight roundabouts (the work did not mention ICD) using surveyors and video recorders. Fieldwork at each roundabout involved three surveyors: one videotaping the traffic flow, one collecting vehicle velocity utilising a radar gun, and one collecting geometry using a laser meter. Compared to the work by Md Diah et al. [9], this research involves about six hours of data collection each day. The researcher adopted Autoscope for volume counting. However, it did not further explain much about the post-recording analysis. Furthermore, this preliminary work did not specify the extraction of weaving data from the recording, although the article aims to develop a weaving parameters capacity model.

A possible reason for the lack of weaving data is the difficulty of collecting lane change data due to the resource required. Typical data collection methods involve using fixed-level recorders or surveyors. Fixed-level video recorders could only provide limited data points due to coverage constraints of equipment. Placing equipment at higher points (e.g., installing equipment at a pole in work by Md Diah et al. [9] or a tall building in work by Wang and Yang [12] could improve the recorder's view. However, an additional instrument is required, or the study site should have tall buildings to upgrade from ground-level to bird-eye views. Otherwise, more workforce and equipment are required to increase ground-level data points, leading to labour and cost-effectiveness issues.

Furthermore, this technique is challenging to collect complex traffic flow (i.e., weaving), which involves multiple lane changes. Besides cost issues, adopting more equipment may cause more prolonged post-recording analysis due to more views to cover. From the workforce perspective, more in-situ surveyors must take care of each small section (i.e.,10 m interval). Otherwise, one surveyor in charge of collecting lane change data over a significant strength will lower data accuracy due to human error caused by quick lane-changing manoeuvres. Internal experiments such as instrumented probe vehicles and driving simulators may not provide sufficient data sample size unless more sessions are involved.

On the other hand, Unmanned Aerial Vehicles (UAVs) have the potential to overcome the limitations of the traditional approaches due to their mobility, flexibility and bird-eye view advantages. One UAV can capture more information (e.g., traffic flow and speed) which generally require at least two surveyors to complete the task. Also, one UAV can provide continuous traffic interaction data (e.g., lane change and car following), which a single video recorder can hardly achieve. Traffic video analysers use image processing techniques to identify and track vehicles from UAV recordings. Most analysers allow offline data editing, enabling users to work offline to extract advanced traffic parameters (e.g., traffic density, vehicle speed, origin–destination count and safety analysis). Such functions ease data analysis by reducing manual work (e.g., hand-count traffic volume) that carries human error. Most analysers also generate output in standard video format (e.g., mp4, mpeg and avi) to ease viewing of the outcome. Given that UAVs and traffic video analysers benefit from data collection and analysis, combining these two could be a promising technique to provide comprehensive traffic data from various road infrastructures. Table 1 summarises some works done using UAVs and traffic video analysers for traffic studies.

2 Methodology

This section describes a methodology for data collection using a UAV and data analysis using a traffic video analyser, Datafromsky.

2.1 Fieldwork Using UAV

The fieldwork was conducted on a sunny Tuesday (12/10/2021) at a multi-lane roundabout with an ICD of 151 m. Two sessions were made available to collect traffic flow for the morning (0630 AM–0830AM) and evening (0430PM–0630PM) peak hours using a UAV. The fieldwork required only one surveyor to fly the UAV. The surveyor must ensure the UAV can capture whole roundabout layouts and traffic flows in and out from the roundabout. Figure 1 shows the study site (the red dot indicates the surveyor's location). The drone could not fly directly above the roundabout due to

Table 1 Some works use UAVs and traffic video analysers for traffic studies

References	Hardware	Software	Findings
[10]	Micro-drone for four propellers	The work adopted Tracker to extract traffic volume and speed data	The work provides a close capacity result with a macroscopic model (−0.5%)
[3]	Radio-controlled quadcopter UAV	The work adopted the Camshift to track vehicles and draw vehicle trajectories	The work provides accurate vehicle trajectories for traffic performance analysis
[7]	UAV	The work adopted Kanade–Lucas–Tomasi (KLT) tracker to extract speed and traffic density	The proposed method has about 96 and 87% accuracy in estimating the traffic stream's average speed and vehicle count
[8]	Argus-One UAV	The work adopted the OpenCV library in C++ to extract vehicle trajectories	The work proposed a streamlined methodology for extracting trajectory from video data
[1]	UAV	The work adopted Wintics to extract traffic counts	The outcome shows low relative error (10%)

signal and wind speed issues. Therefore, the surveyor flew the drone at an angle of 17.02 degrees (Fig. 2) to capture the whole roundabout layout (Fig. 3).

A drone's battery can last 30 min (i.e., 25 min of recording and 5 min of battery change). Therefore, this research utilised five batteries to achieve a two-hours fieldwork session. Table 2 shows the specification of the drone used.

2.2 *Post-recording Analysis Using Traffic Video Analyser*

The raw recordings are first uploaded to the cloud-based platform to obtain vehicle trajectories. With the virtual gate function, vehicle trajectories can provide turning movement and lane change in circulating lanes data. In this project, virtual gates are set at (1) entry and exit lanes to detect vehicle turning movements that lead to weaving in the roundabout and (2) circulating lanes to detect lane change. Lane-based virtual gates are set at roundabout entries and exits (Fig. 4). Procedures below describe virtual gate set up in the circulating lane:

1. Fix the reference point in the central island.
2. Divide the roundabout section into weaving and non-weaving section.
3. Measure the radius from the reference point to the outer edge of the curb.
4. Adopt an arc length of 10 m to represent a 10 m interval.
5. Using the arc length formula, determine the radian angle with an arc length of 10 m.
6. Draw the second line to create the radian angle determined.

Fig. 1 The study site

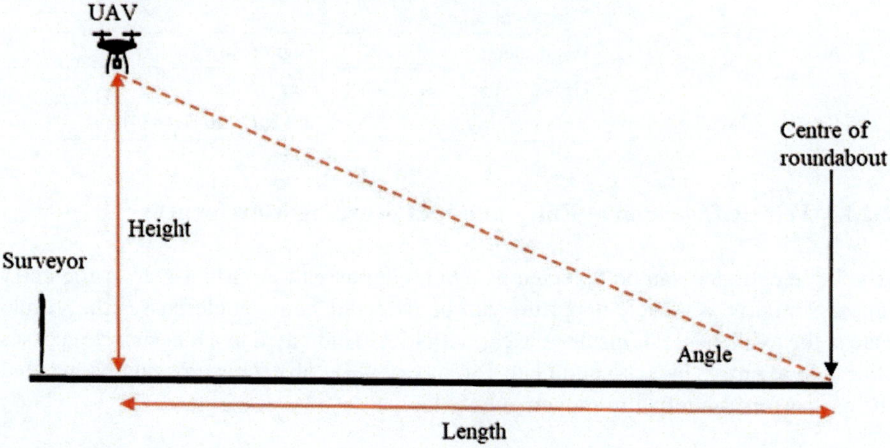

Fig. 2 The flying angle from the centre of a roundabout

Fig. 3 The view captured by the drone

Table 2 Specifications of drone used

Characteristics	Specification
Model	Mavic Mini
Takeoff weight	249 g
Power	LiPo 2S
Max flight time	30 min
Wind speed	Eight m/s
GPS	Yes
Video quality	2.7 K (2720 × 1530)

2.2.1 Vehicle Detection at Entry and Exit (Turning Movement)

A vehicle entrance can be detected as a vehicle passes the virtual gate at the entry lanes. Similarly, a vehicle departure can be detected as a vehicle passes the virtual gate at the exit lanes. For instance, a car with id:39 (indicated in a black circle) passes Gate 145 at entry (Fig. 5a) and Gate 156 at exit (Fig. 5b). This crossing shows that the car performs a left-turn movement.

2.2.2 Lane Change Detection in Circulating Lanes

Detection of lane change could be complex in a three-lane roundabout. A vehicle could change lanes from inner to middle lane, middle to outer lane, inner to outer lane or vice versa. This work adopts virtual gates of a 10 m interval to collect lane changes and their location. As shown in Fig. 6, a lane change can be detected when

Fig. 4 Lane-based virtual gates at entries and exits

a vehicle crosses two consecutive virtual gates over two adjacent circulating lanes (e.g., from Gate 8 to Gate 12).

The collection of lane change is intervals-based rather than a specific point along circulating lanes. Several possible lane-changing scenarios are listed in Table 3, using Fig. 4 as an example (take Gates 4, 5 and 6 as 0 m intervals; Gates 7, 8 and 9 as 10 m and Gates 10, 11 and 12 as 20 m).

Lane-changing scenarios from inner to outer or outer to inner may be less likely to happen within 10 m intervals (e.g., Cases 5 and 9). A driver may find it hard to cross two lanes within 10 m using an acceptable gap. The likely scenario is that a driver may seek a gap and change lanes within the first 10 m, then seek another gap and change lanes to the destination lane, as illustrated in Fig. 7.

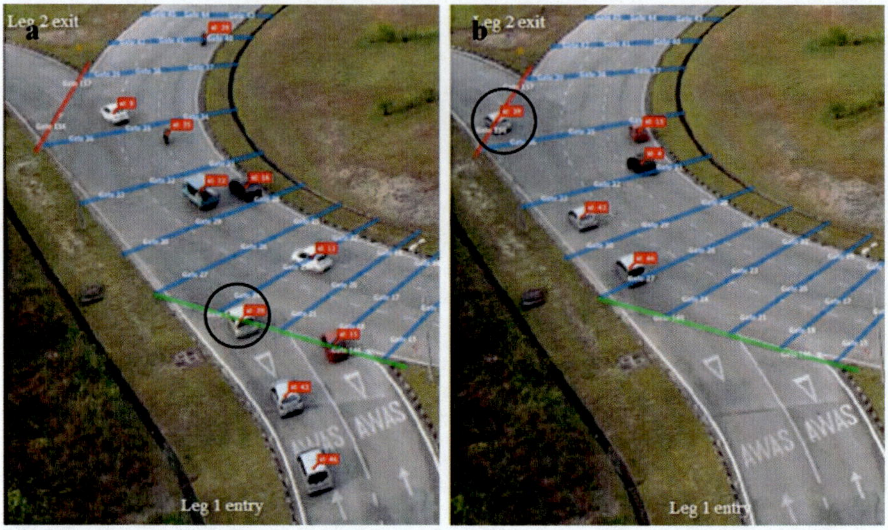

Fig. 5 **a** Vehicle entrance detection, **b** Vehicle departure detection

Table 3 Possible lane-changing scenarios

No	Scenario	Detection
1	A vehicle crosses the gates less than 10 m	Lane change occurs between 0 to 10 m
2	A vehicle crosses the gates more than 10 m	Lane change occurs between 10 to 20 m
3	A vehicle crosses Gate 4 and Gate 7	No lane change at 10 m intervals
4	A vehicle crosses Gate 4 and Gate 8	Vehicles change lanes from the inner to the middle lane at 10 m intervals
5	A vehicle crosses Gate 4 and Gate 9[a]	Vehicles change lanes from the inner to the outer lane at 10 m intervals
6	A vehicle crosses Gate 5 and Gate 7	Vehicles change lanes from the middle to the inner lane at 10 m intervals
7	A vehicle crosses Gate 5 and Gate 8	No lane change at 10 m intervals
8	A vehicle crosses Gate 5 and Gate 9	Vehicles change lanes from the middle to the outer lane at 10 m intervals
9	A vehicle crosses Gate 6 and Gate 7*	Vehicles change lanes from the outer to the inner lane at 10 m intervals
10	A vehicle crosses Gate 6 and Gate 8	Vehicles change lanes from the outer to the middle lane at 10 m intervals
11	A vehicle crosses Gate 6 and Gate 9	No lane change at 10 m intervals

Remarks [a]indicates an unlikely case

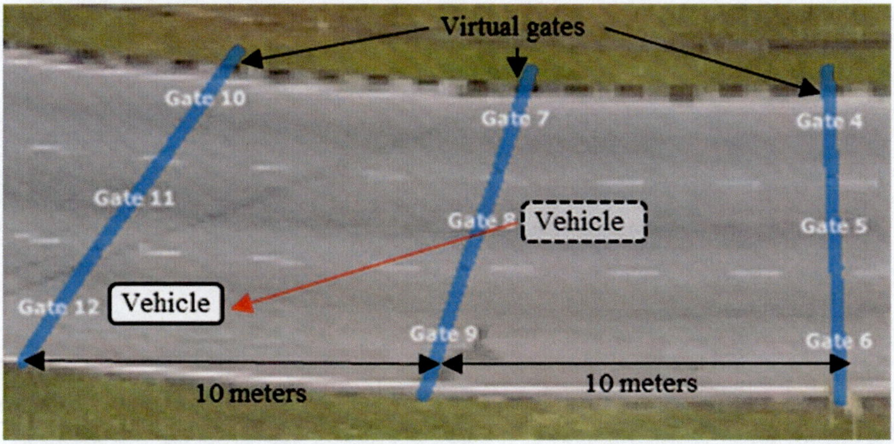

Fig. 6 Detection of a lane change in circulating lanes

3 Results and Discussion

This section discusses weaving in a roundabout followed by lane change location along circulating lanes.

3.1 Weaving in the Roundabout

Weaving occurs when two traffic streams swap paths and cross with each other in a weaving section. For entry flows (i.e., the red line in Fig. 8), through (T), right-turn (R), and U-turn (U) movements from different legs must make at least one require lane change, assuming they enter the weaving section using the middle or inner lane. Similarly, circulating flow (i.e., the blue line in Fig. 8) must change at least one lane, assuming they leave the roundabout using the inner exit lane.

This four-legged roundabout consists of four interconnected weaving sections. A vehicle may pass more than one weaving section if it performs T, R or U movements. Therefore, each weaving section caters to three movements from all legs. Table 4 summarises weaving flows in a roundabout.

Each weaving section (e.g., weaving section between Legs 1 and 2) accommodates all movements from upstream entry (1L, 1T, 1R and 1U), a U movement from the next approach (2U), R and U movements from opposite approach (3R and 3U) and T, R and U movements from previous approach (4T, 4R and 4U). Table 5 shows the weaving section flow of all sections, determined using the technique mentioned under Sect. 2.2.1.

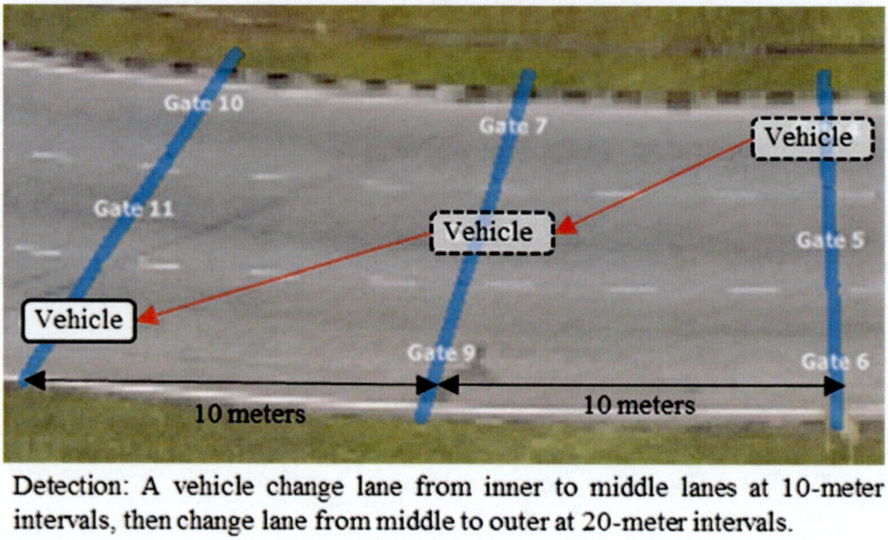

Fig. 7 Detection of vehicle travel from the inner to the outer lane

Table 4 Weaving flow for each weaving section

Weaving section	Flow passing the section	Weaving flow in the section
Between legs 1 & 2	1L, 1T, 1R, 1U, 2U, 3R, 3U, 4T, 4R, 4U	1T, 1R, 1U, 2U, 3R, 4T
Between legs 2 & 3	1T, 1R, 1U, 2L, 2T, 2R, 2U, 3U, 4R, 4U	1T, 2T, 2R, 2U, 3U, 4R
Between legs 3 & 4	1R, 1U, 2T, 2R, 2U, 3L, 3T, 3R, 3U, 4U	1R, 2T, 3T, 3R, 3U, 4U
Between legs 4 & 1	1U, 2R, 2U, 3T, 3R, 3U, 4L, 4T, 4R, 4U	1U, 2R, 3T, 4T, 4R, 4U

Remarks L: Left-turn; T: Through; R: Right-turn; U: U-turn
The number in front movement signifies leg number, e.g., 1L indicates Left-Turn movement from Leg 1

Table 5 Weaving flow in a roundabout

Weaving section	Flow passing the section (pcu/hr)	Weaving flow in the section (pcu/hr)
Between legs 1 & 2	2431	1560
Between legs 2 & 3	2155	1889
Between legs 3 & 4	1849	1157
Between legs 4 & 1	2530	2350

Remarks pcu/hr indicates passenger car unit per hour

3.2 Lane Change and Its Location in the Roundabout

As mentioned in Sect. 3.1, weaving happens as two vehicles swap paths to reach their destination lane. This section reports lane changes and their location in a roundabout. Approximately 86% of lane-changing happens in the weaving section compared to the non-weaving Section (14%). Such observation is related to the locations of weaving and non-weaving sections in the roundabout. As a section adjacent to an entry and exit, most vehicles enter and start changing lanes in the weaving section to leave at the destination exit. In contrast, fewer vehicles change lanes in non-weaving sections because they have chosen the lane they want to travel on in the roundabout.

For the weaving section, lane changes for Legs 1 and 3 are most frequent near the exit (i.e., between 60–70 m and 30–40 m, respectively), as shown in Fig. 9. This finding shows that drivers change lanes when they are about to exit. Conversely, most drivers from Legs 2 and 4 prefer to change lanes once they enter the roundabout. Thus lane changes are more frequent near entry (i.e., between 10–20 m and 0–10 m, respectively).

For the non-weaving section, lane changes for Legs 1 and 2 are most frequent near the entry (i.e., between 30–0 m and 30–40 m, respectively), as shown in Fig. 9. For Leg 3, most lane-changing happens between 20–30 m. Leg 4 has the most lane

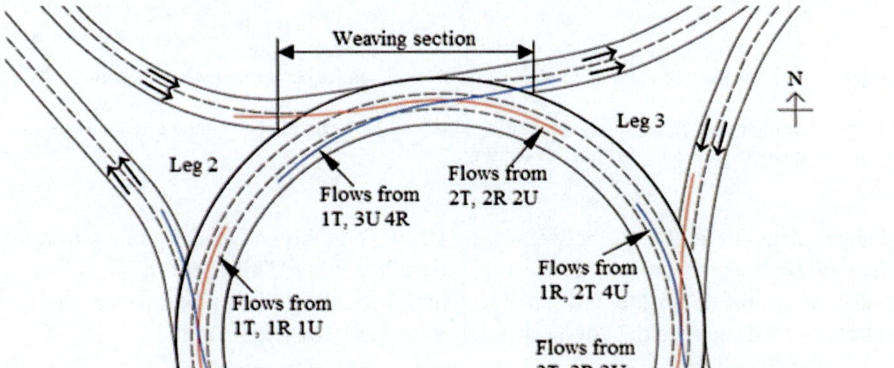

Fig. 8 Weaving pattern in a roundabout

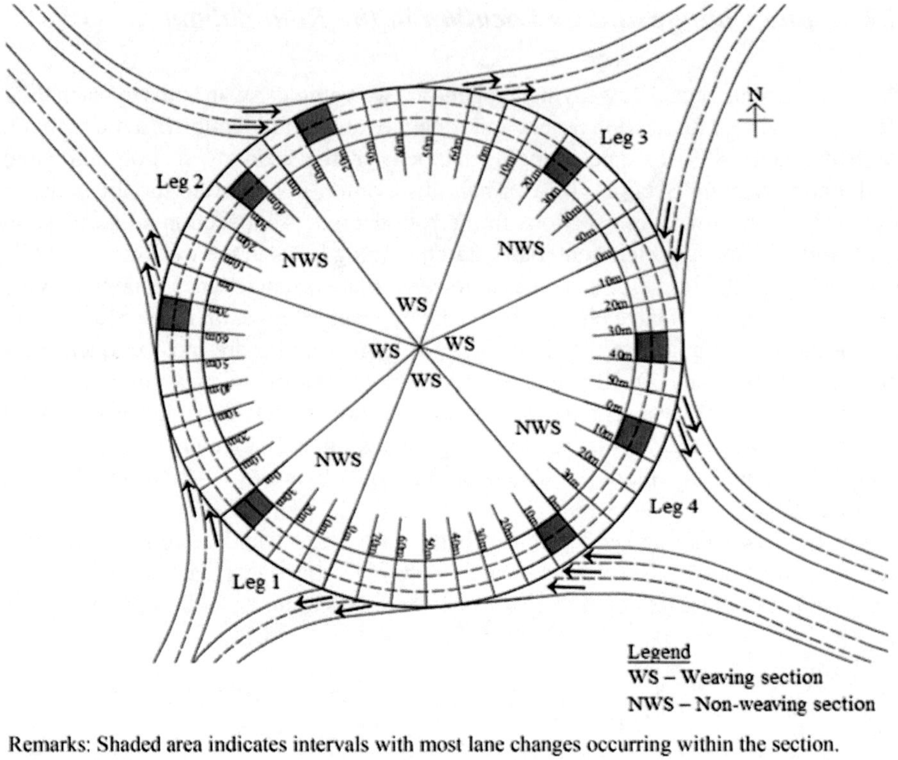

Fig. 9 Lane change location along circulating lanes. *Remarks* Shaded area indicates intervals with most lane changes occurring within the section

changes near the exit (i.e., between 0 to 10 m). Notably, all lane changes in non-weaving sections occur between the inner to middle lanes. This observation indicates that drivers on the middle and outer lanes tend to change lanes before passing the entrance (for Legs 1 and 2) or after passing the exit (for Legs 3 and 4).

4 Conclusion

This paper demonstrated that using UAV and traffic video analyser is feasible to investigate lane change in multi-lane roundabouts with ICDs more than 100 m. UAVs can capture continuous lane-changing data within a single traffic survey session. Using the virtual gate function, the extraction of UAV-based recording by traffic video analyser could provide turning movement and lane change data.

The finding shows 86% of lane change happens in the weaving section. Most drivers make lane change decisions when entering the weaving section, leaving using the nearest exit or travelling along the roundabout. Hence, lane-changing mainly occurs in the weaving section than in the non-weaving section.

This behavioural data can be integrated into the available roundabout capacity models. Generally, circulating flow parameters in all models consider multi-lane circulating flow as one total flow, ignoring interactions between entry and circulating flows (i.e., weaving). After introducing this technique, the circulating flow parameters can now consider weaving as the behavioural data can be obtained using UAVs and traffic video analysers. Therefore, work can be extended study roundabout capacity considering weaving.

Acknowledgements The authors thank the Swinburne University of Technology Sarawak Campus for funding this project.

References

1. Brahimi M, Karatzas S, Theuriot J, Christoforou Z (2020) Drones for traffic flow analysis of urban roundabouts. J Traffic Transp Eng. https://doi.org/10.5923/j.ijtte.20200903.02
2. Gallelli V, Iuele T, Vaiana R (2016) Conversion of a semi-two lanes roundabout into a turbo-roundabout: a performance comparison. Procedia Comput Sci 83:393–400. https://doi.org/10.1016/j.procs.2016.04.201
3. Gao H, Kong SL, Zhou S, Lv F, Chen Q (2014) Automatic extraction of multi-vehicle trajectory based on traffic videotaping from quadcopter model. Appl Mech Mater 552:232–239. https://doi.org/10.4028/www.scientific.net/AMM.552.232
4. HCM (2000) Highway capacity manual 2000, vol 4. Transportation Research Board, National Research Council, Washington, DC
5. HCM (2010) Highway capacity manual 2010, vol 5. Transportation Research Board, National Research Council, Washington, DC
6. HCM (2016) Highway capacity manual 2016, vol 6. Transportation Research Board, National Research Council, Washington, DC
7. Ke R, Li Z, Kim S, Ash J, Cui Z, Wang Y (2017) Real-time bidirectional traffic flow parameter estimation from aerial videos. IEEE Trans Intell Transp Syst 18(4):890–901. https://doi.org/10.1109/TITS.2016.2595526
8. Khan MA, Ectors W, Bellemans T, Ruichek Y, Yasar AUH, Janssens D, Wets G (2018) Unmanned aerial vehicle-based traffic analysis: a case study to analyze traffic streams at urban roundabouts. Procedia Comput Sci 130:636–643. https://doi.org/10.1016/j.procs.2018.04.114
9. Md Diah J, Abdul Rahman MY, Adnan MA, Atan I (2010) Weaving section flow model at the weaving area of Malaysian conventional roundabout. J Transp Eng 136(8):782–792. https://doi.org/10.1061/(ASCE)TE.1943-5436.0000133
10. Salvo G, Caruso L, Scordo A (2014) Urban traffic analysis through an UAV. Procedia Soc Behav Sci 111:1083–1091. https://doi.org/10.1016/j.sbspro.2014.01.143
11. Silva AB, Santos S, Vasconcelos L, Seco Á, Silva JP (2014) Driver behavior characterization in roundabout crossings. Transp Res Procedia 3:80–89. https://doi.org/10.1016/j.trpro.2014.10.093
12. Wang YX (2012) Research on capacity of roundabouts in Beijing. Procedia Soc Behav Sci 43:157–168. https://doi.org/10.1016/j.sbspro.2012.04.088
13. Yang X, Li X, Xue K (2004) A new traffic-signal control for modern roundabouts: method and application. IEEE Trans Intell Transp Syst 5(4):282–287. https://doi.org/10.1109/TITS.2004.838181

Using Unmanned Aerial Vehicles for Traffic Flow Analysis at Large Multi-Lane Roundabout

Elizabeth E. M. Chong and Jit-Boon Bong

Abstract Traditional traffic turning movement survey of large multi-lane roundabout is a daunting task as the traffic surveyors are unable to track the vehicles' movement. The direct method to obtain the turning movement requires the survey equipment to be placed at a higher level, usually at masts or high-rise buildings adjacent to the intersection. Indirect methods such as license plate matching may yield inaccurate results. With the advancement of technology, the use of unmanned aerial vehicles (UAV) in various fields has seen an increase, including traffic and transportation, due to its versatility. This paper compares the method of collecting traffic turning movement data using (i) video camera recording; and (ii) UAV to analyse the traffic flow at a 200 m inscribed diameter roundabout in Kuching, Malaysia. The results give an insight into the capability of a single unit UAV in determining the turning movement. Description of the challenges and limitations of UAV application are documented. The differences in the data obtained from methods were considered small and negligible. Due to the battery limitation, the UAV has to be flown down for battery change, which cause approximately 10 min of data loss for an hour's survey. Continuous data captured with a video camera shows that data loss is within the range of 20% of the total volume. With the expected uptrend of UAV applications in the near future, this kind of research serves as a useful reference for industries and researchers.

Keywords Unmanned aerial vehicle · Traffic analysis · Traffic flow · Multi-lane roundabout

---

E. E. M. Chong (✉) · J.-B. Bong
Faculty of Engineering, Computing and Science, Swinburne University of Technology Sarawak Campus, Kuching, Sarawak, Malaysia
e-mail: echong@swinburne.edu.my

C. S. Choo et al. (eds.), *Proceedings of ASEAN-Australian Engineering Congress (AAEC2022)*, Lecture Notes in Electrical Engineering 1072,
https://doi.org/10.1007/978-981-99-5547-3_23

1 Introduction

With rapid urbanization, cities and towns will usually experience increasing traffic flow and congestion levels. Managing this vehicular growth has been very challenging for authorities worldwide. Traffic congestion is the result of an imbalance in demand and supply in the transportation network. Intersection treatment options range from simple junction widening [5] to complex grade-separated intersections [19]. Roundabout is a popular intersection treatment based on a priority rule where the circulating vehicles have priority over the entering vehicles. Many studies exist on various aspects of roundabout such as critical gap [20, 23], entry capacities [15, 18], weaving sections [17], lane choice [3] and influence of roundabout types [8, 16, 24]. The evaluation of the roundabout design and performance is of utmost importance to ensure smooth traffic flow. Thus, precise and up-to-date traffic data is usually required.

The traditional method of data collection is expensive and a challenging task due to the need for a large amount of equipment or a high number of traffic surveyors in order to cover the survey area [4]. In addition, the administrative tasks such as management of surveyors, equipment and post-fieldwork data can be time-consuming too. In most cases, the costs are often dependent on the level of detail necessary for the project, the size and the type of intersection. Frequently collected data such as traffic flow and turning movement count data are necessary for intersection design and performance analysis.

For large multi-lane roundabout, the conduct of collecting turning movement count data is made even more challenging due to the difficulty in tracking the vehicles' flow either by traffic surveyors at the site or playback of video recording during post-fieldwork data reduction. This is because these techniques only allow limited view at the locations of the surveyor or the video camera, restricted by the ability to see the other parts of the roundabout due to the large diameter size of the roundabout. Indirect methods such as license plate have been used to track the vehicles' movement [3, 19]. However, the use of license plate method may yield inaccurate results due to the difficulty of the surveyors to capture the fast-moving vehicles, particularly motorcycles. Video recording technique provides a higher accuracy as it allows playback of the traffic survey in order to track the vehicles' turning movement. This problem can potentially be solved by using unmanned aerial vehicles (UAV) to record the traffic movement which provides a bird's eye view to easily track the vehicles' trajectory.

This paper discusses the traffic data collection methods of using (i) at-grade video recording technique, and (ii) off-the-shelf UAV aerial traffic flow monitoring; and compares the extracted turning movement data from both methods. The challenges of using UAV will be further discussed. In anticipation that the use of UAV for traffic related studies will increase in the near future, it is hope that this experience will serve as a future reference.

2 Applications of UAV in Transport-Related Projects

The application of UAV has been on an increasing trend in recent years, including transport-related projects [13, 14, 25, 26]. Largely, the studies of UAV in transport can be categorized into two: (i) traffic monitoring, which focused on the tracking of vehicles in the traffic stream with some in real time, and (ii) traffic analysis, where the traffic parameters such as travel behaviour, traffic congestion details, or collision risks assessment are studied with greater depth.

Traffic parameters of roundabout such as traffic volume and critical gaps can be easily obtained through the use of UAV [12]. The advantage of UAV in such studies is that it offers a more naturalistic way of observing drivers' behaviour, i.e. drivers are not influenced by the presence of traffic surveyors or equipment [22]. Brahimi et al. [2] presented a comparison between two counting tools, i.e. UAV and cameras fitted on counting masts, for a low volume roundabout. Results indicated there were large relative errors for some of the movements due to the angle of the videos that did not show a clear view of the legs. They suggested for a mix of equipment to optimize the results.

3 Methodology

The traffic survey was conducted on 22 March 2022 for both morning and evening peak periods, 06:30 AM to 08:30 AM and 04:30 PM to 06:30 PM, respectively. Two methods were used to collect the traffic data at the same time, which were (i) at-grade video camera recording; and (ii) UAV recording. Each method will be discussed further in the following sections.

3.1 Site Selection

In Malaysia, a conventional roundabout has an inscribed circle diameter (ICD) of greater than 50 m [9]. No limit has been set for the maximum ICD. Many countries specify the ICD of roundabouts to be less than 80 m [1, 6, 7].

Typical roundabouts in Kuching city have ICD of greater than 80 m. This has caused difficulty during traffic data collection to determine the turning movement count. The selected site, shown in Fig. 1, has an ICD of 200 m with four approaching legs. Legs 2, 3 and 4 consisted of two entry lanes while Leg 1 has three entry lanes. Circulating lanes consisted of three lanes. The roundabout is a major intersection located within a mixed development area (Fig. 1).

Fig. 1 Satellite view of the selected multi-lane roundabout

3.2 *Traffic Data Collection and Reduction with Video Camera*

One video camera was setup at the splitter island of each leg, which totalled to four video cameras used. Each video camera was adjusted to record the entry and circulating lanes of the leg where it is setup, and the exit lanes of the subsequent leg. This was to enable the journey of each vehicle to be tracked as it passed in front of the video cameras so that the movement of the vehicle is known. The traffic volume was recorded continuously for 2 h as stated in the previous section.

In the post-fieldwork data reduction, the peak hours were determined based on the highest amount of traffic volume for one hour that was summed from four consecutive 15-min intervals, which are 06:45–07:45 AM and 04:30–05:30 PM. Upon determination of the peak hour, the turning movement count was conducted by using vehicle tracing to determine the origin (leg on which the vehicle enters the roundabout) and destination (leg on which the vehicle exits the roundabout) of the vehicle at the roundabout. In the example given in Fig. 2, it can be traced that the black colour vehicle entered the roundabout via Leg 4 and exited via Leg 2, which means the vehicle is making a thorough movement.

3.3 *Traffic Data Collection and Reduction with UAV*

The specifications of the UAV that was used in this study are presented in Table 1. The traffic data collection using UAV followed the proposed universal guiding framework for UAV-based traffic analysis by Khan et al. [11].

Step 1: Scope Definition

The first step involves defining the problem statement with a clear scope of the study, as with any other project. The parameters to be evaluated within the area of study need to be selected together with the characteristics of the UAV used.

Fig. 2 An example of vehicle tracing from video camera recording: a entering the roundabout via Leg 4 and b exiting the roundabout via Leg 2

Table 1 Specifications of the UAV used in this project

Model	DJI Mavic mini
Takeoff weight	249 g
Power	LiPo 2S
Max flight time	30 min
Wind speed	8 m/s
GPS	Yes
Video quality	2.7K (2720 × 1530)

Step 2: Flight Planning

It is important to plan for the implementation of the actual UAV flight during the data collection to avoid any mishaps. The steps involve considerations for safety, environmental influence and route planning aspects. In Malaysia, the Civil Aviation Authority of Malaysia (CAAM) is responsible for regulating the use of unmanned aircraft system (UAS). The UAS can be classified into three categories, namely small unmanned aircraft system (not more than 20 kg), small unmanned surveillance aircraft (for surveillance or data acquisition) and unmanned aircraft system of more than 20 kg. They also dictate the No Fly zones. Environmental influences such as weather and wind conditions, location characteristics (extent of built-up area) and time of study must be given attention to increase the quality of the videos. Due to the sensitivity of UAV to weather elements such as wind and rain [21], the fieldwork was conducted on a fine, sunny day in this study. It was also carried out according to the safety regulations outlined by CAAM.

Step 3: Flight Implementation

During this stage, it is important to capture video that is stable. With a fully stabilized 3-axis gimbal, the UAV used in this study offered a stabilized video when it was hovered at a height of 80 m above ground level. In order to view all the legs of the roundabout clearly, it was necessary to position the UAV approximately 150 m away from the circulating lane at Leg 1, as shown in Fig. 3. The actual view of the roundabout can be seen in Fig. 4. As the traffic survey was conducted for two hours, it was necessary for the single UAV to be flown down every 25 min for a battery change, which resulted in loss of traffic data. During this time, the video footage was captured using video cameras as described in the previous section.

Step 4: Data Acquisition

High-quality data (video recording) needs to be acquired from the UAV. Calibration of the recorded video may be necessary using other data from the sensors (infrared, thermal, ultrasonic, etc.) and flight telemetry data (altitude, horizontal speed, vertical

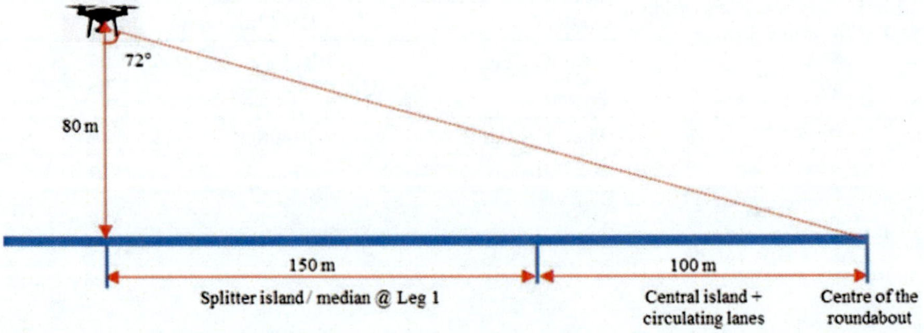

Fig. 3 Position of the UAV in relation to the roundabout

Fig. 4 View of the roundabout taken from the UAV

speed along with the position and orientation data). To ensure the accuracy of the data recorded by the UAV, continuous on-ground video recording was used as described in Sect. 3.2. This calibration technique was also used by Brahimi et al. [2].

The UAV video can be real time or offline, depending on the needs of the project. The offline method is the most often used, including the current study.

Step 5: Data Processing and Analyses

Due to the advancement in traffic data collection with UAV, a lot of data processing techniques have been developed in recent years. The techniques can be broadly classified into (i) semi-automated video processing, which utilizes stabilized and calibrated video to speed up the tracking process, and (ii) automated video analysis, which requires more advanced image processing filters to track the vehicles. For this study, manual data processing was conducted to determine the turning movement count and vehicle classification.

4 Results and Discussion

4.1 Comparison of Video Camera and UAV

It is common for mixed traffic (heterogenous) to be converted into equivalent homogenous traffic. The vehicle count in the current study was converted into passenger car unit (PCU) using the equivalent PCU values shown in Table 2.

Table 2 Conversion factors to PCU [10]

Vehicle type	Equivalent PCU value
Passenger car	1.00
Motorcycle	0.33
Van	1.75
Medium lorry	1.75
Heavy lorry	2.25
Bus	2.25

Due to the battery limitations and having to change the battery during the field survey, only approximately 21–25 min of data could be used for the analysis of both the AM and PM peak hours. In the data extraction process, the video camera's data were extracted to match the duration of useful data from UAV. For example, if a particular UAV flight has useful data from 05:00 to 05:22 PM, data from the video camera were extracted during the same duration. With two flights for each peak hour, the total duration of useful data is approximately 50 min for one hour survey. Hence, for simplicity, the unit 'pcu/50 min' will be used in Table 3 to present the data. High traffic volume could be observed for the intersection, as evidenced by the intersection total of about 6000 and 5500 in the AM and PM peak hours, respectively. Leg 2 and Leg 4 see the highest traffic flows, consistent with the higher road category of the Kuching-Samarahan Expressway. Separate left-turn slip lane facilities were available at Legs 2, 3 and 4, in which the traffic volumes are included in Table 3.

Table 3 Turning movement count results from video camera and UAV recordings for 50 min

Method	From/To	Volume: AM (PM) (pcu/50 min)					
		Leg 1	Leg 2	Leg 3	Leg 4	Leg total	Intersection total
Video camera	Leg 1	14 (5)	414 (168)	268 (397)	175 (269)	871 (839)	6053 (5550)
	Leg 2	869 (381)	41 (6)	720 (663)	1376 (1108)	3006 (215)	
	Leg 3	293 (143)	423 (766)	3 (5)	24 (34)	743 (948)	
	Leg 4	275 (36)	1079 (1376)	40 (119)	39 (74)	1433 (366)	
UAV	Leg 1	14 (5)	416 (165)	270 (391)	178 (271)	878 (832)	5944 (5462)
	Leg 2	876 (373)	40 (6)	706 (654)	1334 (1090)	2956 (212)	
	Leg 3	287 (140)	411 (757)	3 (5)	25 (35)	726 (937)	
	Leg 4	271 (36)	1033 (1345)	41 (117)	39 (72)	1384 (359)	

The data difference between the two methods are presented in Table 4 in PCU and percentage to reflect the differences better. The differences in PCU were obtained by deducting the video camera's data from UAV's data. A negative PCU means the UAV data is lower than the video camera's data. The differences were largely caused by miscount and incorrect vehicle classification. Generally, it can be observed that video camera data are higher than the UAV data. The vehicle types were more difficult to classify when it was viewed from the UAV videos because the plan view images only show the rooftop of the vehicles, making it difficult to differentiate between passenger cars and vans/medium vehicles. Meanwhile, the video camera gave an elevation view, which allows easier identification of the vehicle type. However, the overall differences were relatively small (−1.83% for AM and −1.61% for PM) in comparison to the total volume for the peak hours.

Another issue encountered while processing the video was the difficulty in detecting the trajectories of motorcycles in both methods due to their small size and greater acceleration power. In the video camera's footage, motorcycles could be hidden by larger vehicles, causing missing data. Motorcycles also look quite similar, making the tracing of motorcycle turning movement hard to differentiate.

As the UAV was positioned 150 m outside of the inscribed diameter, the vehicles on the furthest leg away from the UAV (Leg 3) could not be clearly categorized, especially if the data processing was done with a small screen. This is easily corrected by using larger screens to obtain the turning movement count data. However, if any minute details are required, it may not be clear enough for such analysis. There were two trees blocking the view of the circulating and exit lanes but it was not a major issue as the vehicles could still be traced.

Table 4 The differences between traffic volumes analysed from video camera and UAV in PCU and percentage

From/To	Leg 1	Leg 2	Leg 3	Leg 4
Difference in PCU: AM (PM)				
Leg 1	0(0)	2(−3)	2(−6)	3(2)
Leg 2	7(−8)	−1(0)	−14(−9)	−42(−18)
Leg 3	−6(−3)	−12(−9)	0(0)	1(1)
Leg 4	−4(0)	−46(−31)	1(−2)	0(−2)
Difference in percentage: AM (PM)				
Leg 1	0(0)	0(2)	1(2)	2(1)
Leg 2	1(2)	3(0)	2(1)	3(2)
Leg 3	2(2)	3(1)	0(0)	4(3)
Leg 4	1(0)	4(2)	2(2)	0(3)
Overall difference: −1.83 (−1.61)				

4.2 Effect of the Data Loss Due to Battery Change

As was discussed in the previous section, there were some data losses while the single unit UAV was flown down for a quick battery change. The presence of the video camera afforded a continuous recording of the traffic flow. Since it was established that the differences between the two methods were small and negligible, this section examines the effect of a full hour one recording during the peak hours. Table 5 shows the data extracted from the video camera for 50 and 60-min durations. The 50-min data is the same data presented in Table 3 and is included here for ease of reference.

Table 6 presents the differences between the two sets of data in PCU and percentage. For the 10 min loss of data, the video camera accounted for approximately 10–20% of the total volume, although the range of PCU is from 1 to 195 PCU. In practice, if a single unit UAV is used without the video camera, the data acquired by the UAV can be increased by between 10–20% to account for the loss of data during the battery change operation. To avoid the data loss issue, additional UAV is recommended to be flown alternatively so that a continuous traffic stream can be captured.

Table 5 Turning movement count results from video camera for 50 and 60-min durations

Time (minutes)	From/To	Volume: AM (PM)					
		Leg 1	Leg 2	Leg 3	Leg 4	Leg total	Intersection total
50	Leg 1	14 (5)	414 (168)	268 (397)	175 (269)	871 (839)	6053 (5550)
	Leg 2	869 (381)	41 (6)	720 (663)	1376 (1108)	3006 (2158)	
	Leg 3	293 (143)	423 (766)	3 (5)	24 (34)	743 (948)	
	Leg 4	275 (36)	1079 (1376)	40 (119)	39 (74)	1433 (1605)	
60	Leg 1	16 (5)	472 (192)	302 (467)	206 (308)	996 (972)	6832 (6378)
	Leg 2	963 (426)	43 (6)	821 (779)	1571 (1279)	3398 (2490)	
	Leg 3	331 (161)	479 (893)	3 (6)	28 (38)	841 (1098)	
	Leg 4	305 (41)	1203 (1566)	44 (132)	45 (79)	1597 (1818)	

Table 6 Differences between 50 and 60-min data in PCU and percentage

From/To	Leg 1	Leg 2	Leg 3	Leg 4
Difference in PCU: AM (PM)				
Leg 1	2(0)	58(24)	34(70)	31(39)
Leg 2	94(45)	2(0)	101(116)	195(171)
Leg 3	38(18)	56(138)	0(1)	4(4)
Leg 4	30(5)	124(190)	4(13)	6(5)
Difference in percentage: AM (PM)				
Leg 1	14(0)	14(14)	13(18)	18(14)
Leg 2	11(12)	5(0)	14(17)	14(15)
Leg 3	13(13)	13(18)	0(20)	17(12)
Leg 4	11(14)	11(14)	10(11)	15(7)

5 Conclusions

This study gives an insight into the capability of a single UAV as a tool to collect traffic turning movement count volume data at the major large multi-lane roundabout with an inscribed diameter of 200 m, in comparison to a video camera as a typically used data collection tool. Description of the challenges and limitations of UAV application are documented.

The following conclusions can be drawn from this study:

1. With the increasing improvement in UAV technology, an off-the-shelf commercial UAV was found to be suitable for collecting traffic turning movement data with high accuracy because of the ability to view and count every single recorded vehicle.
2. Differences in the data between the two methods were due to miscount and incorrect vehicle classification arising from different perspectives of viewing the vehicles.
3. The challenges of using UAV for data collection include short flight time due to battery limitations, causing the need for the UAV to descend for battery change. During the battery change operation, traffic data was not captured which causes some traffic data loss. This can be corrected by factoring additional 10–20% traffic volume.

As the application of UAV in traffic monitoring and analysis is still in its infancy, improvements are still necessary in future work. Some of the challenges faced in this study, such as the loss of recording time, could be rectified in the immediate future, while the development of automated data processing will require a longer time. It is suggested that a second UAV be deployed alternatively with overlapping timing to triangulate and address the data loss issue. Despite all the challenges, the benefits outweigh the downsides of this technology. Furthermore, there are still many more potentials of UAV that have yet to be fully discovered.

Acknowledgements This research project was partially supported by the Learning and Teaching Grant of Swinburne University of Technology Sarawak Campus.

References

1. Austroads (2021) Guide to road design part 4B: roundabouts. Austroads Ltd., Sydney, Australia
2. Brahimi M, Karatzas S, Theuriot J, Christoforou Z (2020) Drones for traffic flow analysis of urban roundabouts. J Traffic Transp Eng. https://doi.org/10.5923/j.ijtte.20200903.02
3. Chong EEM, Al-Mamari RNS (2020) Characteristics of drivers' lane choices at large multi-lane roundabout. Int J Integr Eng 12(9):176–183
4. Coifman B, McCord M, Mishalani RG, Iswalt M, Ji Y (2006) Roadway traffic monitoring from an unmanned aerial vehicle. IEE Proc-Intell Transp Syst 153(1):11–20
5. Danting Z, Jun M, Bowen W, Ying L (2021) The influence of left-turn lane widening on the usability of guidance system at intersection entrance. In: Proceedings of fifth international conference on traffic engineering and transportation system (ICTETS 2021). SPIE. https://doi.org/10.1117/12.2619961
6. Departemen Pemukiman dan Prasarana Wilayah (2004) Pedoman Perencanaan bundaran untuk persimpangan sebidang Departemen Pemukiman dan Prasarana Wilayah. Indonesia
7. FGSV (2006) Merkblatt für die Anlage von Kreisverkehren (Leaflet for the construction of roundabouts). Forschungsgesellschaft für Straßen- und Verkehrswesen (Road and Transport Association), Cologne
8. Hatami H, Aghayan I (2017) Traffic efficiency evaluation of elliptical roundabout compared with modern and turbo roundabouts considering traffic signal control. Promet-Traffic & Transp 29:1–11. https://doi.org/10.7307/ptt.v29i1.2053
9. Jabatan Kerja Raya (1987a) Arahan Teknik (Jalan) 11/87–A guide to the design of at-grade intersections. Cawangan Jalan, Ibu Pejabat JKR, Malaysia
10. Jabatan Kerja Raya (1987b) Arahan Teknik (Jalan) 13/87–A guide to the design of traffic signals. Cawangan Jalan, Ibu Pejabat JKR, Malaysia
11. Khan MA, Ectors W, Bellemans T, Janssens D, Wets G (2017) UAV-based traffic analysis: a universal guiding framework based on literature survey. Transp Res Procedia 22:541–550. https://doi.org/10.1016/j.trpro.2017.03.043
12. Khan MA, Ectors W, Bellemans T, Ruichek Y, A-u-H Y, Janssens D, Wets G (2018) Unmanned aerial vehicle-based traffic analysis: a case study to analyze traffic streams at urban roundabouts. Procedia Comput Sci 130:636–643. https://doi.org/10.1016/j.procs.2018.04.114
13. Kujawski A, Dudek T (2021) Analysis and visualization of data obtained from camera mounted on unmanned aerial vehicle used in areas of urban transport. Sustain Cities Soc 72:103004. https://doi.org/10.1016/j.scs.2021.103004
14. Luo H, Chu S-C, Wu X, Wang Z, Xu F (2020) Traffic collisions early warning aided by small unmanned aerial vehicle companion. Telecommun Syst 75(2):169–180. https://doi.org/10.1007/s11235-015-0131-5
15. Macioszek E (2020) Roundabout entry capacity calculation—a case study based on roundabouts in Tokyo, Japan, and Tokyo surroundings. Sustainability 12:1533
16. Mauro R, Branco F (2010) Comparative analysis of compact multilane roundabouts and turbo-roundabouts. J Transp Eng 136(4):316–322. https://doi.org/10.1061/(ASCE)TE.1943-5436.0000106
17. Md Diah J, Abdul Rahman MY, Adnan MA, Atan I (2010) Weaving section flow model at the weaving area of Malaysian conventional roundabout. J Transp Eng 136(8):782–792. https://doi.org/10.1061/(ASCE)TE.1943-5436.0000133
18. Mohamed AIZ, Ci Y, Tan Y (2020) A novel methodology for estimating the capacity and level of service for the new mega elliptical roundabout intersection. J Adv Transp 2020:8467152. https://doi.org/10.1155/2020/8467152

19. Ng B, Chong EEM (2019) Implementation of Shallow Underpasses at an Urban Roundabout. International Journal of Advanced Trends in Computer Science and Engineering 8(3):375–382
20. Qu Z, Duan Y, Song X, Hu H, Liu H, Guan K (2014) Capacity prediction model based on limited priority gap-acceptance theory at multilane roundabouts. Math Probl Eng 2014:490280. https://doi.org/10.1155/2014/490280
21. Ranquist E, Steiner M, Argrow B (2017) Exploring the range of weather impacts on UAS operations. In: Proceedings of 18th conference on aviation, range and aerospace meteorology. Seattle, Washington, USA, 22–26 January 2017, p J3.1
22. Salvo G, Caruso L, Scordo A, Guido G, Vitale A (2017) Traffic data acquirement by unmanned aerial vehicle. Eur J Remote Sens 50(1):343–351. https://doi.org/10.1080/22797254.2017.1328978
23. Shaaban K, Hamad H (2018) Group gap acceptance: a new method to analyze driver behavior and estimate the critical gap at multilane roundabouts. J Adv Transp 2018:1350679. https://doi.org/10.1155/2018/1350679
24. Shaaban K, Hamad H (2020) Critical gap comparison between one-, two-, and three-lane roundabouts in Qatar. Sustainability 12(10). https://doi.org/10.3390/su12104232
25. Shan D, Lei T, Yin X, Luo Q, Gong L (2021) Extracting key traffic parameters from UAV video with on-board vehicle data validation. Sensors 21(16). https://doi.org/10.3390/s21165620
26. Zhang H, Liptrott M, Bessis N, Cheng J (2019) Real-time traffic analysis using deep learning techniques and UAV based video. In: 2019 16th IEEE international conference on advanced video and signal based surveillance (AVSS). IEEE, pp 1–5

To Jack or Not to Jack: Inside the Minds of Pipe Jacking Operators

Lit Yen Yeo, Chung Siung Choo, and Sue Han Lee

Abstract Limited available land is leading to the emerging demand for trenchless methods for the installation of buried pipelines. Pipe jacking is one such method. The prediction of jacking forces is crucial in assessing the performance of the pipe jacking process, thus many efforts have been made to predict the jacking forces. Current approaches are based on the negotiated geological conditions, either through empirical or mechanistic approaches. However, such methods are only useful during the planning and design stages. During the construction stage, the microtunnel boring machine (mTBM) operators base their decisions on the operation parameters. This is further complicated when pipe jacking is performed through highly weathered lithology, i.e. 'soft rocks'. There continues to be a gap in this domain of modelling jacking forces for soft rocks based on operation parameters. In the current study, deep learning techniques were applied to pipe jacking field-measured operation parameters obtained from the Kuching Wastewater Management System. The deep learning models helped to relate the time series input operation parameters to the jacking forces. Subsequently, a study on different moving average periods towards the input features was applied to understand the changes in the prediction performance of the jacking forces using the deep learning model. From the study, the results showed that the cumulative parameters which are the drive length, the cumulative days elapsed, and the cumulative lubricant used have the highest moving average periods. This suggested that along the pipe jacking process, these features would have a significant impact on the jacking forces.

Keywords Pipe jacking · Jacking forces · Highly weathered lithology (soft rocks) · Deep learning · Moving average

L. Y. Yeo · C. S. Choo (✉) · S. H. Lee
Faculty of Engineering, Computing and Science, Swinburne University of Technology, Kuching, Sarawak, Malaysia
e-mail: cschoo@swinburne.edu.my

C. S. Choo et al. (eds.), *Proceedings of ASEAN-Australian Engineering Congress (AAEC2022)*, Lecture Notes in Electrical Engineering 1072,
https://doi.org/10.1007/978-981-99-5547-3_24

1 Introduction

The pipe jacking technique is a pipe installation process that comprises the construction of vertical shafts for the jacking of pipes horizontally into the ground. However, this process is challenging due to the combination of uncertainties and complexities arising from the changes in geological conditions. These can cause the malfunction of machines or even lead to geo-hazards like surface subsidence and sinkholes [4].

In recent years, urbanization and limited available land have resulted in the increasing demand for underground space to be utilized for construction [10]. However, many features such as geological conditions, jacking operation parameters, and tunnel geometry will affect the tunnel machine's performance during construction [20]. In this research, pipe jacking through 'soft rocks' of the Tuang Formation underlying the Kuching CBD was used for the construction of trunk sewers. Frictional forces anticipated during construction are critical, as excessive jacking forces could damage the installed pipes [6]. Hence, it was important to understand how pipe jacking operators controlled jacking forces using operation parameters, such as injection of lubrication, jacking speed, cutter torque, slurry pressure, and steering. Moreover, this can be further complicated by complex geological conditions. Knowledge of how geology could affect pipe jacking forces is significant, for example, the highly weathered geology of Kuching, Malaysia [13]. The highly weathered rock masses exhibited rock quality designation (RQD) values of zero, while demonstrating soil-like behaviour during tunnelling [15]. Therefore, this served as the motivation to develop a deep learning model capable of predicting jacking forces using the pipe jacking operation parameters as input features. This study aims to identify the effects of different moving average periods on the prediction of jacking forces. The proposed model will identify the significance of the pipe jacking operation through highly weathered by looking at the increased periods of moving average in obtaining the lowest root mean square error (RMSE).

2 Literature Review

In pipe jacking, the estimation of frictional jacking forces is commonly based on mechanistic approaches, where geomaterial strength parameters with site dimensions are substituted into simple closed-form equations such as those developed by Bennett [1], Chapman and Ichioka [2], Osumi [14], Pellet-Beaucour and Kastner [16], and Staheli [18]. Mission-critical operation parameters (such as jacking speed, cutter torque, lubrication, and slurry pressure) from pipe jacking operations can also be used to estimate jacking forces [12]. However, there is a lack of quantitative relationships between the operation parameters and jacking forces. Furthermore, the influence of operation parameters on jacking forces can directly affect the productivity, speed, and success of pipe jacking projects.

Hence, the decision of "to jack or not to jack" will be dependent on the jacking forces. However, it is challenging to obtain high accuracy in estimating the jacking forces, primarily due to the uncertainty of geological conditions [4, 11]. Chen et al. [3] and Sterling [19] emphasized that pipe jacking operators also tend to base on their prior experience when operating the machine, which leads to the purpose of the current study.

2.1 *Jacking Forces During Pipe Jacking*

Jacking force, F_J is generally derived from two main components which are the frictional force and the face pressure. During the excavation phase, face pressure is caused by lateral earth pressure. In order to advance the pipe jacking machine, a sufficient amount of total jacking forces is required to outweigh the frictional force acting on the soil-shield and soil-pipe interfaces (such as shield friction F_S, pipe friction F_P, and edge resistance F_E) as well as the face pressure F_F. These forces, which act to push the pipe through the ground, can be expressed as in Eq. (1) [9].

$$F_J \geq F_F + (F_E + F_S + F_P) \tag{1}$$

2.2 *Operation and Geological Factors Affecting the Jacking Forces*

During the actual pipe jacking construction, there are many factors that influence jacking forces, such as the geological conditions, the dimensions of the pipe, the depth of burial, the progress drive length for the installation of pipe, the material of the pipe used, overcut, annulus lubrication, roughness outer surface of the pipe and construction technology [17]. Cheng et al. [5] showed the influences of lubrication, stoppage, geological conditions and misalignment towards the jacking forces through the Shulin district sewer network in Taipei Country, Taiwan, of the four pipe jacking drives. Table 1 below summarizes the influence of different factors towards the jacking forces.

2.3 *Types of Deep Learning Algorithms Used in Tunnel Construction*

The purpose of using deep learning algorithms is to imitate the learning sequence of the human brain. By discovering and learning patterns from the given data, it is able

Table 1 Summary of different influence factors towards jacking forces

Influence factors	Jacking forces
Geological condition [5, 7, 17]	Increase in face pressure/resistance
Lubrication [6, 7, 17]	Lower down frictional force
Stoppage [5, 6, 17]	Increase in frictional static resistance
Progress drive length [9, 16]	Increase in frictional force produced throughout the pipeline surface
Jacking speed [5, 16, 18]	Increase in face pressure force

to improve the accuracy of future estimation as well as improve the decision-making process. Deep learning sequential models such as recurrent neural networks (RNN) consisting of long short-term memory (LSTM) and gated recurrent unit (GRU) have been applied to tunnel construction.

Long Short-Term Memory (LSTM)

LSTM is a deep learning network developed to overcome gradient vanishing and exploding issues. The only difference between LSTM and traditional RNN is the inner hidden units in LSTM which consist of cell state, input gate, output gate, and forget gate. All three gates of LSTM were created to monitor the information flow by identifying which to remove and which to remember (add) into the cell state [3]. Therefore, the essential and related information will be remembered and learned throughout the LSTM models.

Gated Recurrent Unit (GRU)

GRU consists of the input layer, hidden layer, and output layer, which has the same structure as LSTM. The historical sequential data is fed through the input layer and flow into the hidden layers by the weighted connections which will be used to update the accuracy of prediction [10]. There are two operational gates used in GRU, which are the reset gate and the update gate to determine significant information from the given historical data.

Table 2 shows the case studies of different deep learning techniques used in tunnel construction.

2.4 Moving Average Technique Used in Prediction

Classical moving average (MA), autoregressive, and autoregressive integrated moving average (ARIMA) models were used to solve the issues of time series forecasting such as improving the prediction of the tunnel boring machine (TBM) penetration rate using the LSTM model [8]. MA improves the prediction performance by using the input of lagged prediction errors that push the model towards the actual values [22].

Table 2 Case studies on different prediction models used in tunnelling

ML/DL techniques used	Purpose
SVM [4]	Identify changes in soil type
LSTM [3]	Predict thrust and torque
RNN, LSTM [8]	Develop a penetration rate (PR) prediction model
RNN [12]	Predict the TBM operation parameters (advance speed, articulation force, cutterhead rotation speed, thrust force)
ANNs, GA-ANNs [20]	Predict the jacking forces using ten influence factors (tunnel geometry, geological conditions, jacking operation)
ANNs, LSTM [11]	Classify rock mass behaviour
ANNs, LSTM, GRU, Conv1d [21]	Predict tunnelling-induced ground settlement

Support vector machine (SVM), Long short-term memory (LSTM), Recurrent neural network (RNN), Artificial neural networks (ANNs), Hybrid genetic algorithm optimized ANNs (GA-ANNs), Gated recurrent unit (GRU), 1d convolutional neural networks (Conv1d)

2.5 *Evaluation of the Performance of Models*

Root mean square error (RMSE) represents the difference between the predicted output with the actual value. Hence, the smaller RMSE values imply better accuracy. RMSE is the square root of mean square error (MSE), where MSE is calculated by averaging the square errors from the individual samples with deviations. The RMSE was used to evaluate the prediction performance of the LSTM model by Gao et al. [8]. Besides, Wei et al. [20] compared the prediction performance between ANN models and GA-ANN models of jacking force in vertical tunnelling projects using the coefficient of determination (R^2), root mean square error (RMSE), and absolute value of relative error (RE). To conclude, the lower the RMSE/RE, the better the predicted performance. In contrast, the higher the R^2, the better the prediction performance.

3 Methodology

The methodology used in the current study is as shown in Fig. 1 with detailed descriptions described, hereinafter.

(1) ***Data Preparation***

Field-measured pipe jacking operation data collected from a drive traversing phyllite was used to perform the parametric study on the effect of prediction performance using different moving average periods. There were a total of 2964 data points used in the developed model.

(B) ***Parametric Study Using Different Moving Average Periods Towards the Selected Dataset***

The selected pipe jacking operation parameters were separated into input features and output prediction. The moving average technique was used to smoothen the noise from the collected dataset. Hence, using for loop in Python to run all the combinations of different moving average periods of two (2), four (4), six (6), and eight (8) towards each input and output feature before proceeding to evaluate the prediction performance of the deep learning model.

(C) ***Evaluate the Prediction Performance of the Deep Learning Model***

The input and output features allocated with different combinations of moving average periods were normalized and split into 80% of training and 20% of testing sets during data pre-processing. The training set was used to train the deep learning model, and the testing set was used to evaluate the prediction performance using RMSE. The results were then stored in the append function of Python coding.

(D) ***Analyse the Findings on Different Moving Average Periods Towards Each Feature***

The evaluated prediction performance results in RMSE were collected and exported to a CSV file. The results were plotted to determine the periods of moving average for each feature in obtaining the lowest RMSE. Finally, the impact of each feature on the prediction performance was identified by looking at the increasing periods of moving average in obtaining the lowest RMSE (Fig. 1).

4 Results and Discussion

Decision-making prediction models using deep learning techniques from Python programming language were used for the following findings. The model was designed to predict the pipe jacking forces using the actual construction operation parameters as input features. RMSE values are shown for the evaluation of the performance of the model.

4.1 Prediction Performance with and Without Using the Moving Average Technique

The results shown in Fig. 2 illustrate three different phyllite drives named Drive A, Drive B, and Drive C which were collected from the micro-tunnelling works in Kuching City. The results consist of ground truth, smoothed data (moving average), and prediction. These six graphs can be separated into upper and lower rows, with the upper row showing the predicted jacking forces with the use of moving average, and the lower row showing the same without moving average. The results are presented to compare the performance of the model with and without the use of a moving average.

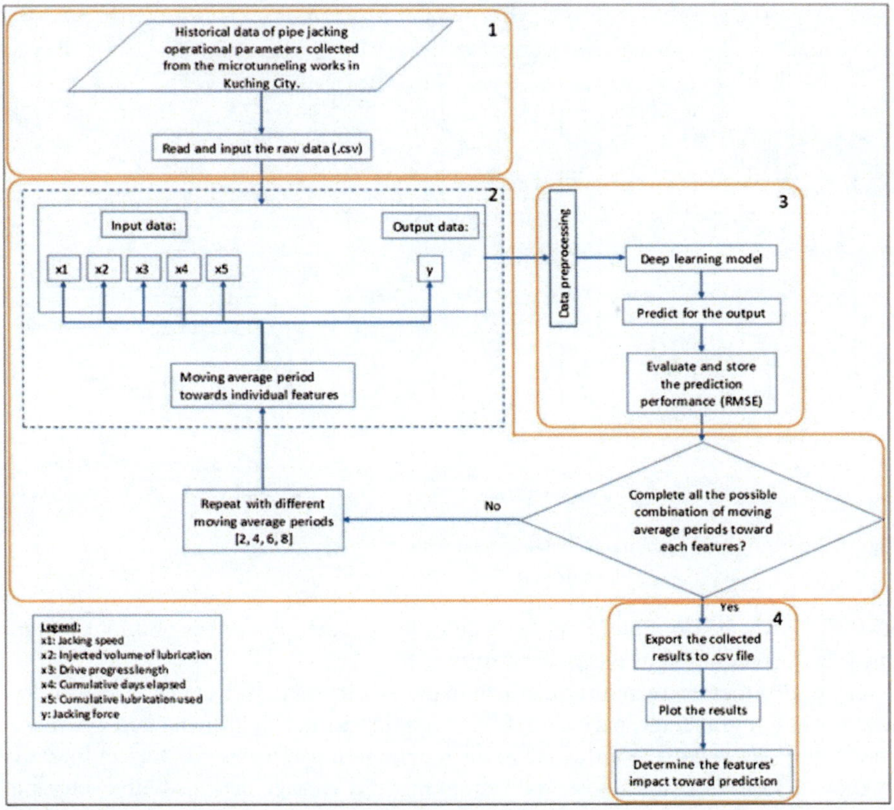

Fig. 1 Proposed methodology used in the current study

The findings explained that using the moving average technique will improve the prediction performance. For instance, the prediction of the jacking forces with moving average periods of 8 shown in Drive A produces 14.31 RMSE which is lower than the prediction without moving average producing 27.42 RMSE. However, after applying the moving average periods of 8 to the plots of the jacking forces, result showed a higher RMSE of 32.87. This indicated that increasing the moving average periods beyond an optimum value could lead to large deviations from the ground-truth data. The moving average period of 1 which resulted in 0 RMSE when compared to the ground-truth data, indicated that no moving average was applied in this case.

Drive B with moving average periods of 8 produced 47.64 RMSE which was lower than the prediction without the moving average, which resulted in 57.24 RMSE. However, the prediction performance is low even after applying moving average periods of 8. This was because of the sudden drop in the jacking forces at approximately 138 m, creating an abnormal situation and poor model prediction.

A moving average period of 8 was applied to Drive C, producing 14.54 RMSE which was lower than the prediction without a moving average, resulting in 24.45

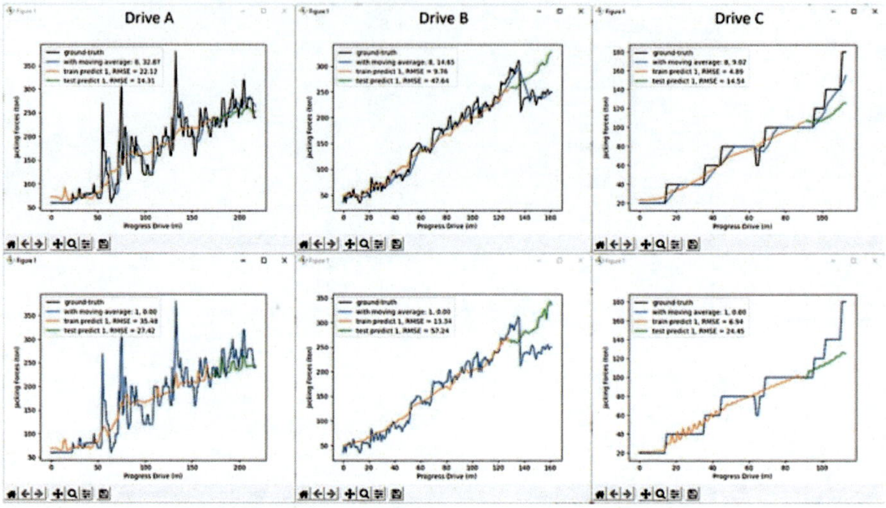

Fig. 2 Prediction of the jacking forces with and without using moving average

RMSE. However, from the plots, we could not see the effect of each input feature towards the prediction of the jacking forces.

Generally, there was an improvement in the prediction performance of the models when a moving average period of 8 was applied to the field-measured operation parameters. However, the influence of each operation parameter on jacking forces is unknown. Therefore, the study on different moving average periods with each input feature against the prediction is needed to identify the impact between the jacking forces with the features.

4.2 Study on Different Moving Average Periods Against the Prediction

The input features of pipe jacking operation parameters consisted of the jacking speed, the volume of injected lubricant, the drive length, the cumulative days elapsed, and the cumulative volume of lubrication used. All the input features were smoothened using moving average periods of two (2), four (4), six (6), and eight (8) to determine the features with different degrees of impact on the prediction performance by looking at the increased periods of moving average in obtaining the lowest RMSE.

Table 3 shows the highest moving average period of eight (8) observed throughout the three different phyllite drives focused on the volume of injected lubricant, the drive length, and the cumulative volume of lubrication used. Furthermore, the top three average periods of the moving average are the drive length, the cumulative

Table 3 Table of different moving average periods against prediction performance (RMSE)

Phyllite drive	Train score (RMSE)	Test score (RMSE)	MA w/ GT score (RMSE)	Moving average periods					
				x1	x2	x3	x4	x5	y
A	19.81201	7.32608	32.86667	4	8	4	6	6	8
	36.56984	16.81122	0	2	2	8	6	4	1
B	10.9903	29.25496	12.90563	6	4	2	6	6	6
	13.89387	35.94509	0	4	6	6	4	6	1
C	5.545867	5.823387	9.019958	2	2	8	2	8	8
	7.921124	17.38158	0	4	2	6	2	2	1

y = 1, output prediction without smoothening

Min	5.823387	Max	6	8	8	6	8
		Total	22	24	34	26	32
		Average	3.666667	4	5.666667	4.333333	5.333333

Legend
x1 = Jacking speed
x2 = Lubricant
x3 = Drive length
x4 = Cum days
x5 = Cum lub
y = Jacking force

volume of lubrication used, and the cumulative days elapsed, with average periods of 5.67, 5.33, and 4.33, respectively.

These features were found to have a significant impact on the prediction performance. The higher the moving average period observed from each feature, the greater the contribution of that feature to the prediction performance by producing the lowest RMSE.

Feature x3 was consistently significant across the three drives. This can be validated in mechanistic jacking force models, which describe the increase in frictional pipe jacking forces in relation to the drive length [6, 7, 18]. Other cumulative features (features x4, and x5) also demonstrated significant influence on frictional jacking forces. These were reasonable responses from the model, since frictional forces increase along with the cumulative length of the pipe string, which are expected to increase with cumulative days and lubricant. These provide a reasonable validation for the predictive model.

5 Conclusion

The findings from the study show an improvement in the prediction after using the moving average technique. However, excessively increasing the moving average periods would cause large deviation from the ground-truth data. The proposed model in this paper shows the increased periods of moving average focused on the cumulative parameters which are the drive length, the cumulative volume of lubricant used, and the cumulative days elapsed in obtaining the lowest RMSE in three different phyllite drives. The increased moving average periods on these features showed the increasing influence of these features on the predicted jacking forces. The findings here show the potential of using deep learning in the prediction of jacking forces from operation parameters, enabling the understanding of how decisions are made at pipe jacking sites. Future work would validate the features of this study through the use of the attention mechanism.

Acknowledgements The authors are very grateful for the support from Swinburne University of Technology, Sarawak Campus and Hock Seng Lee Bhd.

References

1. Bennett RD (1998) Jacking forces and ground deformations associated with microtunneling. University of Illinois at Urbana-Champaign, Illinois
2. Chapman DN, Ichioka Y (1999) Prediction of jacking forces for microtunneling operations. Tunn Undergr Space Technol 14:31–41. https://doi.org/10.1016/S0886-7798(99)00019-X
3. Chen H, Xiao C, Yao Z, Jiang H, Zhang T, Guan Y (2019) Prediction of TBM tunneling parameters through an LSTM neural network. In: Proceeding of the IEEE international conference on robotics and bionimetics
4. Cheng WC, Bai XD, Sheil BB, Li G, Wang F (2020) Identifying characteristics of pipejacking parameters to assess geological conditions using optimisation algorithm-based support vector machines. Tunn Undergr Space Technol 106. https://doi.org/10.1016/j.tust.2020.103592
5. Cheng WC, Ni JC, Shen JSL, Huang HW (2017) Investigation into factors affecting jacking force: a case study. Proc Inst Civ Eng-Geotech Eng 170(4):322–334. https://doi.org/10.1680/jgeen.16.00117
6. Choo CS, Ong DEL (2012) Back-analysis of frictional jacking forces based on shear box testing of excavated spoils. In: Proceedings of the second international conference on geotechnique, construction materials and environment (GEOMATE 2012), pp 462–467
7. Choo CS, Ong DEL (2015) Evaluation of pipe-jacking forces based on direct shear testing of reconstituted tunneling rock spoils. J Geotech Geoenvironmental Eng 141(10). https://doi.org/10.1061/(asce)gt.1943-5606.0001348
8. Gao B, Wang R, Lin C, Guo X, Liu B & Zhang W (2020) TBM penetration rate prediction based on the long short-term memory neural network. Undergr Space. https://doi.org/10.1016/j.undsp.2020.01.003
9. Ji X, Zhao W, Ni P, Barla M, Han J, Jia P, Chen Y, Zhang C (2019) A method to estimate the jacking force for pipe jacking in sandy soils. Tunn Undergr Space Technol 90:119–130. https://doi.org/10.1016/j.tust.2019.04.002
10. Jong SC, Ong DEL, Oh E (2021) State-of-the-art review of geotechnical-driven artificial intelligence techniques in underground soil-structure interaction. Tunn Undergr Space Technol 113. https://doi.org/10.1016/j.tust.2021.103946
11. Marcher T, Erharter GH, Winkler M (2020) Machine learning in tunnelling–capabilities and challenges. Geomech Tunn 13(2):191–198. https://doi.org/10.1002/geot.202000001
12. Nagrecha K, Fisher L, Mooney M, Rodriguez-Nikl T, Mazari M, Pourhomayoun M (2020) As-encountered prediction of tunnel boring machine performance parameters using recurrent neural networks. Transp Res Rec J Transp Res Board 2674(10):241–249. https://doi.org/10.1177/0361198120934796
13. Ong DEL, Choo CS (2016) Back-analysis and finite element modeling of jacking forces in weathered rocks. Tunn Undergr Space Technol 51:1–10. https://doi.org/10.1016/j.tust.2015.10.014
14. Osumi T (2000) Calculating jacking forces for pipe jacking methods. No-Dig Int Res 40–42
15. Peerun MI, Ong DEL, Choo CS, Cheng WC (2020) Effect of interparticle behavior on the development of soil arching in soil-structure interaction. Tunn Undergr Space Technol. 106. https://doi.org/10.1016/j.tust.2020.103610
16. Pellet-Beaucour AL, Kastner R (2002) Experimental and analytical study of friction forces during microtunneling operations. Tunn Undergr Space Technol 17(1):83–97. https://doi.org/10.1016/S0886-7798(01)00044-X

17. Shao B, Ma B, Shi L (2009) A sewer pipeline installation using pipe-jacking in Lang Fang. Am Soc Civ Eng 1413–1424. https://doi.org/10.1061/41073(361)148
18. Staheli K (2006) Jacking force prediction: an interface friction approach based on pipe surface roughness. PhD thesis, School of Civil and Environmental Engineering
19. Sterling RL (2020) Developments and research directions in pipe jacking and microtunneling. Undergr Space 5(1):1–19. https://doi.org/10.1016/j.undsp.2018.09.001
20. Wei XJ, Wang X, Wei G, Zhu CW, Shi Y (2021) Prediction of jacking force in vertical tunneling projects based on neuro-genetic models. J Mar Sci Eng 9(1). https://doi.org/10.3390/jmse9010071
21. Zhang N, Zhou A, Pan Y, Shen SL (2021) Measurement and prediction of tunnelling-induced ground settlement in karst region by using expanding deep learning method. Measurement 183. https://doi.org/10.1016/j.measurement.2021.109700
22. Zhang WQ, Lin Z, Liu XL (2022) Short-term offshore wind power forecasting-a hybrid model based on Discrete Wavelet Transform (DWT), Seasonal Autoregressive Integrated Moving Average (SARIMA), and deep-learning-based Long Short-Term Memory (LSTM). Renew Energy 185:611–628. https://doi.org/10.1016/j.renene.2021.12.100